T0256373

Material Balances for Chemical Reacting Systems

Written for use in the first course in a typical chemical engineering program, *Material Balances for Chemical Reacting Systems* introduces and teaches students a rigorous approach to solving the types of macroscopic balance problems they will encounter as chemical engineers. This first course is generally taken after students have completed their studies of calculus and vector analysis, and these subjects are employed throughout this text. Since courses on ordinary differential equations and linear algebra are often taken simultaneously with the first chemical engineering course, these subjects are introduced as needed.

- Teaches readers the fundamental concepts associated with macroscopic balance analysis of multicomponent, reacting systems
- Offers a novel and scientifically correct approach to handling chemical reactions
- Includes an introductory approach to chemical kinetics
- Features many worked out problems, beginning with those that can be solved by hand and ending with those that benefit from the use of computer software

This textbook is aimed at undergraduate chemical engineering students but can be used as a reference for graduate students and professional chemical engineers as well as readers from environmental engineering and bioengineering. The text features a solutions manual with detailed solutions for all problems, as well as PowerPoint lecture slides, available to adopting professors.

Material Balances for Chemical Reacting Systems

R.L. Cerro, B.G. Higgins, and S. Whitaker

CRC Press
Taylor & Francis Group
Boca Raton   London   New York

CRC Press is an imprint of the
Taylor & Francis Group, an **informa** business

First edition published 2023
by CRC Press
6000 Broken Sound Parkway NW, Suite 300, Boca Raton, FL 33487-2742

and by CRC Press
4 Park Square, Milton Park, Abingdon, Oxon, OX14 4RN

CRC Press is an imprint of Taylor & Francis Group, LLC

The LCCN is 2022013487.

ISBN: 978-1-032-25529-3 (hbk)
ISBN: 978-1-032-25530-9 (pbk)
ISBN: 978-1-003-28375-1 (ebk)

DOI: 10.1201/9781003283751

Access the Support Material: www.routledge.com/Material-Balances-for-Chemical-Reacting-Systems/ Cerro-Higgins-Whitaker/p/book/9781032255293

To our wives
Frances, Sandra, and Suzanne
for their patience and support

Contents

Preface

This text has been written for use in a first course in a rigorous chemical engineering program. That first course is generally taken after students have completed their studies of calculus and vector analysis, and these subjects are employed throughout this text. Since courses on ordinary differential equations and linear algebra are often taken simultaneously with the first chemical engineering course, these subjects are introduced as needed.

Chapter 1 introduces students to the types of macroscopic balance problems they will encounter as chemical engineers, and Chapter 2 presents a review of the types of units (dimensions) they will need to master. While the fundamental concepts associated with units are inherently simple, the practical applications can be complex and chemical engineering students must be experts in this area. Chapter 3 treats macroscopic balance analysis for single component systems and this provides the obvious background for Chapter 4 that deals with the analysis of multicomponent systems in the absence of chemical reactions. Chapter 5 presents the analysis of two-phase systems and equilibrium stages. This requires a brief introduction to concepts associated with phase equilibrium. Chapter 6 deals with stoichiometry and provides the framework for the study of systems with reaction and separation presented in Chapter 7. Chapter 8 treats steady and transient batch systems with and without chemical reactions. Chapter 9 provides a *connection* between stoichiometry as presented in Chapter 6 and reactor design as presented in subsequent courses.

Throughout the text, one will find a variety of problems beginning with those that can be solved by hand and ending with those that benefit from the use of computer software. The problems have been chosen to illustrate concepts and to help develop skills, and many solutions have been prepared as an aid to instructors. Students are encouraged to use the problems to teach themselves the *fundamental concepts* associated with macroscopic balance analysis of multicomponent, reacting systems for this type of analysis will be a recurring theme throughout their professional lives.

Many students and faculty have contributed to the completion of this text, and there are too many for us to identify individually. However, we would be remise if we did not point out that Professor Ruben Carbonell first introduced this approach to teaching material balances at UC Davis in the late 1970s.

R.L. Cerro

B.G. Higgins

S. Whitaker

About the Authors

R.L. Cerro is professor emeritus at the University of Alabama in Huntsville. He studied undergraduate chemical engineering at the Universidad del Litoral in Santa Fe, Argentina and completed MS and PhD degrees at the University of California, Davis. After two years post-doctoral work at the University of Minnesota, where he team-taught chemical engineering courses, he returned to the University of Litoral where he was a professor and later a fellow of the National Research Council of Argentina and a member of the National Academy of Engineering. He was a professor of chemical engineering at The University of Tulsa, OK and later chair of the Chemical and Materials Engineering Department at the University of Alabama in Huntsville. He had sabbatical leaves at the University of Minnesota, the Universidad de Salamanca in Spain, and the University of Litoral where he taught undergraduate and graduate courses. His area of research is transport phenomena, and he was an industrial consultant for Allied-Signal (Catoosa, OK), Corning Research (Fontainebleu, France), and Koch-Glitsch (Wichita, KS).

B.G. Higgins is professor emeritus at the University of California, Davis. He studied undergraduate chemical engineering at the University of the Witwatersrand, Johannesburg, South Africa, and completed his graduate studies at the University of Minnesota in chemical engineering. After three years at the Institute of Paper Chemistry in Appleton, WI, he joined the faculty at the University of California. He was chair of the department from 1988–1996. His principal area of research is the fluid mechanics and stability of thin film coating processes, and he has consulted with US and Japanese companies on coating technology as related to optical displays, printable electronics, and drying technology for thin films. He was also a software developer from 2009–2011 for Wolfram Alpha. From 2008–2018, Prof. Higgins was visiting professor at the University of Tokyo and Tokyo University of Science. He is currently a visiting professor of chemical engineering at the Hanoi University of Mining and Geology.

S. Whitaker is professor emeritus at the University of California, Davis. He studied undergraduate chemical engineering at the University of California, Berkeley and completed his graduate studies at the University of Delaware. After three years at the DuPont Experimental Station in Delaware, he joined the faculty at Northwestern University. Three years later, he moved to the newly established Chemical Engineering Department at the University of California, Davis. His main area of research is multi-phase transport phenomena, and he is the author of four textbooks: *Introduction of Fluid Mechanics* (Prentice-Hall, 1968), *Elementary Heat Transfer Analysis* (Pergamon Press, 1976), *Fundamental Principles of Heat Transfer* (Pergamon Press, 1977), and *The Method of Volume Averaging* (Kluwer Academic Publishers, 1999).

Nomenclature

A	area, m²; surface area of the control volume V, m²; absorption factor
$A_a(t)$	surface area of an arbitrary, moving control volume $V_a(t)$, m²
$A(t)$	surface area of a specific moving control volume $V(t)$, m²
A_e	area of the entrances and exits for the control volume V, m²
$A_e(t)$	area of the entrances and exits for the control volume $V_a(t)$, m²
$A_m(t)$	surface area of a body, m²
AW_J	atomic mass of the Jth atomic species, g/mol
$\mathbf{A}$	atomic matrix, also identified as $[N_{JA}]$
$\mathbf{A}^*$	row reduced echelon form of the atomic matrix, also identified as $[N_{JA}]^*$
$\mathbf{B}$	Bodenstein matrix that maps r onto $\mathbf{R}_B$
c_A	ρ_A/MW_A, molar concentration of species A, mol/m³
c	$\Sigma_{A=1}^{A=N} c_A$, total molar concentration, mol/m³
C	conversion
D	diameter, m
$\mathbf{f}$	force vector, N
f	magnitude of the force vector, N
$\mathbf{g}$	gravity vector, m/s²
g	magnitude of the gravity vector, m/s²
h	height, m
$\mathbf{i, j, k}$	unit base vectors
$\mathbf{I}$	unit matrix
$K_{eq,A}$	equilibrium coefficient for species A
L	length, m
m	mass, kg
$\dot{m}$	mass flow rate, kg/s
m_A	mass of species A, kg
$\dot{m}_A$	mass flow rate of species A, kg/s
$\dot{M}_A$	molar flow rate of species A, mol/s
$\dot{M}$	$\Sigma_{A=1}^{A=N} \dot{M}_A$, total molar flow rate, mol/s
$\mathbf{M}$	dimensionless molar flow rate
M_A	dimensionless molar flow rate of species A
MW_A	molecular mass of species A, g/mol
$\mathbf{M}$	mechanistic matrix that maps r onto $\mathbf{R}_M$
$\mathbf{n}$	outwardly directed unit normal vector
n_A	number of moles of species A
n	$\Sigma_{A=1}^{A=N} n_A$, total number of moles
N	number of molecular species in a multicomponent system
N_{JA}	number of J-type atoms associated with molecular species A
$[N_{JA}]$	atomic matrix, also identified as $\mathbf{A}$
$[N_{JA}]^*$	row reduced echelon form of the atomic matrix, also identified as $\mathbf{A}^*$
p	$\Sigma_{A=1}^{A=N} p_A$, total pressure, N/m²

p_A	partial pressure of species A, N/m^2
$p_{A,vap}$	vapor pressure of species A, N/m^2
p_g	$p - p_o$, gauge pressure, N/m^2
p_o	reference pressure (usually atmospheric), N/m^2
P	pivot matrix that maps $\mathbf{R}_P$ on to $\mathbf{R}_{NP}$
Q	volumetric flow rate, m^3/s
r	radial position, m
r	rank of a matrix
r	column matrix of elementary reaction rates, mol/m^{3}s
R	universal gas constant, see Table 5-1 for units
r_A	net mass rate of production of species A per unit volume, kg/m^{3}s
R_A	r_A/MW_A, net molar rate of production of species A per unit volume, mol/m^{3}s
$\mathcal{R}_A$	*global* net molar rate of production of species A, mol/s
R	column matrix of net molar rates of production, mol/m^{3}s
$\mathbf{R}_{NP}$	column matrix of non-pivot species net molar rates of production, mol/m^{3}s
$\mathcal{R}_{NP}$	column matrix of non-pivot species *global* net molar rates of production, mol/s
$\mathbf{R}_P$	column matrix of pivot species net molar rates of production, mol/m^{3}s
$\boldsymbol{R}_P$	????column matrix of pivot species *global* net molar rates of production, mol/s
$\mathbf{R}_M$	column matrix of all net rates of production, mol/m^{3}s
$\mathbf{R}_B$	column matrix of net rates of production of Bodenstein products, mol/m^{3}s
S	stoichiometric matrix
S	selectivity
T	temperature, K
t	time, s
$\mathbf{u}_A$	$\mathbf{v}_A - \mathbf{v}$, mass diffusion velocity of species A, m/s
$\mathbf{u}_A^*$	$\mathbf{v}_A - \mathbf{v}^*$, molar diffusion velocity of species A, m/s
$\mathbf{v}_A$	species A velocity, m/s
v	$\sum_{A=1}^{A=N} \omega_A \mathbf{v}_A$, mass average velocity, m/s
$\mathbf{v}^*$	$\sum_{A=1}^{A=N} x_A \mathbf{v}_A$, molar average velocity, m/s
v	magnitude of velocity vector, m/s
$\mathbf{v}_r$	relative velocity, m/s
V	volume, m^3; volume of a fixed control volume, m^3
$V_a(t)$	volume of an arbitrary moving control volume, m^3
$V(t)$	volume of a specific moving control volume, m^3
$V_m(t)$	volume of a body also referred to as a material volume, m^3
v_x, v_y, v_z	components of the velocity vector, $\mathbf{v} = \mathbf{i}v_x + \mathbf{j}v_y + \mathbf{k}v_z$, m/s
$\mathbf{w} \cdot \mathbf{n}$	speed of displacement of the surface of the arbitrary moving control volume $V_a(t)$, m/s
x_A	c_A/c, mole fraction of species A in the liquid phase
y_A	c_A/c, mole fraction of species A in the gas phase

| x, y, z | rectangular Cartesian coordinates, m |
| Y | yield |

GREEK LETTERS

λ	unit tangent vector.
γ	ρ/ρ_{H_2O}, specific gravity
θ	angle, radians
ν	μ/ρ, kinematic viscosity, m²/s
ρ_A	mass density of species A, kg/m³
ρ	$\Sigma_{A=1}^{A=N}\rho_A$, total mass density, kg/m³
ω_A	ρ_A/ρ, mass fraction of species A
ε	void fraction, volume of void space per unit volume
μ	viscosity, P
μ	specific growth rate, $(m^3 s)^{-1}$
σ	surface tension, N/m
τ	residence time, s

1 Introduction

1.1 INTRODUCTION

This text has been prepared for use in what is normally the first chemical engineering course in a typical chemical engineering program. There are essentially two major objectives associated with this text. The *first objective* is to carefully describe and apply the axioms for the conservation of mass in multicomponent, reacting systems. Sometimes these ideas are stated as

mass is conserved

or

mass is neither created nor destroyed

and in this text, we will replace these vague comments with definitive mathematical statements of the axioms for the conservation of mass. Throughout the text, we will use these axioms to analyze the *macroscopic transport of molecular species* and their production or consumption owing to chemical reactions. The macroscopic mass and mole balances presented in this text are often referred to as *material balances*. A course on material balances is generally taken after students have completed courses in calculus, vector analysis, and ordinary differential equations, and these subjects will be employed throughout the text. Since a course on linear algebra is often taken simultaneously with the first chemical engineering course, the elements of linear algebra required for problem solving will be introduced as needed.

The *second objective* of this text is to introduce students to the types of problems that are encountered by chemical engineers and to use modern computing tools for the solution of these problems. To a large extent, chemical engineers are concerned with the transport and transformation (by chemical reaction) of various molecular species. Although it represents an oversimplification, one could describe chemical engineering as the business of *keeping track of molecular species*. As an example of the problem of "keeping track of molecular species," we consider the coal combustion process illustrated in Figure 1.1. While coal combustion is a steadily diminishing source of energy, it still serves as a good example for this discussion. Coal fed to the combustion chamber may contain sulfur, and this sulfur may appear in the stack gas as SO_2 or in the ash as $CaSO_3$. In general, the calcium sulfite in the ash does not present a problem; however, the sulfur dioxide in the stack gas represents an important contribution to air pollution.

DOI: 10.1201/9781003283751-1

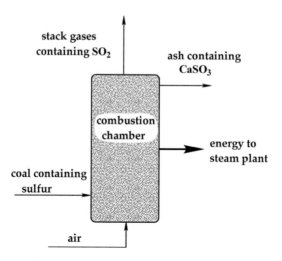

FIGURE 1.1   Coal combustion.

The sulfur dioxide in the stack gas can be removed by contacting the gas with a limestone slurry (calcium hydroxide) in order to affect a conversion to calcium sulfite. This process takes place in a gas-liquid contacting device as illustrated in Figure 1.2. There we have shown the stack gas bubbling up through a limestone slurry in which SO_2 is first absorbed as suggested by

$$(SO_2)_{gas} \rightleftarrows (SO_2)_{liquid} \tag{1.1}$$

The absorbed sulfur dioxide then reacts with water to form sulfurous acid

$$H_2O + SO_2 \rightleftarrows H_2SO_3 \tag{1.2}$$

which subsequently reacts with calcium hydroxide according to

$$Ca(OH)_2 + H_2SO_3 \rightleftarrows CaSO_3 + 2H_2O \tag{1.3}$$

Here, we have used Eq. 1.1 to represent the process of *gas absorption*, while Eqs. 1.2 and 1.3 are *stoichiometric representations* of the two reactions involving sulfurous acid. The situation is not as simple as we have indicated in Eq. 1.3 for the sulfurous acid may react either *homogeneously* with calcium hydroxide or *heterogeneously* with the limestone particles. This situation is also illustrated in Figure 1.2 where we have indicated that *homogeneous reaction* takes place in the fluid surrounding the limestone particles and *heterogeneous reaction* occurs at the fluid-solid interface between the particles and the fluid.

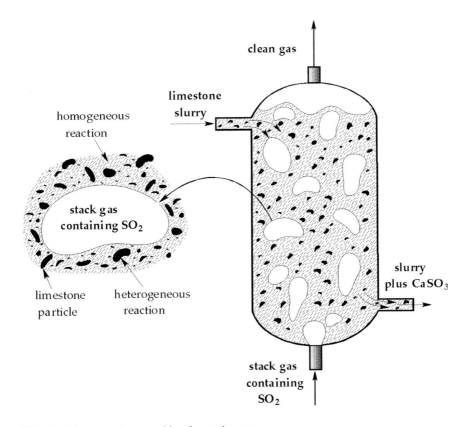

FIGURE 1.2 Limestone scrubber for stack gases.

There are other mass balance problems that are less complicated than those illustrated in Figures 1.1 and 1.2, and these are problems associated with the study of a *single chemical component* in the absence of chemical reaction. Consider, for example, a water balance on Mono Lake which is illustrated in Figure 1.3. It is not difficult to see that the *sources* of water for the lake are represented by the average rainfall and snowfall in the Mono Lake watershed. Over time, these sources must be balanced by the two "sinks" i.e., evaporation and shipments to Los Angeles[*]. There are two questions to answer concerning the impact of Los Angeles on Mono Lake:

1. What will be the final configuration of the lake?
2. When will this configuration occur?

The water balance for Mono Lake can be analyzed in a relatively simple manner, and both of these questions will be discussed in Chapter 3.

---

[*] For details see the Mono Lake Committee at https://www.monolake.org.

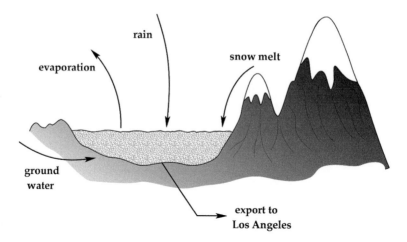

FIGURE 1.3   Water balance on Mono Lake.

The biological processes that occur in Mono Lake are altered by the changing level and the changing chemical composition of the lake. The analysis of these natural processes is very complex; however, commercial biological reactors, such as the chemostat illustrated in Figure 1.4, can be analyzed using the techniques that are developed in this text. In a chemostat, nutrients and oxygen enter a well-mixed tank containing a cell culture, and biological reactions generate new cells that are harvested in the product stream. This biological process will be analyzed in Chapter 8.

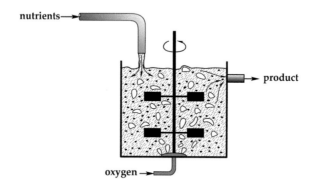

FIGURE 1.4   Continuous cell growth in a chemostat.

1.2   ANALYSIS VERSUS DESIGN

In this text, we will generally concern ourselves with the *analysis* of systems of the type illustrated in Figures 1.1–1.4. For example, if we know how much SO_2 is *entering* the scrubber shown in Figure 1.2 and we can measure the amount of $CaSO_3$ *leaving* with the slurry, then we can use material balance techniques to determine the amount of SO_2 leaving the scrubber in the "clean" gas. The *design* of a limestone slurry scrubber is a much more difficult problem. In that problem, we would be given the amount and composition of the stack gas to be treated, and the allowable amount of SO_2 in the clean gas would be specified. The task of the chemical engineer would be to determine the size of the equipment and the flow rate of the limestone slurry required to produce the desired clean gas. The design of a stack gas scrubber is not a trivial problem because the *rate of transfer* of SO_2 from the gas to the liquid is influenced by both the homogeneous and heterogeneous chemical reactions. In addition, this *rate of transfer* depends on the bubble size and velocity, the viscosity of the slurry, and a number of other parameters. Because of this, there are *many possible designs* that will provide the necessary concentration of SO_2 in the stack gas; however, it is the *responsibility of the chemical engineer* to develop the least expensive design that minimizes environmental impact, protects the health and safety of plant personnel, and assures a continuous and reliable operation of the chemical plant.

1.3 REPRESENTATION OF CHEMICAL PROCESSES

Chemical processes are inherently complex. In a continuous chemical plant[†], as we have illustrated in Figure 1.5, raw materials are prepared, heated or cooled, and reacted with other raw materials.

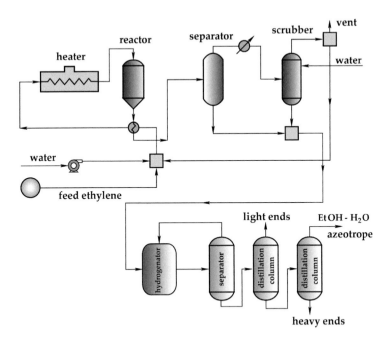

FIGURE 1.5   Simplified flowsheet for the manufacture of ethyl alcohol from ethylene.

[†] *Shreve's Chemical Process Industries*, 1998, 5th Edition, edited by J. Saeleczky and R. Margolies, McGraw-Hill Professional, New York.

The products are heated or cooled and separated according to specifications. A chemical plant, including its utilities, has many components such as chemical reactors, distillation towers, heat exchangers, compressors, and pumps. These components are connected to each other by pipelines or other means of transportation for carrying gases, liquids, and solids. To describe these complex systems, chemical engineers use two fundamental elements:

a. **Structure**: This is the manner in which the components of a plant are connected to each other with pipelines or other means of transportation. The *structure* is unique to a plant. Two components connected in different sequences can completely alter the nature of the products. Structure is represented using flowsheets. A complete version of a flowsheet, including all utilities, control, and safety devices is known as the Piping and Instrumentation Diagram (P+ID). Figure 1.5 is a pictorial representation of a simple flowsheet.

b. **Performance**: This is the *duty* or *basic operating specifications* of the individual units. The duty is described using specification sheets for all units of the process and by listing the properties of the streams connecting the units. The properties of the streams include flow rates, compositions, pressures, and temperatures. In relatively simple systems, a single document includes the flowsheet and the properties of the streams. To describe complex systems, one needs several flowsheets as well as a collection of specification sheets.

To perform material balances for complex systems, one uses information about the *structure* of the flowsheet and the *performance* of the units to determine the properties of the connecting streams. The processes illustrated in Figures 1.1–1.5 appear to be dramatically different; however, the *fundamental concepts* used to analyze these systems are the same. Hidden behind the complexity of these processes is a simplicity that we will describe in subsequent chapters. To make this point very clear, we consider the complex system illustrated in Figure 1.5, and we identify the scrubber as the object of a separate analysis as illustrated in Figure 1.6. In this text, most of our efforts will be directed toward the analysis of single units such as scrubbers, distillation columns, chemical, and biological reactors, in addition to systems such as Mono Lake. After establishing the framework for the analysis of single units, we will move on to a study of more complex systems in Chapter 7. Transient processes are examined briefly in Chapter 8, and an introduction to reaction kinetics is provided in Chapter 9.

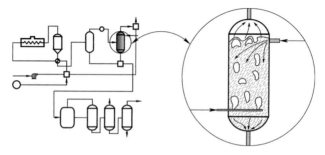

FIGURE 1.6   Analysis of an individual unit.

The concept of analyzing small parts of a problem and then assembling the small parts into a comprehensive representation of the whole problem is extremely important. In addition, the concept of studying the whole problem and then breaking it apart into smaller problems is also extremely important. For example, a chemical complex, such as the one shown in Figure 1.7, can be seen in terms of a sequence of progressively smaller elements. The entire *production system* consists of a natural gas plant that provides feed for an ethylene plant which, in turn, produces feedstocks for a vinyl chloride plant. Within the vinyl chloride plant, there are various units such as reactors and distillation columns. In addition, there are various feed preparation units, secondary distillation units, utilities, and flares used to warn of emergencies. Similar units exist in the natural gas plant and the ethylene plant, and we have not shown those details in Figure 1.7. However, we have shown the details of the mass transfer unit that must be analyzed as part of the design of the vinyl chloride plant. In addition, we have illustrated the details of the gas-liquid mass transfer process that lies at the *foundation* of the design of the mass transfer unit. Clearly, there is a series of length scales associated with the production of vinyl chloride and there is important analysis to be done at each length scale.

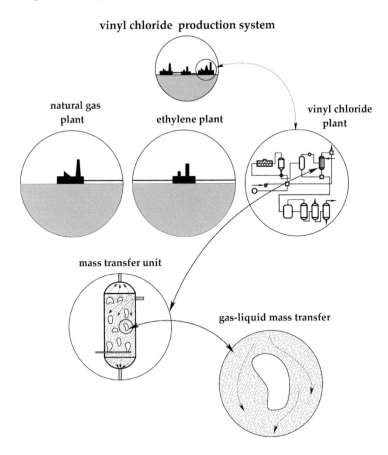

FIGURE 1.7 Hierarchy of length scales associated with the production of vinyl chloride.

The circles illustrated in Figure 1.7 represent *control volumes* that we use for accounting purposes, i.e., we want to know what goes in, what goes out, and what is accumulated or depleted. In some cases, we do not need to know what is happening inside the control volume and we are only concerned with the inputs and outputs of the control volume. This situation is shown in Figure 1.8 where we have shown only the inputs and outputs for the vinyl chloride plant. If both the steady state and dynamic behavior of the systems associated with the natural gas plant, the ethylene plant, and the vinyl chloride plant are known, the behavior of the *vinyl chloride production system* is also known. However, to learn how those systems behave, we need to move down the length scales to determine the details of the various processes. This is illustrated in Figure 1.7 where we have shown a mass transfer unit that is one element of the vinyl chloride plant, and we have shown a bubble at which mass transfer takes place within the mass transfer unit. In Figure 1.7, we have illustrated the concept that we must be able to keep track of molecular forms at a variety of length scales.

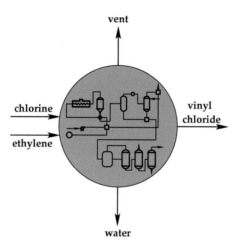

FIGURE 1.8   Control volume representation of the vinyl chloride plant.

As another example of the importance of keeping track of molecular species in both large and small regions, we consider the problem of *lead contamination* in California (see Figure 1.9). The title of the article by Stedling, Dunlap, and Flegal[‡] suggests that we should *keep track of lead* in the San Francisco Bay estuary system; however, the lead that appears in the estuary comes from several sources. Endless weathering of granite in the Sierra Nevada mountains releases lead that is transported by streams and rivers and eventually arrives in the bay. Other lead comes from hydraulic mine sediments transported across the Central Valley and into the bay during the 19th century. Finally, the lead generated by the earlier use of leaded gasoline has made its way into the estuary by a variety of paths.

[‡] Steding, D.J., Dunlap, C.E. and Flegal, A.R. 2000, New isotopic evidence for chronic lead contamination in the San Francisco Bay estuary system: Implications for the persistence of past industrial lead emissions in the biosphere, Proc. Natl Acad. Sci. **97** (19), 11181–11186.

Within the estuary itself, the impact of lead contamination varies. In the shallow salt marshes, seasonal floods and daily tidal flows have a small effect on the transport of lead, and the local bio-reactors are confronted with an unhealthy diet. Clearly, the study of lead contamination in the San Francisco Bay estuary requires *keeping track of lead* over a variety of length scales as we have indicated in Figure 1.9. The analysis of this lead contamination process in Northern California has some of the same characteristics as the analysis of water conservation in Mono Lake, of stack gas scrubbing in a coal-fired power plant, of cell growth in a chemostat, and of vinyl chloride production. In this text, we will develop a framework that allows us to analyze all of these systems from a single perspective based on the axioms for the mass of multicomponent systems. This will allow us to solve mass balance problems associated with a wide range of phenomena; however, chemical engineers must remember that in addition to these physical problems, there are *economic*, *environmental*, and *safety* concerns associated with every process and these concerns must be addressed.

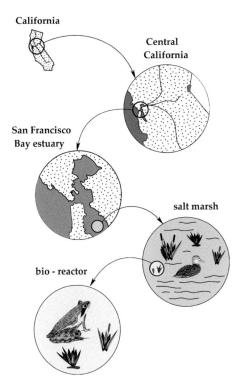

FIGURE 1.9 Multiple regions associated with lead contamination.

1.4 PROBLEMS

1.1. Read the MSDS (Material Safety Data Sheet) and write a one-paragraph description of the hazards associated with: dimethyl mercury (students with last names ending in (A–E), methanol (F–J), ethyl ether (K–O), benzene (P–T), phosgene (U–Z). Information is available on the Internet, at your campus library, and on the Web page of your local department of environmental health and safety.

2 Units

There are three things that every engineer should know about units. First, the fundamental significance of units must be understood. Second, the conversion from one set of units to another must be a routine matter. Third, one must learn to use units to help prevent the occurrence of algebraic and conceptual errors. The second of these is emphasized by the following:

> In the Sacramento Bee, November 11, 1999 one finds the headline:
> **Training faulted in loss of $125 million Marsprobe**.
> In the article that follows one reads, "The immediate cause of the spacecraft's Sept. 23 disappearance as it entered Mars orbit was a failure by a young engineer……to make a simple conversion from English units to metric…."

Physics is a quantitative science. By this we mean that the physicist attempts to compare *measured observables* with values predicted from theory. There is basically only *one measuring process* and that is the process of counting. For example, the distance between two points is determined by counting the number of times that a *standard length* fits between the two points. Often we call this length a *unit length*. The business of measuring began with the Egyptians, but we are generally more familiar with the work of the Greek geometers such as Pythagoras. In physics, the process of performing experiments *and* measuring observables is often attributed to Galileo (1564–1642). The process of measuring by counting standard units can be described as[*]:

> Since the measurement process is one of counting multiples of some chosen standard, it is reasonable to ask how many standards we need. If we need a standard for each *observable*, we will need a large Bureau of Standards. As a matter of fact, we need only *four standards*: a standard of *length*, a standard of *mass*, a standard of *time*, and a standard of *electric charge*. This is an extraordinary fact. It means that if one is equipped with a set of these four standards and the ability to count, one can (*in principle*) assign a numerical value to any observable, be it distance, velocity, viscosity, temperature, pressure, etc.

Here we find that our confrontation with units begins with a great deal of simplicity since we require only the following *four fundamental standards*:

LENGTH
MASS
TIME
ELECTRIC CHARGE

---

[*] Hurley, J.P. and Garrod, C. 1978, page 1, *Principles of Physics*, Houghton Mifflin Co., Boston.

DOI: 10.1201/9781003283751-2

The reason for this simplicity is that observables, in one way or another, must satisfy the laws of physics, and these laws can be quantified in terms of length, mass, time, and electric charge.

Although the concept of a standard is simple, the matter is complicated by the fact that the *choice* of a standard is arbitrary. For example, a football player prefers the yard as a *standard of length* because one yard is significant and 100 yards represents an upper bound for the domain of interest. The carpenter prefers the foot as a *standard of length* since one foot is significant and the distance of 100 feet spans the domain of interest for many building projects. For the same reasons, a truck driver prefers the mile as a *standard of length*, i.e., one of them is significant and 100 of them represent a certain degree of accomplishment. It is a fact of life that people like to work in terms of standards that give rise to counts somewhere between 1 and 100, and we therefore change our standards to fit the situation. While the football player thinks in terms of *yards* on Saturday, his Sunday chores are likely to be measured in *feet* and the distance to the next game will surely be thought of in terms of *miles*. Outside of the United States, a football player (soccer) thinks in terms of *meters* on Saturday, perhaps *centimeters* on Sunday, and the distance to the next match will undoubtedly be determined in *kilometers*.

2.1 INTERNATIONAL SYSTEM OF UNITS

In 1960, a conference was held in Paris to find agreement on a set of standards. From that conference there arose what are called the SI (Système International) system of *basic units* which are listed in Table 2.1. Note that the SI system does not use the electric charge as a standard, but rather the *electric charge per unit time* or the *electric current*. In addition, the SI system of basic units includes three additional units, the thermodynamic temperature, the mole, and the luminous intensity. These three additional units are *not necessary* to assign numerical values to any observable; thus, their role is somewhat different than the four fundamental standards identified by Hurley and Garrod. For example, a mole consists of $6.02209\ldots\times10^{23}$ entities such as atoms, molecules, photons, etc., while the *basic unit* associated with *counting entities* is one.

This point has been emphasized by Feynman *et al.*[†] who commented that: "We use 1 as a unit, and the chemists use 6×10^{23} as a unit!" Nevertheless, a mole is a *convenient unit* for engineering calculations because one of them is significant, and we will use moles to count atoms and molecules throughout this text. Sometimes chemical engineers make use of the "pound-mole" as a unit of measure; however, in this text we will be consistent with chemists, physicists and biologists and use only the mole as a unit of measure.

2.1.1 MOLECULAR MASS

Here we follow the SI convention concerning the definition of molecular mass which is:

$$\text{molecular mass} = \frac{\text{mass of the substance}}{\text{amount of the substance}} \tag{2.1}$$

[†] Feynman, R.P., Leighton, R.B. and Sands, M. 1963, *The Feynman Lectures on Physics*, Addison-Wesley Pub. Co., New York.

TABLE 2.1

SI Basic Units

Quantity	Name	Symbol	Definition
Length	Meter	m	The meter is the length of the path traveled by light in vacuum during a time interval of 1/299,792,458 of a second.
Mass	Kilogram	kg	The kilogram is the unit of mass equal to the international prototype of the kilogram.
Time	Second	s	The second is the duration of 9,192,631,770 periods of the radiation corresponding to the transition between the two hyperfine levels of the ground state of the cesium 133 atom.
Electric current	Ampere	A	The ampere is that constant current which, if maintained in two straight parallel conductors of infinite length, of negligible circular cross section, and placed 1 meter apart in vacuum, would produce between these conductors a force equal to 2×10^{-7} newton per meter of length.
Temperature	Kelvin	K	The Kelvin, unit of thermodynamic temperature, is the fraction of 1/273.16 of the thermodynamic temperature of the triple point of water.
Elemental entities	Mole	mol	The mole is the amount of substance of a system which contains as many elementary entities as there are atoms in 0.012 kilogram of carbon-12.
Luminous intensity	Candela	cd	The candela is the luminous intensity, in a given direction, of a source that emits monochromatic radiation of frequency 540×10^{12} hertz and that has a radiant intensity in that direction of 1/683 watt per steradian.

Continuing with the SI system, we represent the *mass* of a substance in kilograms and the *amount* of the substance in moles. For the case of carbon-12 identified in Table 2.1, this leads to

$$\text{molecular mass of carbon-12} = \frac{0.012 \text{ kilogram}}{\text{mole}} \tag{2.2}$$

Using the compact notation indicated in Table 2.1, we express this result as

$$MW_{C^{12}} = \frac{0.012 \text{ kg}}{\text{mol}} \tag{2.3}$$

in which the symbol *MW* is based on the historical use of *molecular weight* to describe the *molecular mass*. The molecular mass of carbon-12 can also be expressed in terms of grams leading to

$$MW_{C^{12}} = \frac{12 \text{ g}}{\text{mol}} \tag{2.4}$$

While Eq. 2.3 represents the molecular mass in the preferred SI system of units, the form given by Eq. 2.4 is extremely common, and we have used this form to list atomic masses and molecular masses in Tables A1 and A2 of Appendix A.

Energy can be described in units of kg m^2/s^2; however, the thermodynamic temperature represents an extremely *convenient unit* for the description of energy and many engineering calculations would be quite cumbersome without it. The same comment applies to the luminous intensity which is an observable that can be assigned a numerical value in terms of the four fundamental standards of length, mass, time, and electric charge. One of the attractive features of the SI system is that alternate units are created as multiples and submultiples of powers of 10, and these are indicated by prefixes such as *giga* for 10^9, *centi* for 10^2, *nano* for 10^{-9}, etc. Some of these alternate units are listed in Table 2.2 for the meter. NIST (National Institute of Standards and Technology) provides a more extensive list of prefixes. In other systems of units, multiples of 10 are not necessarily used in the creation of alternate units, and this leads to complications which in turn leads to errors.

2.1.2 SYSTEMS OF UNITS

If we focus our attention on the *fundamental standards* and ignore the electric charge, we can think of the SI system as dealing with *length*, *mass*, and *time* in terms of *meters*, *kilograms*, and *seconds*. At one time, this was known as the MKS-system to distinguish it from the CGS-system in which the fundamental units were expressed as *centimeters*, *grams* and *seconds*. Another well-known system of units is referred to as the British (or English) system in which the fundamental units are expressed in terms of *feet*, *pounds-mass*, and *seconds*. Even though there was general agreement in 1960 that the SI system was preferred, and is now required in most scientific and technological applications, one must be prepared to work with the CGS and the British system, in addition to other systems of units that are associated with specific technologies.

2.2 DERIVED UNITS

In addition to using some alternative units for length, time, mass, and electric charge, we make use of many *derived units* in the SI system and a *few* are listed in Table 2.3. Some derived units are sufficiently notorious so that they are named after famous scientists and represented by specific symbols. For example, the unit of kinematic viscosity is the stokes (St), named after the British mathematician Sir George G. Stokes[‡], while the equally important molecular and thermal diffusivities are known only by their generic names and represented by a variety of symbols. The key point to remember concerning units is that the *basic units* represented in Table 2.1 are sufficient to describe all physical phenomena, while the *alternate units* illustrated Table 2.2 and the *derived units* listed in Table 2.3 are used as a matter of convenience.

While the existence of derived units is simply a matter of convenience, this *convenience* can lead to *confusion*. As an example, we consider the case of Newton's second law which can be stated as

$$\left\{ \begin{array}{c} \text{force acting} \\ \text{on a body} \end{array} \right\} = \left\{ \begin{array}{c} \text{time rate of change} \\ \text{of linear momentum} \\ \text{of the body} \end{array} \right\} \tag{2.5}$$

[‡] Rouse, H. and Ince, S. 1957, *History of Hydraulics*, Dover Publications, Inc., New York.

TABLE 2.2
Alternate Units of Length

1 *kilo*meter (km) = 10^3 meter (m)
1 *deci*meter (dm) = 10^{-1} m
1 *centi*meter (cm) = 10^{-2} m
1 *milli*meter (mm) = 10^{-3} m
1 *micro*meter (μm) = 10^{-6} m
1 *nano*meter (nm) = 10^{-9} m

In terms of mathematical symbols, we express this axiom as

$$\mathbf{f} = \frac{d}{dt}(m\mathbf{v}) \tag{2.6}$$

Here we adopt a nomenclature in which a lower case, boldface Roman font is used to represent *vectors* such as the force and the velocity. Force and velocity are quantities that have both *magnitude and direction* and we need a special notation to remind us of these characteristics.

Let us now think about the use of Eq. 2.6 to calculate the force required to accelerate a mass of 7 kg at a rate of 13 m/s². From Eq. 2.6 we determine the magnitude of this force to be

$$f = (7 \text{ kg})(13 \text{ m/s}^2) = 91 \text{ kg m/s}^2 \tag{2.7}$$

where f is used to represent the magnitude of the vector $\mathbf{f}$. Note that the force is expressed in terms of *three* of the four fundamental standards of measure, i.e., mass, length, and time. There is no real need to go beyond Eq. 2.7 in our description of force; however, our intuitive knowledge of force is rather different from our

---

TABLE 2.3
Derived SI Units

Physical Quantity	Unit (Symbol)	Definition
Force	Newton (N)	kg m/s²
Energy	Joule (J)	kg m²/s²
Power	Watt (W)	J/s
Electrical potential	Volt (V)	J/(A s)
Electric resistance	Ohm (Ω)	V/A
Frequency	Hertz (Hz)	cycle/second
Pressure	Pascal (Pa)	N/m²
Kinematic viscosity	Stokes (St)	cm²/s
Thermal diffusivity	Square meter/second	m²/s
Molecular diffusivity	Square meter/second	m²/s

knowledge of mass, length and time. Consider for example, pushing against a wall with a "force" of 91 kg m/s². This is simply not a satisfactory description of the event. What we want here is a unit that describes the *physical nature* of the event, and we obtain this unit by *defining* a unit of force as

$$1 \text{ newton} = (1 \text{ kg}) \times (1 \text{ m/s}^2) \qquad (2.8)$$

When pushing against the wall with a force of 91 kg m/s² we feel comfortable describing the event as

$$\text{force} = 91 \text{ kg m/s}^2 = (91 \text{ kg m/s}^2) \times (1)$$

$$= (91 \text{ kg m/s}^2) \times \left(\frac{1 \text{ newton}}{\text{kg m/s}^2} \right) = 91 \text{ newtons} \qquad (2.9)$$

Here we have arranged Eq. 2.8 in the form

$$1 = \frac{\text{newton}}{\text{kg m/s}^2} = \frac{N}{\text{kg m/s}^2} \qquad (2.10)$$

and multiplied the quantity 91 kg m/s² by *one* in order to affect the change in units. Note that in our definition of the unit of force we have made use of a *one-to-one* correspondence represented by Eq. 2.8. This is a characteristic of the SI system and it is certainly one of its attractive features. One must keep in mind that Eq. 2.8 is *nothing more than a definition* and if one wished it could be replaced by the *alternate definition* given by Truesdell[§]

$$1 \text{ euler} = (17.07 \text{ kg}) \times (1 \text{ m/s}^2) \qquad (2.11)$$

However, there is nothing to be gained from this second definition, aside from honoring Euler, and it is clearly less attractive than the definition given by Eq. 2.8.

In the British system of units, the *one-to-one* correspondence is often lost and confusion results. In the British system, we choose our standards of mass, length and time as

$$\text{standard of mass} = \text{lb}_m$$
$$\text{standard of length} = \text{ft}$$
$$\text{standard of time} = \text{s}$$

and we define the pound-force according to

$$1 \text{ lb}_f = (1 \text{ lb}_m) \times (32.17 \text{ ft/s}^2) \qquad (2.12)$$

[§] Truesdell, C. 1968, *Essays in the History of Mechanics*, Springer-Verlag, New York.

This definition was chosen so that mass and force would be numerically equivalent when the mass was acted upon by the earth's gravitational field. While this may be convenient under certain circumstances, the definition of a unit of force given by Eq. 2.12 is certainly less attractive than that given by Eq. 2.8.

In summary, we note that there are only four standards needed to assign numerical values to all observables, and the choice of standard is arbitrary, i.e. the standard of length could be a foot, an inch or a centimeter. Once the standard is chosen, i.e. the meter is the standard of length, other *alternate units* can be constructed such as those listed in Table 2.2. In addition to a variety of *alternate units* for mass, length and time, we construct for our own convenience a series of *derived units*. Some of the derived units for the SI system are given in Table 2.3. Finally, we find it convenient to tabulate *conversion factors* for the *derived units* for the various different systems of units and some of these are listed in Table 2.4 and in Table 2.5. An interesting history of the SI system is available from the Bureau International des Poids et Mesures[**].

TABLE 2.4
Basic Conversion Factors

Mass

1 pound mass (lb_m) = 453.6 gram (g)

1 gram = 10^{-3} kilogram (kg)

1 ton (short) = 2,000 lb_m

1 ton (long) = 2,240 lb_m

Force

Pound-force (lb_f) = 32.17 lb_m ft/s^2

Dyne (dyn) = g cm/s^2

1 newton (N) = 10^5 dyne

1 dyne = 2.248×10^{-6} pound force (lb_f)

1 poundal = 3.108×10^{-2} lb_f

Pound-force (lb_f) = 4.448 newton (N)

Pressure

1 atmosphere (atm) = 14.7 lb_f /in^2

1 lb_f/in^2 = 6.89×10^5 N/m^2

1 atm = 1.013×10^6 dyne/cm^2

1 pascal = 1 N/m^2

1 atmosphere (atm) = 1.01325 bar

1 atm → 760 mmHg

Length

1 inch = 2.54 centimeter (cm)

1 angstrom (Å) = 10^{-8} centimeter

1 foot (ft) = 0.3048 meter (m)

1 mile (mi) = 1.609 kilometer (km)

1 yard (yd) = 0.914 meter (m)

1 nanometer (nm) = 10^{-9} meter (m)

Time

1 minute (min) = 60 second (s)

1 hour (h) = 60 minute (min)

1 day = 24 hours (h)

Volume

1 in^3 = 16.39 cm^3

1 ft^3 = 2.83×10^{-2} m^3

1 gallon (US) (gal) = 231 in^3

1 liter = 1 L = 10^{-3} m^3

[**] Bureau International des Poids et Mesures, http://www.bipm.fr/enus/3_SI/si-history.html

TABLE 2.5

Additional Conversion Factors

Energy

1 calorie (cal) = 4.186 joule (J)

1 British Thermal Unit (Btu) = 252 cal

1 erg = 10^{-7} joule (J)

1 Btu = 1,055 watt second

1 ft lb$_f$ = 1.356 joule (J)

Power

1 horsepower (hp) = 745.7 watt

1 hp = 42.6 Btu/min

1 watt = 9.51×10^{-4} Btu/s

1 ft lb$_f$ /s = 1.356 watt (W)

Temperature

$1°C = 1.8°F$

$K = °C + 273.16$

$R = F + 459.60$

$F = 1.8°C + 32$

Area

1 in^2 = 6.452 cm^2

1 in^2 = 9.29×10^{-2} m^2

1 acre = 4.35×10^4 ft^2

Flow Rate

1 gal/min = gpm 6.309×10^{-5} m^3/s

1 ft^3/min = 47.19×10^{-5} m^3/s

Viscosity

1 poise (P) = 1 g/cm s

1 poise (P) = 0.10 N s/m^2

1 poise (P) = 1 g/cm s

1 centipoise (cP) = 10^{-3} kg/m s

Kinematic Viscosity

1 m^2/s = 3.875×10^4 ft^2/h

1 cm^2/s = 10^{-4} m^2/s

1 stokes (St) = 1 cm^2/s

2.3   DIMENSIONALLY CORRECT AND DIMENSIONALLY INCORRECT EQUATIONS

In defining a unit of force on the basis of Eq. 2.6, we made use of what is known as the *law of dimensional homogeneity*. This law states that natural phenomena proceed with no regard for man-made units, thus the basic equations describing physical phenomena must be valid for all systems of units. It follows that each term in an equation based on the laws of physics must have the same units. This means that the units of **f** in Eq. 2.6 must be the same as the units of $d(m\mathbf{v})/dt$. It is this fact which leads us to the definition of a unit of force such as that given by Eq. 2.8.

Equations that satisfy the *law of dimensional homogeneity* are sometime referred to as *dimensionally correct* in order to distinguish them from equations that are *dimensionally incorrect*. While the law of dimensional homogeneity requires that all terms in an equation have the same units, this constraint is often ignored in the construction of empirical equations found in engineering practice. For example, in the sixth edition of *Perry's Chemical Engineers' Handbook*[††] we find an equation for the drop size produced by a certain type of atomizer that takes the form

---

[††] Perry, R.H., Green, D.W., and Maloney, J.O., 1984, *Perry's Chemical Engineer' Handbook*, 6th Edition, New York, McGraw-Hill Books.

$$\bar{X}_{vs} = \frac{1920\sqrt{\alpha}}{V_a\sqrt{\rho_1}} + 597\left(\frac{\mu}{\sqrt{\alpha\rho_1}}\right)^{0.45}\left(\frac{1000\,Q_L}{Q_a}\right)^{1.5} \tag{2.13}$$

In order that this expression produce correct results, it is *absolutely essential* that the quantities in Eq. 2.13 be specified as follows:

$\bar{X}_{vs}$ = average drop diameter, μm (a drop with the same volume to surface ratio as the total sum of all drops formed)

α = surface tension, dyne/cm

μ = liquid viscosity, P

V_a = relative velocity between air and liquid, ft/s

ρ_1 = liquid density, g/cm³

Q_L = liquid volumetric flow rate, any units

Q_a = air volumetric flow rate, same units as Q_L

When the numbers associated with these quantities are inserted into Eq. 2.13, one obtains the average drop diameter, $\bar{X}_{vs}$, in *micrometers*. Such an equation must always be used with great care for any mistake in assigning the values to the terms on the right-hand side will obviously lead to an *undetectable error* in $\bar{X}_{vs}$. In addition to being *dimensionally incorrect*, Eq. 2.13 is an *empirical* representation of the process of atomization. This means that the range of validity is limited by the range of values for the parameters used in the experimental study. For example, if the liquid density in Eq. 2.13 tends toward infinity, $\rho_1 \to \infty$, we surely *do not* expect that $\bar{X}_{vs}$ will tend to zero. This indicates that Eq. 2.13 is only valid for some finite range of densities. In addition, it is well known that when the relative velocity between air and liquid is zero ($V_a = 0$) the drop size is *not infinite* as predicted by Eq. 2.13, but instead it is on the order of the diameter of the atomizer jet. Once again, this indicates that Eq. 2.13 is only valid for some range of values of $\bar{X}_{vs}$; however, the range is known only to those persons who examine the original experimental data. Clearly, a dimensionally incorrect empiricism carries its own warning: beware!

The dimensionally incorrect result given by Eq. 2.13 can be used to construct a *dimensionally correct equation* by finding the units that should be associated with the coefficients 1,920 and 597. For example, the correct form of the first term should be expressed as

$$\left(\begin{array}{c}\text{correct form needed}\\ \text{to produce micrometers}\end{array}\right) = \frac{1920\,(\text{units})\,\sqrt{\alpha}}{V_a\,\sqrt{\rho_1}} \tag{2.14}$$

in which the correct units are determined by

$$(\text{units}) = (\text{micrometers})\left[\frac{(\text{ft/s})\,\sqrt{\text{g/cm}^3}}{\sqrt{\text{dyne/cm}}}\right] = 10^{-4}\,\text{cm}\left[\frac{(\text{ft/s})\,\sqrt{\text{g/cm}^3}}{\sqrt{\text{g/s}}}\right]$$

$$= 30.48\times10^{-4}\,\sqrt{\text{cm}} \tag{2.15}$$

Thus the dimensionally correct form of the first term in Eq. 2.13 is given by

$$\frac{1920\sqrt{\alpha}}{V_a\sqrt{\rho_1}} \Rightarrow \left(5.85\sqrt{cm}\right)\frac{\sqrt{\alpha}}{V_a\sqrt{\rho_1}} \tag{2.16}$$

and it is an exercise for the student to determine the correct form of the second term in Eq. 2.13.

2.4 CONVENIENCE UNITS

Occasionally, one finds that certain quantities are represented by terms that do not have the appropriate units. The classic example of this situation is associated with the use of mercury barometers to measure the pressure. It is a straightforward matter to use the laws of hydrostatics to show that the atmospheric pressure measured by the barometer shown in Figure 2.1 is given by

$$p_o = \rho_{Hg}\, g\, h + p_{Hg}^o \tag{2.17}$$

Here p_{Hg}^o represents the vapor pressure of mercury and under normal circumstances, this pressure is extremely small compared to p_o.

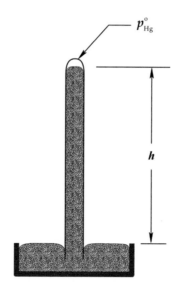

FIGURE 2.1 Mercury barometer.

This allows us to write Eq. 2.17 as

$$p_o = \rho_{Hg}\, g\, h \tag{2.18}$$

Since the density of mercury and the gravitational constant are essentially constant, the atmospheric pressure is often reported in terms of h, i.e. millimeters of mercury. While this is convenient, it can lead to errors if units are not used carefully. The message here should be clear: beware of convenience units!

The pressure over and above the constant ambient pressure is often a convenient quantity to use in engineering calculations. For example, if one is concerned about the possibility that a tank might rupture because of an excessively high pressure, it would be the pressure *difference* between the inside and outside that one would want to know. The pressure over and above the surrounding ambient pressure is usually known as the *gauge pressure* and is identified as p_g. The gauge pressure is defined by

$$p_g = p - p_o \tag{2.19}$$

where p_o is the ambient pressure. One should note that the gauge pressure may be negative if the pressure in the system is less than p_o.

2.5 PROBLEMS

SECTION 2.1

2.1. The following prefixes are officially approved for various multiples of ten:

$$10 \rightarrow deca \rightarrow D, \qquad 100 \rightarrow hecto \rightarrow H, \qquad 1000 \rightarrow kilo \rightarrow K$$
$$10^6 \rightarrow mega \rightarrow M, \qquad 10^9 \rightarrow giga \rightarrow G, \qquad 10^{12} \rightarrow tera \rightarrow T$$

Indicate how to express the following quantities in terms of a prefix and a symbol using the appropriate SI units.

(a) 100,000 watt,   (b) 100 meter,   (c) 300,000 meter
(d) 100,000 hertz,   (e) 200,000 kg,   (f) 2,000 ampere

2.2. If a 1-inch nail has a mass of 2.2 g, what will be the mass of one mole of 1-inch nails?

SECTION 2.2

2.3. Convert the following quantities as indicated:

a. 5,000 cal to Btu
b. 5,000 cal to watt-s
c. 5,000 cal to N-m

2.4. At a certain temperature, the viscosity of a lubricating oil is 0.136×10^{-3} lb_f s/ft^2. What is the kinematic viscosity in m^2/s if the density of the oil is $\rho = 0.936$ g/cm^3?

2.5. The density of a gas mixture is $\rho = 1.3$ kg/m^3. Calculate the density of the gas mixture in the following list of units:

a. lb_m/ft^3
b. g/cm^3
c. g/L
d. lb_m/in^3

2.6. Write an expression for the *volume per unit mass*, $\hat{V}$, as a function of the *molar volume*, $\tilde{V}$, (that is the volume per mole) and the molecular mass, MW. Write an expression for the molar volume, $\tilde{V}$, as a function of the density of the component, ρ, and its molecular mass, MW.

2.7. The CGS system of units was once commonly used in science. What is the unit of force in the CGS system? Find the conversion factor between this unit and a Newton. Find the conversion factor between this unit and lb_f.

2.8. In rotating systems one uses angular velocity in radians per second. How do you convert revolutions per minute (rev/min) to rad/s?

2.9. Platinum is used as a catalyst in many chemical processes and in automobile catalytic converters. If a troy ounce of platinum costs $100 and a catalytic converter has 5 grams of platinum, what is the value in dollars of the platinum in a catalytic converter? To solve this problem, one needs to determine the relation between a troy ounce and a gram, and this conversion factor is not listed in Table 2.4. While the troy ounce originated in 16th century Britain, it has largely been replaced by the measures of mass indicated in Table 2.4. However, it is still retained today for the measure of precious stones and metals such as platinum, gold, etc.

2.10. In the textile industry, filament and yarn sizes are reported in denier which is defined as the mass in grams of a length of 9,000 meters. If a synthetic fiber has an average specific gravity of 1.32 and a filament of this material has a denier of 5.0, what is the mass per unit length in pounds-mass per yard? What is the cross-sectional area of this fiber in square inches? The specific gravity is defined as

$$\text{specific gravity} = \frac{\text{density}}{\text{density of water}}$$

2.11. (Adopted from Safety Health and Loss Prevention in Chemical Processes by AIChE). The level of exposure to hazardous materials for personnel of chemical plants is a very important safety concern. The Occupational Safety and Health Act (OSHA) defines as a hazardous material any substance or mixture of substances capable of producing adverse effects on the health and safety of a human being. OSHA also requires the Permissible Exposure Limit, or PEL, to be listed on the Material Safety Data Sheet (MSDS) for the particular component. The PEL is defined by the OSHA authority and is usually expressed in *volume parts per million* and abbreviated as ppm. Vinyl chloride is believed to be a human carcinogen that is an agent which causes or promotes the initiation of cancer. The PEL for vinyl chloride in air is 1 ppm, i.e., one liter of vinyl chloride per one million liters of mixture. For a dilute mixture of a gas in air at ambient pressure and temperature, one can show that volume fractions are equivalent to molar fractions. Compute the PEL of VC in the following units:

   a. moles of VC/m^3
   b. grams of VC/m^3
   c. moles VC/mole of air

2.12. This problem is adopted from Safety Health and Loss Prevention in Chemical Processes by the AIChE. Trichloroethylene (TCE) has a molecular mass of 131.5 g/mol so the vapors are much more dense than air. The density of air at 25°C and 1 atm is $\rho_{air} = 1.178$ kg/m³, while the density of TCE is $\rho_{TCE} = 5.37$ kg/m³. Being much denser than air, one would expect TCE to descend to the floor where it would be relatively harmless. However, gases easily mix under most circumstances, and at toxic concentrations the difference in density of a toxic mixture with respect to air is negligible. Assume that the gas mixture is ideal (see Sec. 5.1) and compute the density of a mixture of TCE and air at the following conditions:

a. The time-weighted average of PEL (see Problem 2.11) for 8 hours exposure, 100 ppm.
b. The 15-minute ceiling exposure, 200 ppm.
c. The 5-minute peak exposure, 300 ppm.
   Here ppm represents "parts per million" and in this particular case it means moles per million moles. Answer (a): $\rho_{mix} = 1.17842$ (kg/m³)

2.13. A liquid has a specific gravity of 0.865. What is the density of the liquid at 20°C in the following units:

(a) kg/m³, (b) lb_m/ft³, (c) g/cm³, (d) kg/L

SECTION 2.3

2.14. In order to develop a dimensionally correct form of Eq. 2.13, the appropriate units must be included with the numerical coefficients, 1,920 and 597. The units associated with the first coefficient are given by Eq. 2.15, and in this problem you are asked to find the units associated with 597.

2.15. In the literature, you have found an empirical equation for the pressure drop in a column packed with a particular type of particle. The pressure drop is given by the *dimensionally incorrect* equation

$$\Delta p = 4.7 \left(\frac{\mu^{0.15} H \rho^{0.85} v^{1.85}}{d_p^{1.2}} \right)$$

which requires the following units:

Δp = pressure drop, lb_f/ft²
μ = fluid viscosity, lb_m/ft s
H = height of the column, ft
ρ = density, lb_m/ft³
v = superficial velocity, ft/s
d_p = effective particle diameter, ft

Imagine that you are given data for μ, H, ρ, v, and d_p in SI units and you wish to use it directly to calculate the pressure drop in lb_f/ft². How would you change the empirical equation for Δp to obtain another empirical equation suitable for use with SI units? Note that your objective here is to replace the coefficient 4.7 with a new coefficient. Begin by putting the equation in dimensionally correct form, i.e., find the units associated with the coefficient 4.7, and then set up the empiricism so that it can be used with SI units.

2.16. The ideal gas heat capacity can be expressed as a power series in terms of temperature according to

$$C_p = A_1 + A_2T + A_3T^2 + A_4T^3 + A_5T^4$$

The units of C_p are joule/(mol K), and the units of temperature are degrees Kelvin. For chlorine, the values of the coefficients are: $A_1 = 22.85$, $A_2 = 0.06543$, $A_3 = -1.2517 \times 10^{-4}$, $A_4 = 1.1484 \times 10^{-7}$, and $A_5 = -4.0946 \times 10^{-11}$. What are the units of the coefficients? Find the values of the coefficients to compute the heat capacity of chlorine in cal/g C, using temperature in degrees Rankine.

Section 2.4

2.17. A *standard cubic foot*, or scf, of gas represents one cubic foot of gas at one atmosphere and 273.16 K. This means that a standard cubic foot is a *convenience unit* for moles. This is easy to see in terms of an ideal gas for which the equation of state is given by (see Sec. 5.1)

$$pV = nRT$$

The number of moles in *one standard cubic foot* of an ideal gas can be calculated as

$$n = pV/RT \begin{cases} p = \text{one atmosphere} \\ V = \text{one cubic foot} \\ T = 273.16 \text{ K} \end{cases}$$

and for a *non-ideal gas* one must use an appropriate equation of state[‡‡]. In this problem, you are asked to determine the number of moles that are equivalent to one scf of an ideal gas (see Sec. 5.1).

2.18. Energy is sometimes expressed as $v^2/2g$ although this term does not have the units of energy. What are the units of this term and why would it be used to represent energy? Think about the fact that ρgh represents the gravitational potential energy per unit volume of a fluid and that h is often used as a *convenience unit* for gravitational potential energy. Remember that $\frac{1}{2}\rho v^2$ represents the kinetic energy per unit volume where v is determined by $v^2 = \mathbf{v} \cdot \mathbf{v}$ and consider the case for which the fluid density is a constant.

[‡‡] Sandler, S.I. 2006, *Chemical, Biochemical, and Engineering Thermodynamics*, 4th edition, John Wiley and Sons, New York.

3 Conservation of Mass for Single Component Systems

In Chapter 1, we pointed out that much of chemical engineering is concerned with keeping track of *molecular species* during processes in which chemical reactions and mass transport take place. However, before attacking the type of problems described in Chapter 1, we wish to consider the special case of single component systems. The study of single component systems will provide an opportunity to focus attention on the concept of *control volumes* without the complexity associated with multi-component systems. We will examine the *accumulation* of mass and the *flux* of mass under relatively simple circumstances, and this provides the foundation necessary for our subsequent studies.

There is more than one way in which the principle of conservation of mass for single component systems can be stated. Provided that a body is not moving at a velocity close to the speed of light[*], one attractive form is:

the mass of a
body is constant

however, we often express this idea in the *rate form* leading to an equation given by

$$\left\{ \begin{array}{c} \textit{time rate of} \\ \textit{change of the} \\ \textit{mass of a body} \end{array} \right\} = 0 \tag{3.1}$$

The principle of conservation of mass is also known as the *axiom* for the conservation of mass. In physics, one uses the word axiom to describe an accepted principle that cannot be derived from a more general principle. Axioms are based on specific experimental observations, and from those *specific observations* we construct the *general statement* given by Eq. 3.1.

As an example of the application of Eq. 3.1, we consider the motion of the cannon ball illustrated in Figure 3.1. Newton's second law requires that the force acting on the cannon ball be equal to the time rate of change of the linear momentum of the solid body as indicated by

$$\mathbf{f} = \frac{d}{dt}(m\mathbf{v}) \tag{3.2}$$

---

[*] Hurley, J.P. and Garrod, C. 1978, *Principles of Physics*, Houghton Mifflin Co., Boston.

DOI: 10.1201/9781003283751-3

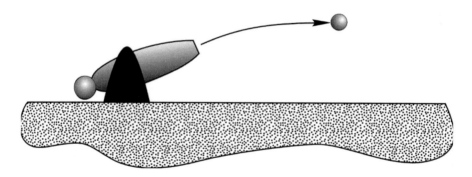

FIGURE 3.1   Cannon ball.

We now apply Eq. 3.1 in the form

$$\frac{dm}{dt} = 0 \tag{3.3}$$

to find that the force is equal to the mass times the acceleration.

$$\mathbf{f} = m\frac{d\mathbf{v}}{dt} = m\,\mathbf{a} \tag{3.4}$$

Everyone is familiar with this result from previous courses in physics and perhaps a course in engineering mechanics.

3.1   CLOSED AND OPEN SYSTEMS

While Eq. 3.1 represents an attractive statement for conservation of mass when we are dealing with *distinct bodies* such as the cannon ball illustrated in Figure 3.1, it is not particularly useful when we are dealing with a *continuum* such as the water jet shown in Figure 3.2. In considering Eq. 3.1 and the water jet, we are naturally led to ask the question, "Where is the body?" Here we can identify a body, illustrated in Figure 3.3, in terms of the famous *Euler cut principle* that we state as:

> Not only do the laws of continuum physics apply to *distinct bodies*, but they also apply to any *arbitrary body* that one might imagine as being *cut out* of a distinct body.

The idea that the laws of physics, laboriously deduced by the observation of distinct bodies, can also be applied to bodies *imagined* as being *cut out of* distinct bodies is central to the engineering analysis of continuous systems. The validity of the Euler cut principle for bodies rests on the fact that the governing equations developed on the basis of this principle are consistent with experimental observation. Because of this, we can apply this principle to the liquid jet and *imagine* the moving, deforming fluid body illustrated in Figure 3.3. There we have shown a fluid body at $t = 0$ that moves and deforms to a new configuration, $V_m(t)$, at a later time, t. The nomenclature here deserves some attention, and we begin by noting that we have used the

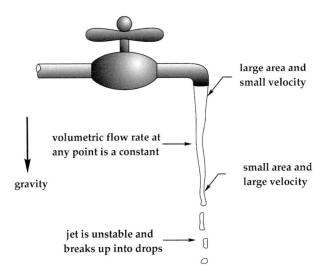

large area and
small velocity

volumetric flow rate at
any point is a constant

small area and
large velocity

gravity

jet is unstable and
breaks up into drops

FIGURE 3.2 Water jet.

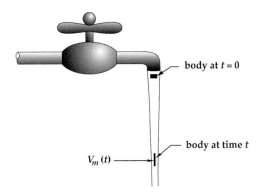

body at $t = 0$

body at time t

$V_m(t)$

FIGURE 3.3 Moving deforming body.

symbol V to represent a *volume*. This particular volume always contains the same *material*, thus we have added the subscript m. In addition, the position of the volume changes with *time*, and we have indicated this by adding (t) as a descriptor. It is important to understand that the fluid body, which we have constructed on the basis of the Euler cut principle, always contains the same fluid since no fluid crosses the surface of the body.

In order to apply Eq. 3.1 to the moving, deforming fluid body shown in Figure 3.3, we first illustrate the fluid body in more detail in Figure 3.4. There we have shown the volume occupied by the body, $V_m(t)$, along with a differential volume, dV. The mass, dm, contained in this differential volume is given by

$$dm = \rho\, dV \qquad (3.5)$$

in which ρ is the density of the water leaving the faucet. The total mass contained in $V_m(t)$ is determined by summing over all the differential elements that make up the body to obtain

$$\left\{\begin{array}{c} \text{mass of} \\ \text{the body} \end{array}\right\} = \int\limits_{V_m(t)} \rho\, dV \tag{3.6}$$

Use of this representation for the mass of a body in Eq. 3.1 leads to

Axiom: $$\frac{d}{dt} \int\limits_{V_m(t)} \rho\, dV = 0 \tag{3.7}$$

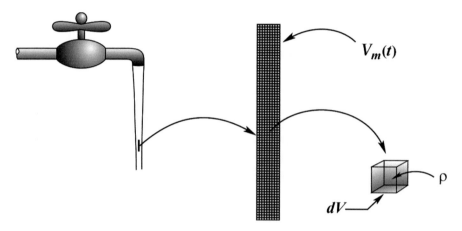

FIGURE 3.4   Integration over $V_m(t)$ to obtain the mass of a body.

It should be clear that Eq. 3.7 provides *no information* concerning the velocity and diameter of the jet of water leaving the faucet. For such systems, the knowledge that the mass of a fluid body is constant is not very useful, and instead of Eq. 3.7 we would like an axiom that tells us something about the *mass contained within some specified region in space*. We base this more general axiomatic statement on an extension of the Euler cut principle that we express as:

> Not only do the laws of continuum physics apply to *distinct bodies*, but they also apply to any *arbitrary region* that one might imagine as being *cut out* of Euclidean 3-space.

We refer to this arbitrary region in space as a *control volume*, and a *more general alternative* to Eq. 3.1 is given by

Axiom: $\left\{\begin{array}{c} \text{time rate of change} \\ \text{of the mass contained} \\ \text{in } any \text{ control volume} \end{array}\right\} = \left\{\begin{array}{c} \text{rate at which} \\ \text{mass } enters \text{ the} \\ \text{control volume} \end{array}\right\} - \left\{\begin{array}{c} \text{rate at which} \\ \text{mass } leaves \text{ the} \\ \text{control volume} \end{array}\right\}$ (3.8)

To illustrate how this more general axiom for the mass of single component systems is related to Eq. 3.1, we consider the *control volume* to be the space occupied by the *fluid body* illustrated in Figure 3.4. This control volume moves with the body, thus no mass enters or leaves the control volume and the two terms on the right-hand side of Eq. 3.8 are zero as indicated by

$$\left\{ \begin{array}{c} \text{rate at which} \\ \text{mass } enters \text{ the} \\ \text{control volume} \end{array} \right\} = \left\{ \begin{array}{c} \text{rate at which} \\ \text{mass } leaves \text{ the} \\ \text{control volume} \end{array} \right\} = 0 \qquad (3.9)$$

Under these circumstances, the axiomatic statement given by Eq. 3.8 takes the *special form*

$$\left\{ \begin{array}{c} \text{time rate of change} \\ \text{of the mass contained} \\ \text{in the control volume} \end{array} \right\} = \frac{d}{dt} \int_{V_m(t)} \rho \, dV = 0 \qquad (3.10)$$

This indicates that Eq. 3.8 *contains* Eq. 3.7 as a special case. Another special case of Eq. 3.8 that is especially useful is the *fixed control volume* illustrated in Figure 3.5. There we have identified the control volume by V to clearly indicate that it represents a

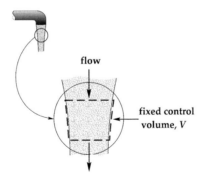

flow

fixed control volume, V

FIGURE 3.5   Control volume fixed in space.

volume fixed in space. This fixed control volume can be used to provide useful information about the velocity of the fluid in the jet and the cross-sectional area of the jet. As an application of the control volume formulation of the principle of conservation of mass, we consider the production of a polymer fiber in the following example:

EXAMPLE 3.1 OPTICAL FIBER PRODUCTION

The use of fiber optics is essential to high-speed Internet communication, thus the production of optical fibers is extremely important. As an example of the application of Eq. 3.8, we consider the production of an optical fiber from molten glass. In Figure 3.1a, we have illustrated a stream of glass extruded through a small tube, or *spinneret*.

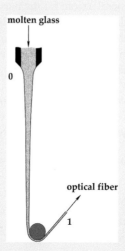

molten glass

0

optical fiber

1

FIGURE 3.1a Optical fiber production.

The surrounding air is at a temperature below the solidification temperature of the glass which is a solid when the fiber reaches the take-up wheel or *capstan*. A key quantity of interest in the fiber spinning operation is the *draw ratio*[†] which is the ratio of the area of the of the spinneret hole (indicated by A_o) to the area of the optical fiber leaving the take-up wheel (indicated by A_1). Application of the macroscopic mass balance given by Eq. 3.8 will show us how the draw ratio is related to the parameters of the spinning operation.

We begin our analysis by assuming that the process operates at *steady-state* so that Eq. 3.8 reduces to

$$\left\{ \begin{array}{c} \text{rate at which} \\ \text{mass } \textit{enters} \text{ the} \\ \text{control volume} \end{array} \right\} = \left\{ \begin{array}{c} \text{rate at which} \\ \text{mass } \textit{leaves} \text{ the} \\ \text{control volume} \end{array} \right\} \qquad (1)$$

In order to apply this result, we need to carefully specify the control volume, and this requires some judgment concerning the particular process under consideration. In this case, it seems clear that the surface of the control volume should *cut* the glass at both the entrance and exit, and that these two cuts should be joined by a surface that is coincident with the interface between the glass and the surrounding air. This leads to the control volume that is illustrated in Figure 3.1b where we have shown portions of the control surface at the *entrance*, at the *exit*, and at the glass-air *interface* where we neglect any mass transfer that may occur by *diffusion*. Because of this, we are concerned only with the rate at which mass enters and leaves the control volume at the entrance and exit.

[†] Denn, M.M. 1980, Continuous drawing of liquids to form fibers, Ann. Rev. Fluid. Mech. 12, 365–387.

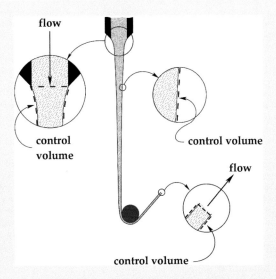

FIGURE 3.1b   Control volume for the fiber optic spinning process.

The rate at which mass *enters* the control volume is given by

$$\left\{ \begin{array}{c} \text{rate at which} \\ \text{mass } \textit{enters} \text{ the} \\ \text{control volume} \end{array} \right\} = \left\{ \begin{array}{c} \text{density of} \\ \text{the glass} \end{array} \right\} \left\{ \begin{array}{c} \text{volumetric flow} \\ \text{rate of the glass} \end{array} \right\} = \rho_o Q_o \quad (2)$$

Here, we have used Q_o to represent the volumetric flow rate of the glass at the entrance, and ρ_o to represent the density of the glass at the entrance. Thus, the units associated with Eq. 2 are represented by

$$\left\{ \begin{array}{c} \text{units of density} \\ \text{times volumetric} \\ \text{flow rate} \end{array} \right\} = \left\{ \frac{kg}{m^3} \right\} \left\{ \frac{m^3}{s} \right\} = \frac{kg}{s} \quad (3)$$

and $\rho_o Q_o$ has units of mass per unit time. At the exit of the control volume, we designate the volumetric flow rate by Q_1 and the density of the solidified glass as ρ_1. This leads to an expression for the rate at which mass leaves the control volume given by

$$\left\{ \begin{array}{c} \text{rate at which} \\ \text{mass leaves the} \\ \text{control volume} \end{array} \right\} = \left\{ \begin{array}{c} \text{density of} \\ \text{the glass} \end{array} \right\} \left\{ \begin{array}{c} \text{volumetric flow} \\ \text{rate of the glass} \end{array} \right\} = \rho_1 Q_1 \quad (4)$$

Use of Eqs. 3 and 4 in the macroscopic mass balance given by Eq. 1 leads to

$$\rho_o Q_o = \rho_1 Q_1 \qquad (5)$$

This result tells us how the volumetric flow rates at the entrance and exit are related to the densities at the entrance and exit; however, it does not tell us what we want to know, i.e., the *draw ratio*, A_o/A_1. In order to extract this ratio from Eq. 5, we need to express the volumetric flow rate in terms of the average velocity and the cross-sectional area. At the entrance, this expression is given by

$$\left\{ \begin{array}{c} \text{volumetric flow} \\ \text{rate of the glass} \end{array} \right\} = \left\{ \begin{array}{c} \text{average} \\ \text{velocity} \end{array} \right\} \left\{ \begin{array}{c} \text{cross} \\ \text{sectional} \\ \text{area} \end{array} \right\} = \langle v \rangle_o A_o \qquad (6)$$

Here, we have used $\langle v \rangle_o$ to represent the average velocity of the glass at the entrance, and A_o to represent the cross-sectional area, thus the units associated with Eq. 6 are given by

$$\left\{ \begin{array}{c} \text{units of velocity} \\ \text{times cross} \\ \text{sectional area} \end{array} \right\} = \left\{ \frac{m}{s} \right\} \left\{ m^2 \right\} = \frac{m^3}{s} \qquad (7)$$

Use of an analogous representation for the volumetric flow rate at the exit allows us to express Eq. 5 in the form

$$\rho_o \langle v \rangle_o A_o = \rho_1 \langle v \rangle_1 A_1 \qquad (8)$$

From this form of the macroscopic mass balance, we find the draw ratio to be given by

$$\left\{ \begin{array}{c} \text{draw} \\ \text{ratio} \end{array} \right\} = \frac{A_o}{A_1} = \frac{\rho_1 \left. \langle v \rangle_1 \right|_{\substack{\text{take-up} \\ \text{wheel}}}}{\rho_o \langle v \rangle_o} \qquad (9)$$

Here, we have clearly indicated that the average velocity at the exit of the control volume is specified by the speed of the take-up wheel. Normally, this result would be arranged as

$$\left\{ \begin{array}{c} \text{draw} \\ \text{ratio} \end{array} \right\} = \frac{A_o}{A_1} = \left(\frac{\rho_1 A_o}{\rho_o Q_o} \right) \left. \langle v \rangle_1 \right|_{\substack{\text{take-up} \\ \text{wheel}}} \qquad (10)$$

with the thought that the velocity at the take-up wheel is the most convenient parameter used to control the draw ratio.

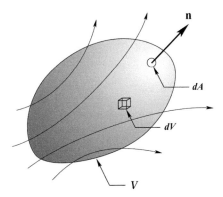

FIGURE 3.6   Control volume fixed in space.

In the previous example, we have shown how Eq. 3.8 can be used to determine the velocity at the take-up wheel in order to produce a glass fiber of a specified cross-sectional area. In order to prepare for more complex problems, we need to translate the *word statement* given by Eq. 3.8 into a precise *mathematical statement*. We begin by considering the *fixed control volume* illustrated in Figure 3.6. This volume is identified by V, the surface area of the volume by A, and the outwardly directly unit normal vector by **n**. The surface A may contain *entrances* and *exits* where fluid flows into and out of the control volume, and it may contain *interfacial areas* where mass transfer may or may not take place.

We begin our analysis of Eq. 3.8 for the control volume illustrated in Figure 3.6 by considering the differential volume, dV, and denoting the mass contained in this differential volume by dm. In terms of the mass density ρ, we have

$$dm = \rho \, dV \tag{3.11}$$

and the mass contained in the control volume can be represented as

$$\left\{ \begin{array}{c} \text{mass contained} \\ \text{in the control} \\ \text{volume} \end{array} \right\} = \int_V \rho \, dV \tag{3.12}$$

This allows us to express the first term in Eq. 3.8 as

$$\left\{ \begin{array}{c} \text{time rate of change} \\ \text{of the mass contained} \\ \text{in the control volume} \end{array} \right\} = \frac{d}{dt} \int_V \rho \, dV \tag{3.13}$$

The determination of the rate at which *mass leaves the control volume* illustrated in Figure 3.6 requires the use of the projected area theorem[‡] and before examining the *general case* we consider the *special case* illustrated in Figure 3.7. In Figure 3.7,

---

[‡] Stein, S.K. and Barcellos, A. 1992, Sec. 17.1, *Calculus and Analytic Geometry*, McGraw-Hill, Inc., New York.

we have shown a control volume, V, that can be used to analyze the flow rate at the entrance and exit of a tube. In that *special case*, the velocity vector, $\mathbf{v}$, is *parallel* to the unit normal vector, $\mathbf{n}$, at the exit. Thus, the volume of the fluid, ΔV, leaving a differential area dA in a time Δt is given by

$$\Delta V = v\,\Delta t\,dA = \mathbf{v}\cdot\mathbf{n}\,\Delta t\,dA, \qquad \text{flow orthogonal to an exit} \qquad (3.14)$$

From this, we determine that the volume of fluid that that flows across the surface dA per unit time is given by

$$\frac{\Delta V}{\Delta t} = \mathbf{v}\cdot\mathbf{n}\,dA, \qquad \left\{\begin{array}{c}\text{volumetric flow rate orthogonal}\\ \text{to an area } dA\end{array}\right\} \qquad (3.15)$$

Following the same development given by Eq. 4 in Example 3.1, we express the mass flow rate as

$$\frac{\rho\,\Delta V}{\Delta t} = \frac{\Delta m}{\Delta t} = \rho\mathbf{v}\cdot\mathbf{n}\,dA, \qquad \left\{\begin{array}{c}\text{mass flow rate orthogonal}\\ \text{to an area } dA\end{array}\right\} \qquad (3.16)$$

In order to determine the total rate at which mass leaves the exit of the faucet shown in Figure 3.7, we simply integrate this expression for the mass flow rate over the area A_{exit} to obtain

$$\left\{\begin{array}{c}\text{rate at which mass}\\ \text{flows out of the faucet}\end{array}\right\} = \int_{A_{exit}} \rho\mathbf{v}\cdot\mathbf{n}\,dA \qquad (3.17)$$

When the velocity vector and the unit normal vector are *parallel*, the flow field has the form illustrated in Figure 3.7 and the mass flow rate is easily determined. When these two vectors are *not parallel*, we need to examine the flow more carefully and this is done in the following paragraphs.

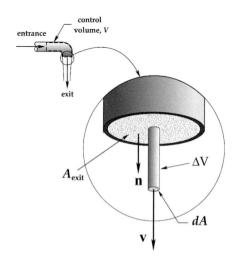

FIGURE 3.7   Mass flow at the exit of a faucet.

3.1.1 GENERAL FLUX RELATION

In order to determine the rate at which mass leaves a control volume when $\mathbf{v}$ and $\mathbf{n}$ are *not parallel*, we return to the differential surface area element illustrated in Figure 3.6. A more detailed version is shown in Figure 3.8 where we have included the unit vector that is *normal to the surface*, $\mathbf{n}$, and the unit vector that is *tangent to the velocity vector*, λ. The unit tangent vector, λ, is defined by

$$\mathbf{v} = v\lambda \qquad (3.18)$$

in which $\mathbf{v}$ represents the fluid velocity vector and v is the magnitude of that vector. In Figure 3.8, we have "marked" the fluid at the surface area element, and in order to determine the rate at which mass leaves the control volume through the area dA, we need to determine the *volume of fluid* that crosses the surface area dA per unit time. In Figure 3.9, we have illustrated this volume which is bounded by the vectors $v\Delta t$ that are parallel to the unit vector λ. One should imagine an observer who is *fixed relative to the surface* and who can determine the velocity $\mathbf{v}$ as the fluid crosses the surface A. The magnitude of $v\Delta t$ is given by $v\Delta t$ and this is the length of the cylinder that is swept out in a time Δt. The volume of the cylinder shown in Figure 3.9 is equal to the length, $v\Delta t$, times the cross-sectional area, dA_{cs}, and that concept is illustrated in Figure 3.10. We express the volume that is *swept out* of the control volume in a time Δt as

$$\Delta V = (v\Delta t)\,dA_{cs} \qquad (3.19)$$

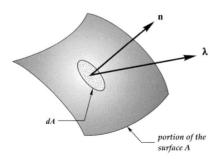

FIGURE 3.8 Surface area element.

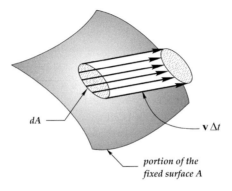

FIGURE 3.9 Volume of fluid crossing the surface dA in a time Δt.

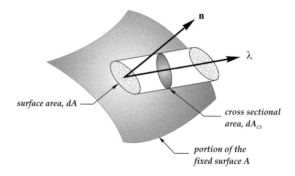

FIGURE 3.10   Volume leaving at the control surface during a time Δt.

In order to relate this volume to the surface area of the control volume, we need to make use of the projected area theorem[§].

This theorem allows us to express the cross-sectional area at an *exit* according to

$$dA_{cs} = \lambda \cdot \mathbf{n}\, dA\,, \quad \text{projected area theorem} \tag{3.20}$$

and from this we see that the differential volume takes the form

$$\Delta V = v\, \Delta t\, \lambda \cdot \mathbf{n}\, dA \tag{3.21}$$

Since the fluid velocity vector is given by,

$$\mathbf{v} = v\lambda \tag{3.22}$$

we see that the volume of fluid that crosses the surface area element *per unit time* is given by

$$\frac{\Delta V}{\Delta t} = \mathbf{v} \cdot \mathbf{n}\, dA\,, \quad \left\{ \begin{array}{c} \text{volumetric flow rate across} \\ \text{an } \textit{arbitrary} \text{ area } dA \end{array} \right\} \tag{3.23}$$

This expression is identical to that given earlier by Eq. 3.15; however, in this case, we have used the projected area theorem to demonstrate that this is a generally valid result. From Eq. 3.23, we determine that the rate at which mass crosses the differential surface area is

$$\frac{\rho\, \Delta V}{\Delta t} = \frac{\Delta m}{\Delta t} = \rho\, \mathbf{v} \cdot \mathbf{n}\, dA\,, \quad \left\{ \begin{array}{c} \text{mass flow rate across} \\ \text{an } \textit{arbitrary} \text{ area } dA \end{array} \right\} \tag{3.24}$$

This representation for $\Delta m / \Delta t$ is identical *in form* to that given for the *special case* illustrated in Figure 3.7 where the velocity vector and the unit normal vectors were parallel. The result given by Eq. 3.24 is *entirely general* and it indicates that $\rho\mathbf{v} \cdot \mathbf{n}$ represents the *mass flux* (mass per unit time per unit area) at any exit.

[§] Whitaker, S. 1968, Sec. 2.6, *Introduction to Fluid Mechanics*, Prentice Hall, Inc., Englewood Cliffs, N.J.

We are now ready to return to Eq. 3.8 and express the rate at which mass *leaves* the control volume according to

$$\left\{ \begin{array}{c} \text{rate at which mass} \\ \text{\textit{flows out of} the} \\ \text{control volume} \end{array} \right\} = \int_{A_{exit}} \rho \mathbf{v} \cdot \mathbf{n} \, dA \qquad (3.25)$$

It should be clear that over the *exits* we have the condition $\mathbf{v} \cdot \mathbf{n} > 0$ since $\mathbf{n}$ is always taken to be the outwardly directed unit normal. At the *entrances*, the velocity vector and the normal vector are related by $\mathbf{v} \cdot \mathbf{n} < 0$, and this requires that the rate at which mass enters the control volume be expressed as

$$\left\{ \begin{array}{c} \text{rate at which mass} \\ \text{\textit{flows into} the} \\ \text{control volume} \end{array} \right\} = - \int_{A_{entrance}} \rho \mathbf{v} \cdot \mathbf{n} \, dA \qquad (3.26)$$

Use of Eqs. 3.13, 3.25, and 3.26 in Eq. 3.8 yields a precise, mathematical statement of the principle of conservation of mass for a control volume that contains *only exits and entrances* as they are described by Eqs. 3.25 and 3.26. This precise mathematical statement takes the form

$$\frac{d}{dt} \int_V \rho \, dV = - \int_{A_{entrances}} \rho \mathbf{v} \cdot \mathbf{n} \, dA - \int_{A_{exits}} \rho \mathbf{v} \cdot \mathbf{n} \, dA \qquad (3.27)$$

In general, any control volume will contain entrances, exits, and *interfacial areas* where mass transfer may or may not occur. For example, the fixed control volume illustrated in Figure 3.11 contains an *entrance*, an *exit*, and an *interfacial area* which

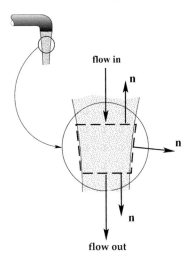

FIGURE 3.11   Entrances, exits and interfacial areas.

is the air-water interface. If we assume that there is negligible mass transfer at the air-water interface, Eq. 3.27 is applicable to the fixed control volume illustrated in Figure 3.11 where no mass transfer occurs at the air-water interface. This is precisely the type of simplification that was made in Example 3.1 where we neglected any mass transfer at the glass-air interface. For the more general case where mass transfer can take place at interfacial areas, we need to express Eq. 3.27 in terms of the fluxes at the entrances, exits, and *interfacial areas* according to

$$\frac{d}{dt} \int_V \rho \, dV = - \int_{A_{entrances}} \rho \mathbf{v} \cdot \mathbf{n} \, dA - \int_{A_{exits}} \rho \mathbf{v} \cdot \mathbf{n} \, dA - \int_{A_{interface}} \rho \mathbf{v} \cdot \mathbf{n} \, dA \qquad (3.28)$$

In our description of the control volume, V, illustrated in Figure 3.6, we identified the surface area of the volume as A and this leads us to express Eq. 3.28 in the compact form given by

$$\frac{d}{dt} \int_V \rho \, dV + \int_A \rho \mathbf{v} \cdot \mathbf{n} \, dA = 0 \qquad (3.29)$$

We should think of this expression as a *precise statement* of the principle of conservation of mass for a fixed control volume.

3.1.2 CONSTRUCTION OF CONTROL VOLUMES

The macroscopic mass balance indicated by Eq. 3.29 represents a law of physics that is valid for all non-relativistic processes. It is a powerful tool and its implementation requires that one pay careful attention to the construction of the control volume, V. There are four important rules that should be followed during the construction of control volumes, and we list these rules as:

Rule I. Construct a cut (a portion of the surface area A) where information is *given*.
Rule II. Construct a cut where information is *required*.
Rule III. Join these cuts with a surface located where $\mathbf{v} \cdot \mathbf{n}$ is *known*.
Rule IV. Be sure that the surface specified by Rule III encloses regions in which volumetric information is either *given* or *required*.

In addition to macroscopic mass balance analysis, these rules also apply to macroscopic momentum and energy balance analysis that students will encounter in subsequent courses.

3.2 MASS FLOW RATES AT ENTRANCES AND EXITS

There are many problems for which the control volume, V, contains only a *single entrance* and a *single exit* and the flux at the interfacial area is zero. This is the type of problem considered in Example 3.1 in terms of the word statement given by Eq. 3.8, and we need to be certain that no confusion exists concerning Eq. 3.8 and the rigorous form given by Eq. 3.29. For systems that contain only a single entrance and a single exit, it is convenient to express Eq. 3.29 as

$$\frac{d}{dt} \int_V \rho \, dV = \int_{A_{entrance}} \rho |\mathbf{v} \cdot \mathbf{n}| \, dA - \int_{A_{exit}} \rho |\mathbf{v} \cdot \mathbf{n}| \, dA \qquad (3.30)$$

in which we have made use of the relations

$$\mathbf{v} \cdot \mathbf{n} = -|\mathbf{v} \cdot \mathbf{n}|, \quad \text{at the entrance}, \qquad \mathbf{v} \cdot \mathbf{n} = |\mathbf{v} \cdot \mathbf{n}|, \quad \text{at the exit} \qquad (3.31)$$

The form of the macroscopic mass balance given by Eq. 3.30 is analogous to the word statement given by Eq. 3.8 which we repeat here as

$$\left\{ \begin{array}{l} \text{time rate of change} \\ \text{of the mass contained} \\ \text{in } any \text{ control volume} \end{array} \right\} = \left\{ \begin{array}{l} \text{rate at which} \\ \text{mass } enters \text{ the} \\ \text{control volume} \end{array} \right\} - \left\{ \begin{array}{l} \text{rate at which} \\ \text{mass } leaves \text{ the} \\ \text{control volume} \end{array} \right\} \qquad (3.32)$$

There are several convenient representations for the terms that appear in Eqs. 3.30 and 3.32, and we present these forms in the following paragraph.

3.2.1 CONVENIENT FORMS

The mass in a control volume can be expressed in terms of the volume V and the *volume average density*, $\langle \rho \rangle$, which is defined by

$$\langle \rho \rangle = \frac{1}{V} \int_V \rho \, dV \qquad (3.33)$$

In terms of $\langle \rho \rangle$, the accumulation in a *fixed control volume* takes the form

$$\frac{d}{dt} \int_V \rho \, dV = V \frac{d \langle \rho \rangle}{dt} \qquad (3.34)$$

To develop a convenient form for the mass flux, we make use of Eq. 3.23 to express the volumetric flow rate at an exit as

$$\int_{A_{exit}} |\mathbf{v} \cdot \mathbf{n}| \, dA = Q_{exit} \qquad (3.35)$$

Given this result, we can define an average density at an exit, $\langle \rho \rangle_{b,exit}$, according to

$$\langle \rho \rangle_{b,exit} = \frac{1}{Q_{exit}} \int_{A_{exit}} \rho |\mathbf{v} \cdot \mathbf{n}| \, dA = \frac{\displaystyle\int_{A_{exit}} \rho |\mathbf{v} \cdot \mathbf{n}| \, dA}{\displaystyle\int_{A_{exit}} |\mathbf{v} \cdot \mathbf{n}| \, dA} \qquad (3.36)$$

This average density is sometimes referred to as the "bulk density", and that is the origin of the subscript b in this definition. In addition, $\langle\rho\rangle_{b,exit}$ is sometimes referred to as the "cup mixed density" since it is the density that one would measure by collecting a cup of fluid at the outlet of a tube and dividing the mass of fluid by the volume of fluid. Eqs. 3.35 and 3.36 also apply to an entrance, thus we can use these results to express Eq. 3.30 in the form

$$V\frac{d\langle\rho\rangle}{dt} = \langle\rho\rangle_{b,entrance}\,Q_{entrance} - \langle\rho\rangle_{b,exit}\,Q_{exit} \tag{3.37}$$

If the process is steady, the volume average density will be independent of time and this result simplifies to

$$\langle\rho\rangle_{b,entrance}\,Q_{entrance} = \langle\rho\rangle_{b,exit}\,Q_{exit} \tag{3.38}$$

If the density is constant over the surfaces of the entrance and exit, the average values in this result can be replaced with the constant values and one recovers the mass balance given by Eq. 5 in Example 3.1.

In addition to expressing the mass flux in terms of a density and a volumetric flow rate, there are many problems in which it is convenient to work directly with the mass flow rate. We designate the mass flow rate by $\dot{m}$ and represent the terms on the right-hand side of Eq. 3.30 according to

$$\int_{A_{entrance}} \rho|\mathbf{v}\cdot\mathbf{n}|\,dA = \dot{m}_{entrance}, \qquad \int_{A_{exit}} \rho|\mathbf{v}\cdot\mathbf{n}|\,dA = \dot{m}_{exit} \tag{3.39}$$

This leads to the following form of the macroscopic mass balance for one entrance and one exit

$$V\frac{d\langle\rho\rangle}{dt} = \dot{m}_{entrance} - \dot{m}_{exit} \tag{3.40}$$

When the process is steady, this form simplifies to

$$\dot{m}_{entrance} = \dot{m}_{exit} \tag{3.41}$$

When systems have more than one entrance and more than one exit, the mass flow rates can be expressed in precisely the manner that we have indicated by Eqs. 3.37–3.39; however, one must be careful to include the flows at *all entrances and all exits*.

EXAMPLE 3.2 POLYMER COATING

As an example of the application of Eq. 3.29, we consider the process of coating a polymer optical fiber with a polymer film. The process is similar to that discussed in Example 3.1; however, in this case, we wish to determine the coating thickness rather than the fiber diameter. To determine the thickness of the polymer film, we make use of the control volume illustrated in Figure 3.2a. In constructing this control volume, we have followed the rules given earlier which are listed here along with the commentary that is appropriate for this particular problem.

Rule I. Construct a cut (a portion of the surface area A) where information is *given*.

In this case, we have constructed a cut at a position downstream from the take-
up wheel where the thickness of the polymer coating will normally be *specified*
and the velocity of the polymer film will be *equal* to the velocity of the optical
fiber. Thus, we have created a cut where information is normally *given*.

Rule II. Construct a cut where information is *required*.

In this case, we have created a *cut* at the entrance where the coating polymer
comes in contact with the optical polymer since we need to know the volumetric flow rate of the polymer that will produce the desired thickness of the
coating. Thus, we have created a cut where information is *required*.

Rule III. Join these cuts with a surface located where $\mathbf{v} \cdot \mathbf{n}$ is *known*.

In this case, we have joined the two cuts along the polymer-polymer interface
on the basis of the assumption that $\mathbf{v} \cdot \mathbf{n} = 0$ at this interface. Thus, we have
joined these cuts with a surface located where $\mathbf{v} \cdot \mathbf{n}$ is *known*.

Rule IV. Be sure that the surface specified by Rule III encloses regions
in which volumetric information is either *given* or *required*.

For this particular problem, there is no volumetric information either given or
required, and this rule is automatically satisfied.

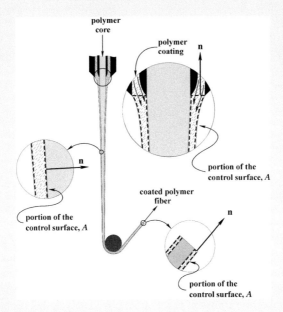

FIGURE 3.2a   Polymer coating process.

We begin our analysis of the coating operation by assuming that the process operates at steady-state so that Eq. 3.29 reduces to

$$\int_A \rho \mathbf{v} \cdot \mathbf{n} \, dA = 0, \qquad \textit{steady state} \tag{1}$$

When no mass transfer occurs at the portion of A located at the glass-polymer interface, we need to only evaluate the area integral at the entrance and exit and Eq. 1 reduces to

$$\int_{A_{entrance}} \rho \mathbf{v} \cdot \mathbf{n} \, dA + \int_{A_{exit}} \rho \mathbf{v} \cdot \mathbf{n} \, dA = 0 \tag{2}$$

At this point, we can follow Example 3.1 to obtain

$$\left(\rho_o Q_o \right)_{\text{polymer}} = \left(\rho_1 Q_1 \right)_{\text{polymer}} \tag{3}$$

At the exit, the velocity of the polymer coating is determined by the take-up wheel and is identical to the velocity of the optical fiber. This leads to

$$\left(Q_1 \right)_{\text{polymer}} = \left(A_1 \right)_{\text{polymer}} \langle v \rangle_1 \Big|_{\substack{\text{take-up} \\ \text{wheel}}} \tag{4}$$

in which $(A_1)_{\text{polymer}}$ represents the area of the *annular region* occupied by the coating polymer. Use of Eq. 4 in Eq. 3 allows one to determine the cross-sectional area of the polymer coating to be

$$\left(A_1 \right)_{\text{polymer}} = \frac{\left(\rho_o Q_o \right)_{\text{polymer}}}{\left(\rho_1 \right)_{\text{polymer}} \langle v \rangle_1 \Big|_{\substack{\text{take-up} \\ \text{wheel}}}} \tag{5}$$

This expression can be used to determine the *coating thickness* in terms of the operating variables, and this will be left as an exercise for the student.

In both Examples 3.1 and 3.2, we illustrated applications of the macroscopic mass balance in order to determine the characteristics of a steady fiber spinning and coating process. Many practical problems are transient in nature and can be analyzed using the complete form of Eq. 3.29. The consumption of propane gas in rural areas represents an important transient problem, and we analyze this problem in the next example.

EXAMPLE 3.3   DELIVERY SCHEDULE FOR PROPANE

In rural areas where natural gas is not available by pipeline, compressed propane gas is used for heating purposes. During the winter months, one must
pay attention to the pressure gauge on the tank in order to avoid running out
of fuel. Deliveries, for which there is a charge, are made periodically and
must be scheduled in advance. The system under consideration is illustrated
in Figure 3.3a. The volume of the tank is 250 gallons, and when full it contains 750 kg of propane at a pressure of 95 psi (gauge). If the gas is consumed
at a rate of 400 scf/day, we need to know whether a monthly or a bi-monthly
delivery is required. Here the abbreviation, scf, represents *standard cubic feet*,
and the word *standard* means that the pressure is equivalent to 760 mm of
mercury and the temperature is 273.16 K. A standard cubic foot represents a
convenience unit (see Sec. 2.4) for the number of moles and this is easily dem-
onstrated for the case of an ideal gas. Given the pressure ($p =$ one atmosphere),
the temperature ($T = 273.16\,\mathrm{K}$), and the volume ($V =$ one cubic foot), one can
use the ideal gas law to determine the number of moles according to

$$n = pV/RT$$

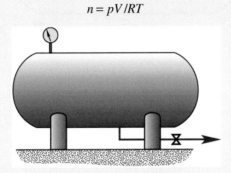

FIGURE 3.3a   Tank of compressed propane.

If the gas under consideration is not an ideal gas, one must use the appropriate
equation of state[**] to determine what is meant by a standard cubic foot.

The propane inside the tank illustrated in Figure 3.3a is an equilibrium mixture
of liquid and vapor, and we have illustrated this situation in Figure 3.3b. There we
have shown a control volume that has been constructed on the following basis:

  I. A cut has been made at the exit of the tank where information is *given*.
 II. There is no *required* information that would generate another cut.
III. The surface of the control volume is located where $\mathbf{v} \cdot \mathbf{n}$ is *known*.
IV. The surface specified by III encloses the region for which volumetric
     information is *required*.

[**] Sandler, S.I. 2006, *Chemical, Biochemical, and Engineering Thermodynamics*, 4th edition,
John Wiley and Sons, New York

The proportions of liquid and vapor inside the tank are not known; however, we do know the initial mass of propane in the tank and we are given (indirectly) the average mass flow rate leaving the tank. Our analysis of the transient behavior of this system is based on the macroscopic mass balance given by

$$\frac{d}{dt} \int_V \rho \, dV + \int_A \rho \mathbf{v} \cdot \mathbf{n} \, dA = 0 \tag{1}$$

for which the fixed control volume, V, is illustrated in Figure 3.3b.

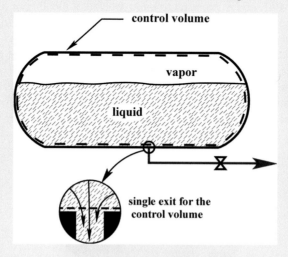

FIGURE 3.3b   Control volume for analysis of the propane tank.

Regardless of how the propane is distributed between the liquid and vapor phase, the mass of propane in the tank is given by

$$\left\{ \begin{array}{c} \text{mass of} \\ \text{propane} \\ \text{in the tank} \end{array} \right\} = \int_V \rho \, dV = m \tag{2}$$

and Eq. 1 takes the form

$$\frac{dm}{dt} + \int_{A_{exit}} \rho \mathbf{v} \cdot \mathbf{n} \, dA = 0 \tag{3}$$

On the basis of the analysis in Sec. 3.2, we can express this result as

$$\frac{dm}{dt} + \rho_{exit} \, Q_{exit} = 0 \tag{4}$$

We are given that the volumetric flow rate is 400 scf per day and this can be converted to SI units to obtain

$$Q_{exit} = 400 \frac{ft^3}{day} \times \left(2.83 \times 10^{-2}\right) \frac{m^3}{ft^3} = 11.32 \frac{m^3}{day} \tag{5}$$

In order to determine the density of the propane vapor leaving the tank, we first recall that one mole of gas at *standard conditions* occupies a volume of 22.42 liters. This leads to the molar concentration of propane given by

$$c_{propane} = \frac{mol}{22.42 \ L} \times \frac{L}{10^3 \ cm^3} \times \left(\frac{100 \ cm}{m}\right)^3 = 44.60 \frac{mol}{m^3} \tag{6}$$

The molecular mass (see Sec. 2.1.1) of propane is given in Table A2 of Appendix A, and use of that molecular mass allows us to determine the mass density of the propane as

$$\rho_{propane} = c_{propane} \ MW_{propane} = 44.60 \frac{mol}{m^3} \times 44.097 \frac{g}{mol} = 1967 \frac{g}{m^3} \tag{7}$$

The mass flow rate at the exit of the control volume illustrated in Figure 3.3b can now be calculated as

$$\dot{m}_{exit} = \rho_{exit} \ Q_{exit} = 22,266 \ g/day = 22.27 \ kg/day \tag{8}$$

Use of this result in Eq. 4 leads to the governing differential equation for the mass in the tank

$$\frac{dm}{dt} = -22.27 \ kg/day \tag{9}$$

This equation is easily integrated, and the initial condition imposed to obtain,

$$m(t) = m(t=0) - \left(22.27 \ kg/day\right) t \tag{10}$$

For an initial mass of 750 kg, we find that the tank will be empty ($m = 0$) when $t = 33.7$ days. For these circumstances, we require one delivery per month to ensure that the propane tank will never be empty.

3.3 MOVING CONTROL VOLUMES

In Figure 3.12, we have illustrated the transient process of a liquid draining from a cylindrical tank, and we want to be able to predict the depth of the fluid in the tank as a function of time. When gravitational and inertial effects dominate the flow process, the volumetric flow rate from the tank can be expressed as

$$Q = C_D \ A_o \sqrt{2gh} \tag{3.42}$$

Here, Q represents the volumetric flow rate, A_o is the cross-sectional area of the orifice through which the water is flowing, and C_D is the *discharge coefficient* that must be determined experimentally or by using the concepts presented in a subsequent course on fluid mechanics. Eq. 3.42 is sometimes referred to as Torricelli's efflux principle[tt] in honor of the Italian scientist who discovered this result in the 17th century. We would like to use Torricelli's law, along with the macroscopic mass balance to determine the height of the liquid as a function of time. The proper control volume to be used in this analysis is illustrated in Figure 3.12 where we see that the top portion of the control surface is *moving with the fluid*, and the remainder of the control surface is *fixed relative to the tank*.

In order to develop a general method of attacking problems of this type, we need to explore the form of Eq. 3.8 for an *arbitrary moving control volume*. In Sec. 3.1, we illustrated how Eq. 3.8 could be applied to the *special case* of a moving, deforming fluid body, and our analysis was quite simple since no fluid crossed the boundary of the control volume as indicated by Eq. 3.9. To develop a general mathematical representation for Eq. 3.8, we consider the *arbitrary moving control volume* shown in Figure 3.13. The speed of displacement of the surface of this control volume is $\mathbf{w} \cdot \mathbf{n}$ which need not be a constant, i.e., our control volume may be moving, deforming, accelerating or decelerating. An observer moving with the control volume determines the rate at which fluid crosses the boundary of the moving control volume, $V_a(t)$, and thus observes the *relative velocity*.

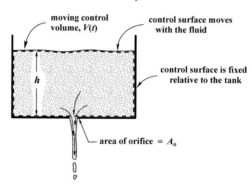

FIGURE 3.12   Draining tank.

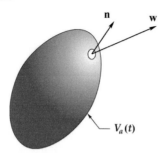

FIGURE 3.13   Arbitrary moving control volume.

[tt] Rouse, H. and Ince, S. 1957, page 61, *History of Hydraulics*, Dover Publications, Inc., New York.

Previously, we used a word statement of the principle of conservation of mass to develop a precise representation of the macroscopic mass balance for a fixed control volume. However, the statement given by Eq. 3.8 is *not limited to fixed control volumes* and we can apply it to the *arbitrary moving control volume* shown in Figure 3.13. The mass within the moving control volume is given by

$$\left\{ \begin{array}{c} \text{mass contained in} \\ any \text{ control volume} \end{array} \right\} = \int_{V_a(t)} \rho \, dV \tag{3.43}$$

and the time rate of change of the mass in the control volume is expressed as

$$\left\{ \begin{array}{c} \text{time rate of change} \\ \text{of the mass contained} \\ \text{in } any \text{ control volume} \end{array} \right\} = \frac{d}{dt} \int_{V_a(t)} \rho \, dV \tag{3.44}$$

In order to determine the *net* mass flow leaving the moving control volume, we simply repeat the development given by Eqs. 3.18–3.29 noting that the velocity of the fluid determined by an observer on the surface of the control volume illustrated in Figure 3.13 is the *relative velocity* $\mathbf{v}_r$. This concept is illustrated in Figure 3.14 where we have shown the volume of fluid *leaving* the surface element dA during a time Δt. On the basis of that representation, we express the macroscopic mass balance for an arbitrary moving control volume as

$$\frac{d}{dt} \int_{V_a(t)} \rho \, dV + \int_{A_a(t)} \rho \mathbf{v}_r \cdot \mathbf{n} \, dA = 0 \tag{3.45}$$

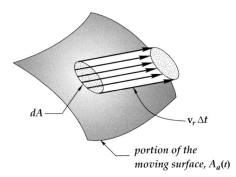

FIGURE 3.14 Volume of fluid crossing a moving surface dA during a time Δt.

The relative velocity is given explicitly by

$$\mathbf{v}_r = \mathbf{v} - \mathbf{w} \tag{3.46}$$

Thus, the macroscopic mass balance for an arbitrary moving control volume takes the form

$$\frac{d}{dt}\int_{V_a(t)}\rho\,dV + \int_{A_a(t)}\rho(\mathbf{v}-\mathbf{w})\cdot\mathbf{n}\,dA = 0 \qquad (3.47)$$

Here, it is helpful to think of $\mathbf{w}$ as the velocity of an observer moving with the surface of the control volume illustrated in Figure 3.13. It is important to recognize that this result contains our previous results for a *fluid body* and for a *fixed control volume*. This fact is illustrated in Figure 3.15 where we see that the arbitrary velocity can be set equal to zero, $\mathbf{w}=0$, in order to obtain Eq. 3.29, and we see that the arbitrary velocity can be set equal to the fluid velocity, $\mathbf{w}=\mathbf{v}$, in order to obtain Eq. 3.7.

$$\frac{d}{dt}\int_{V_a(t)}\rho\,dV + \int_{A_a(t)}\rho(\mathbf{v}-\mathbf{w})\cdot\mathbf{n}\,dA = 0$$

$$\mathbf{w}=0 \qquad\qquad \mathbf{w}=\mathbf{v}$$

$$\frac{d}{dt}\int_{V}\rho\,dV + \int_{A}\rho\mathbf{v}\cdot\mathbf{n}\,dA = 0 \qquad\qquad \frac{d}{dt}\int_{V_m(t)}\rho\,dV = 0$$

FIGURE 3.15   Axiomatic forms for the conservation of mass.

Clearly, Eq. 3.47 is the most general form of the principle of conservation of mass since it can be used to obtain directly the result for a fixed control volume and for a body.

EXAMPLE 3.4   WATER LEVEL IN A STORAGE TANK

A cylindrical API open storage tank, 2m in diameter and 3m in height, is used as a water storage tank. The water is used as a cooling fluid in a batch distillation unit and then sent to waste. The distillation unit runs for six hours and consumes 5 gal/min of water. At the beginning of the process the tank is full of water, and since we do not want to empty the tank during the distillation process, we need to know the liquid level in the tank at the end of the process. The system under consideration is illustrated in Figure 3.4a and we have constructed the control volume on the following basis:

    I. A cut has been made at the exit of the tank where information is *given*.

    II. There is no *required* information that would generate another cut.

    III. The surface of the control volume is located where $(\mathbf{v}-\mathbf{w})\cdot\mathbf{n}$ is *known*.

    IV. The control volume encloses the region for which information is *required*.

For the moving control volume illustrated in Figure 3.4a, the macroscopic mass balance takes the form

$$\frac{d}{dt}\int_{V(t)}\rho\,dV + \int_{A(t)}\rho(\mathbf{v}-\mathbf{w})\cdot\mathbf{n}\,dA = 0 \tag{1}$$

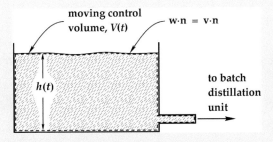

FIGURE 3.4a Cooling water storage tank.

Here, we have replaced $V_a(t)$ with $V(t)$ and $A_a(t)$ with $A(t)$ since the control volume is no longer *arbitrary*. A little thought will indicate that $\rho(\mathbf{v}-\mathbf{w})\cdot\mathbf{n}$ is zero everywhere on the surface of the control volume *except* at the exit of the tank. This leads to the representation given by

$$\int_{A(t)}\rho(\mathbf{v}-\mathbf{w})\cdot\mathbf{n}\,dA = \int_{A_{exit}}\rho\mathbf{v}\cdot\mathbf{n}\,dA = \rho Q_{exit} \tag{2}$$

in which we have assumed that the density can be treated as a constant. Use of Eq. 2 in Eq. 1 provides

$$\frac{d}{dt}\int_{V(t)}\rho\,dV + \rho Q_{exit} = 0 \tag{3}$$

and we again impose the condition of a constant density to obtain

$$\rho\frac{dV(t)}{dt} + \rho Q_{exit} = 0 \tag{4}$$

The control volume is computed as the product of the cross-sectional area of the tank multiplied by the level of water in the tank, i.e., $V(t) = A_T\,h(t)$. Use of this relation in Eq. 4 and canceling the density leads to

$$A_T\frac{dh}{dt} = -Q_{exit} \tag{5}$$

This equation is easily integrated to give

$$h(t) = h(t=0) - \frac{Q_{exit}\, t}{A_T} \tag{6}$$

For consistency, all the variables are computed in SI units (see Table 2.4) leading to

$$A_T = \frac{\pi D^2}{4} = 3.14\,\text{m}^2$$

$$Q_{exit} = \frac{(5\,\text{gal/min})(3.78 \times 10^{-3}\,\text{m}^3/\text{gal})}{60\,\text{s/min}} = 3.15 \times 10^{-4}\,\text{m}^3/\text{s} \tag{7}$$

At the end of the process, $t = 6\,\text{h} = 2.16 \times 10^4\,\text{s}$, and Eq. 6 indicates that the water level in the tank will be

$$h(t=6\text{h}) = 3\,\text{m} - 2.17\,\text{m} = 0.83\,\text{m}$$

EXAMPLE 3.5   WATER LEVEL IN A STORAGE TANK WITH AN INLET AND OUTLET

In Figure 3.5a, we have shown a tank into which water enters at a volumetric rate Q_1 and leaves at a volumetric rate Q_0 that is given by

$$Q_0 = 0.6\, A_0 \sqrt{2gh} \tag{1}$$

Here, A_0 is the area of the orifice in the bottom of the tank. The tank is initially empty when water begins to flow into the tank, thus we have an initial condition of the form

I.C. $$h = 0, \quad t = 0 \tag{2}$$

The parameters associated with this process are given by

$$Q_1 = 10^{-4}\,\text{m}^3/\text{s}, \quad A_0 = 0.354\,\text{cm}^2, \quad D = 1.5\,\text{m}, \quad h_0 = 2.78\,\text{m} \tag{3}$$

and in this example, we want to derive an equation that can be used to predict the height of the water at any time. That equation requires a numerical solution and we illustrate one of the methods (see Appendix B) that can be used to obtain numerical results.

The control volume used to analyze this process is illustrated in Figure 3.5a where we have shown a moving control volume for which the "cuts" at the entrance and exit are joined by a surface at which $(\mathbf{v} - \mathbf{w}) \cdot \mathbf{n}$

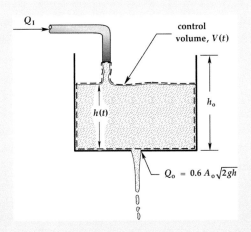

FIGURE 3.5a Tank filling and overflow process.

is zero. If the water overflows, a *second exit* will be created; however, we are only concerned with the process that occurs prior to overflow. Our analysis is based on the macroscopic mass balance for a moving control volume given by

$$\frac{d}{dt} \int_{V(t)} \rho \, dV + \int_{A(t)} \rho (\mathbf{v} - \mathbf{w}) \cdot \mathbf{n} \, dA = 0 \tag{4}$$

and the assumption that the density is constant leads to

$$\frac{d\,V(t)}{dt} + \int_{A(t)} (\mathbf{v} - \mathbf{w}) \cdot \mathbf{n} \, dA = 0 \tag{5}$$

The normal component of the *relative velocity*, $(\mathbf{v} - \mathbf{w}) \cdot \mathbf{n}$, is zero everywhere except at the entrance and exit, thus Eq. 5 simplifies to

$$\frac{dV(t)}{dt} + Q_o - Q_1 = 0 \tag{6}$$

The volume of the control volume is given by

$$V(t) = \frac{\pi D^2}{4} h(t) + V_{jet}(t) \tag{7}$$

and use of this expression in Eq. 6 provides

$$\frac{\pi D^2}{4} \frac{dh}{dt} + \frac{dV_{jet}}{dt} + Q_o - Q_1 = 0 \tag{8}$$

If the area of the jet is much, much smaller than the cross-sectional area of the tank, we can impose the following constraint on Eq. 8

$$\frac{dV_{jet}}{dt} \ll \frac{\pi D^2}{4} \frac{dh}{dt} \tag{9}$$

however, one must keep in mind that there are problems for which this inequality might not be valid. When Eq. 9 is valid, we can simplify Eq. 8 and arrange it in the compact form

$$\frac{dh}{dt} = \alpha - \beta\sqrt{h} \tag{10}$$

where α and β are constants given by

$$\alpha = \frac{4Q_1}{\pi D^2}, \quad \beta = \frac{0.6 \times 4 \times A_o \sqrt{2g}}{\pi D^2} \tag{11}$$

The initial condition for this process takes the form

I.C. $$\qquad\qquad\qquad\qquad h = 0, \quad t = 0 \tag{12}$$

and we need to integrate Eq. 10 and impose this initial condition in order to determine the fluid depth as a function of time. Separating variables in Eq. 10 leads to

$$\frac{dh}{\alpha - \beta\sqrt{h}} = dt \tag{13}$$

and the integral of this result is given by

$$\frac{2}{\beta^2}\left[\left(\alpha - \beta\sqrt{h}\right) - \alpha\ln\left(\alpha - \beta\sqrt{h}\right)\right] = t + C \tag{14}$$

The constant of integration, C, can be evaluated by means of the initial condition given by Eq. 12 and the result is

$$C = \frac{2}{\beta^2}\left(\alpha - \alpha\ln\alpha\right) \tag{15}$$

This allows us to express Eq. 14 in the form

$$\frac{2\alpha}{\beta^2}\left[\left(-\beta\sqrt{h}/\alpha\right) - \ln\left(1 - \beta\sqrt{h}/\alpha\right)\right] = t \tag{16}$$

and this result can be used to predict the height of the water at any time. Since Eq. 16 represents an implicit expression for $h(t)$, some computation is necessary in order to produce a curve of the fluid depth versus time.

Before we consider a numerical method that can be used to solve Eq. 16 for $h(t)$, we want to determine if the tank will overflow. To explore this question, we note that Eq. 16 provides the result

$$\left(1 - \beta\sqrt{h_\infty}/\alpha\right) = 0, \quad t = \infty \tag{17}$$

and this can be used to determine whether $h(t)$ is greater than H as time tends to infinity. One can also obtain Eq. 17 directly from Eq. 10 by imposing the steady-state condition that $dh/dt = 0$. We can arrange Eq. 17 in the form

$$h_\infty = (\alpha/\beta)^2, \quad t = \infty \tag{18}$$

and in terms of Eq. 11 we obtain

$$h_\infty = \left(Q_1/0.6 \times A_o\sqrt{2g}\right)^2, \quad t = \infty \tag{19}$$

from the values given in the problem statement, we find that $h_\infty = 2.26\,\text{m}$ at steady state, thus the tank will not overflow.

At this point, we would like to use Eq. 16 to develop a general solution for the fluid depth in the tank as a function of time, i.e., we wish to know $h(t)$ for the specific set of parameters given in this example. It is convenient to represent this problem in dimensionless variables so that

$$H(x,\theta) = -\left[x + \ln(1-x)\right] - \theta \qquad (20)$$

in which x is the dimensionless dependent variable representing the depth and defined by

$$x = \beta\sqrt{h}/\alpha \qquad (21)$$

In Eq. 20, we have used θ to represent the dimensionless time defined by

$$\theta = \beta^2 t/2\alpha \qquad (22)$$

and our objective is to find the root of the equation

$$H(x,\theta) = 0, \qquad x = x^* \qquad (23)$$

Here, we have identified the solution as x^* and it is clear from Eq. 20 that the solution will require that $x^* < 1$. In Appendix B, we have described several methods for solving implicit equations, such as Eq. 23, and in this example, we will use the simplest of these methods.

BISECTION METHOD

Here, we wish to find the solution to Eq. 23 when the parameter θ is equal to 0.10. A sketch of $H(x,\theta)$ is shown in Figure 3.5b where we see that x^* is located between zero and one. The bisection method begins by locating values of x that produce positive and negative values of the function $H(x,\theta)$ and these values are identified as $x_o = 0.90$ and $x_1 = 0.10$ in Figure 3.5b. The next step in the bisection method is to bisect the distance between x_o and x_1 to produce the value indicated by $x_2 = (x_1 + x_o)/2$. Next one evaluates $H(x_2,\theta)$ in order to determine whether it is positive or negative. From Figure 3.5b we see that $H(x_2,\theta) > 0$, thus, the next estimate for the solution is given by $x_3 = (x_2 + x_1)/2$. This type of geometrical construction is not necessary to carry out the bi-section method. Instead, one only needs to evaluate $H(x_2,\theta)\, H(x_1,\theta)$ to determine whether it is positive or negative. If $H(x_2,\theta)\, H(x_1,\theta) < 0$ the next estimate is given by $x_3 = (x_2 + x_1)/2$. However, if $H(x_2,\theta)\, H(x_1,\theta) > 0$ the next estimate is given by $x_3 = (x_2 + x_o)/2$. This procedure is repeated to achieve a converged value of $x^* = 0.3832$ as indicated in Table 3.5. Given the solution for the dimensionless depth defined by Eq. 21, and given the specified dimensionless time defined by Eq. 22, we can determine that the fluid depth will be 0.435 meters at 11 hours and 9.3 minutes after the start time. The results tabulated in Table 3.1 can be extended to a range of dimensionless times in order to produce a curve of x versus θ, and these results can be transformed to produce a curve of h versus t for any value of α and β.

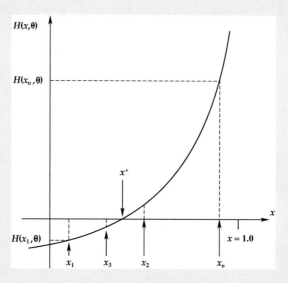

FIGURE 3.5b   Graphical iterative solution for $x = \beta\sqrt{h}/\alpha$.

TABLE 3.5
Converging Values for $x = \beta\sqrt{h}/\alpha$ at $\theta = \beta^2 t/2\alpha = 0.10$

n	x_n	$H(x_n)$
2	0.5000	0.09315
3	0.3000	−0.04333
4	0.4000	0.01083
5	0.3500	−0.01922
6	0.3750	−0.00500
7	0.3875	0.00271
8	0.3812	−0.00120
9	0.3844	0.00074
10	0.3828	−0.00023
11	0.3836	0.00026
12	0.3832	0.00001
13	0.3830	−0.00011
14	0.3831	−0.00005
15	0.3832	−0.00002
16	**0.3832**	**0.00000**

In Examples 3.4 and 3.5, we have illustrated how one can develop solutions to transient macroscopic mass balances. For the system analyzed in Example 3.5, an iterative method of solution was required to find the root of an implicit equation for the fluid depth. More information about iterative methods is provided in Appendix B.

3.4   PROBLEMS

3.4.1   SECTION 3.1

3.1. In Figure 3-1, we have illustrated a body in the shape of a sphere located in the center of the tube. The flow in the tube is laminar and the velocity profile is parabolic as indicated in the figure. Indicate how the shape of the sphere will change with time as the body is transported from left to right. Base your sketch on a cut through the center of the sphere that originally has the form of a circle. Keep in mind that the body does not affect the velocity profile.

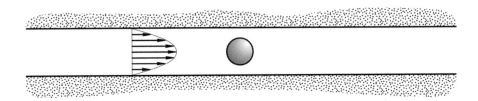

FIGURE 3-1   Body flowing and deforming in a tube.

3.2a. If the straight wire illustrated in Figure 3-2a has a uniform *mass per unit length* equal to ξ_o, the total mass of the wire is given by

$$mass = \xi_o\, L$$

If the mass per unit length is given by $\xi(x)$, the total mass is determined by the following *line integral*:

$$mass = \int_{x=0}^{x=L} \xi(x)\,dx$$

For the following conditions

$$\xi(x) = \xi_o + \alpha\left(x - \frac{1}{2}L\right)^2$$

$$\xi_o = 0.0065 \text{ kg/m}, \quad \alpha = 0.0017 \text{ kg/m}^3, \quad L = 1.4 \text{ m}$$

determine the total mass of the wire.

FIGURE 3-2a   Wire having a uniform or non-uniform mass density.

3.2b. If the flat plate illustrated in Figure 3-2b has a uniform *mass per unit area* equal to ψ_o, the total mass of the plate is given by

$$mass = \psi_o\, A = \psi_o\, L_1\, L_2$$

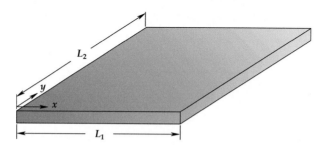

FIGURE 3-2b   Flat plate having a uniform or non-uniform mass density.

If the mass per unit area is given by $\psi(x, y)$, the total mass is determined by the *area integral* given by

$$mass = \int_A \psi\, dA = \int_{y=0}^{y=L_2} \int_{x=0}^{x=L_1} \psi(x, y)\, dx\, dy$$

For the following conditions

$$\psi(x, y) = \psi_o + \alpha xy$$

$$\psi_o = 0.0065 \text{ kg/m}^2, \quad \alpha = 0.00017 \text{ kg/m}^4, \quad L_1 = 1.4 \text{ m}, \quad L_2 = 2.7 \text{ m}$$

determine the total mass of the plate.

3.2c. If the density is a function of position represented by

$$\rho = \rho_o + \alpha\left(x - \frac{1}{2}L_1\right) + \beta\left(y - \frac{1}{3}L_2\right)^2$$

develop a general expression for the mass contained in the region indicated by

$$0 \le x \le L_1 \qquad 0 \le y \le L_2 \qquad 0 \le z \le L_3$$

3.4.2 SECTION 3.2

3.3. To describe flow of natural gas in a pipeline, a utility company uses *mass* flow rates. In a 10-inch internal diameter pipeline, the flow is 20,000 lb_m/h. The average density of the gas is estimated to be 10 kg/m³. What is the volumetric flow rate in ft³/s? What is the average velocity inside the pipe in m/s?

3.4. For the coating operation described in Example 3.2, we have produced an optical fiber having a diameter of 125 micrometers. The speed of the coated fiber at the take-up wheel is 4.5 meters per second and the desired thickness of the polymer coating is 40 micrometers. Assume that there is no change in the polymer density and determine the volumetric flow rate of the coating polymer that is required to achieve a thickness of 40 micrometers.

3.5. Slide coating is one of several methods for continuously depositing a thin liquid coating on a moving web. A schematic of the process is shown in Figure 3-5. In slide coating, a liquid film flows down an inclined plate (called the slide) owing to a gravitational force that is balanced by a viscous force. In a subsequent course in fluid mechanics, it will be shown that the velocity profile upstream on the slide is given by

$$v_x = \frac{\rho g h^2 \sin\theta}{\mu}\left[y/h - \frac{1}{2}(y/h)^2 \right], \qquad \text{on the slide} \qquad (1)$$

Here, y is the distance perpendicular to the slide surface and h is the thickness of the liquid film. Variations of the velocity, v_x, across the width of the slide can be ignored. In a steady operation, all the liquid flowing down the slide is picked up by a vertical web moving at a constant speed, U_o. Far downstream on the moving web, the velocity profile is given by

$$v_x = U_o - \frac{\rho g b^2}{\mu}\left[y/b - \frac{1}{2}(y/b)^2 \right], \qquad \text{on the moving web} \qquad (2)$$

and this velocity profile is illustrated in Figure 3-5. Here, one must note that the coordinate system used in Eq. 2 is different from that associated with Eq. 1. In slide coating operations, the system is operated in a manner such that

$$\frac{\rho g b^2}{\mu} \ll U_o \qquad (3)$$

and Eq. 2 and be replaced with the approximation given by

$$v_x = U_o, \qquad \text{far downstream on the moving web} \qquad (4)$$

In this problem, you are asked to carry out the following steps in the analysis of the slide coating process.

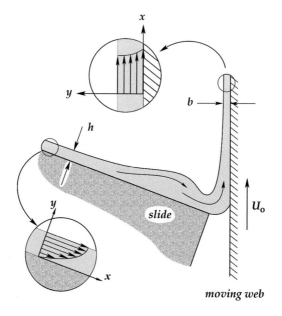

FIGURE 3-5   Slide coating.

a. Demonstrate that the flow on the slide can be expressed as

$$v_x = 3 \langle v_x \rangle \left[y/h - \frac{1}{2}(y/h)^2 \right] \tag{5}$$

To accomplish this, make use of Eq. 3.35 in the form

$$Q_{entrance} = \int_{A_{entrance}} |\mathbf{v} \cdot \mathbf{n}| dA, \qquad |\mathbf{v} \cdot \mathbf{n}| = v_x \tag{6}$$

and apply Eq. 6 of Example 3.1.

b. Construct an appropriate control volume and develop a macroscopic balance that will allow you to determine the thickness of the liquid film, b, on the moving web during steady operation.

3.6. One method for continuously depositing a thin liquid coating on a moving web is known as *slot die coating* and the process is illustrated in Figure 3-6. In slot-die coating, the liquid is forced through a slot of thickness d and flows onto a vertical web moving at a constant speed, U_o. The velocity profile in the slot is given by

$$v_x = 6 \langle v_x \rangle \left[y/d - (y/d)^2 \right] \tag{1}$$

and this profile is illustrated in Figure 3-6. Variations of the velocity, v_x, across the width, w, of the slot can be ignored. In a steady operation, all the feed liquid to the slot die is picked up by the web. As in the case of slide coating (see Problem 3.5), the fluid velocity on the moving web can be approximated by

$$v_x = U_o, \qquad \text{far downstream on the moving web} \qquad (2)$$

a. Select an appropriate control volume and construct the macroscopic balance that will allow you to determine the thickness, b, of the liquid film on the moving web during steady operation.

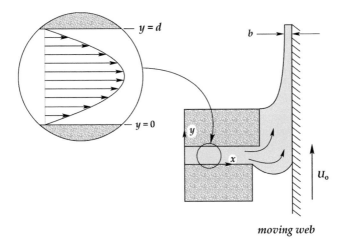

FIGURE 3-6   Slot die coating.

b. If far downstream on the moving web all the liquid in the coated film of thickness, b, moves at the velocity of the web, U_o, determine the thickness of the coated film.

c. If the gap, d, of the die slot changes by 10% and the average velocity remains constant, how much will the final thickness, b, of the coated film change?

3.7. If the delivery charge for the propane tank described in Example 3.3 is $37.50, and the cost of the next largest available tank is $2,500 (for a 2.2 cubic meter tank), how long will it take to recover the cost of a larger tank?

3.8. The steady-state average residence time of a liquid inside a holding tank is determined by the ratio of the volume of the tank to the volumetric flow rate of liquid into and out of the tank, $\tau = V / Q$. Consider a cylindrical tank with volume $V = 3 \, m^3$ and an input mass flow rate of water of 250 kg/minute. At steady-state, the output flow rate is equal to the input flow rate. What is the average residence time of water in the tank? What would be the average residence time if the mass flow rate of water is increased to 300 kg/minute? What would be the average residence time if a load of 1.2 m³ of stones is dropped into the holding tank?

3.9. Mono Lake is located at about 6,000 ft above sea level on the eastern side of the Sierra Nevada mountains, and a simple model of the lake is given in Figure 3-9a. The environment is that of a high, cold desert during the Winter, a thirsty well during the Spring runoff, and a cornucopia of organic and avian life during the Summer. Mono Lake is an important resting place for a variety of birds traveling the flyway between Canada and Mexico and was once the nesting place of one-fourth of the world's population of the California gull.

The decline of Mono Lake began in 1941 when Los Angeles diverted all the water from four of the five creeks flowing into the lake. This sent 56,000 acre-feet per year into the Owens River and on to the Los Angeles aqueduct. Details concerning the fight to save Mono Lake from a diminishing influx of water are available at http://www.monolake.org, and another web site located at http://www.worldsat.ca/image_gallery/aral_sea.html will provide information about a similar problem in the Aral Sea. The demise of Mono Lake was apparently secured in 1970 with the completion of a second barrel of the already-existing Los Angeles aqueduct from the southern Owens Valley. This allowed for a 50% increase in the flow, and most of this water was supplied by increased diversions from the Mono Basin. To be definitive, assume that the export of water from the Mono Basin was increased to 110,000 acre-feet per year in 1970. Given the conversion factor

$$1 \text{ acre-foot} = 43,560 \text{ ft}^3$$

one finds that $4.79 \times 10^9 \text{ ft}^3$ of water are being removed from the Mono Basin each year. In 1970, the surface area of the lake was $185 \times 10^6 \text{ m}^2$ and the maximum depth was measured as 50 m. If the lake is assumed to be circular with the configuration illustrated in Figure 3-9a, we can deduce that the tangent of θ is given by $\tan\theta = h(t)/r(t) = 6.52 \times 10^{-3}$.

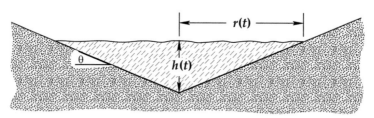

$r_0 = \text{initial radius} \quad r_1 = \text{final radius}$

$h_0 = \text{initial depth} \quad h_1 = \text{final depth}$

FIGURE 3-9a Assumed Mono Lake profile.

In this problem, you are asked to determine the final or steady-state condition of the lake, taking into account the flow of water to Los Angeles. The control volume to be used in this analysis is illustrated in Figure 3-9b and one needs to know the rate of evaporation in order to solve this problem. The rate of evaporation from the lake depends on a number of factors such as water temperature, salt concentration, humidity, and wind velocity, and it varies considerably throughout the year. It appears that the rate of evaporation from Mono Lake is about 36 inches per year. This represents a *convenience unit* and in order to determine the actual mass flux, we write

$$\dot{m}_2 = \left\{ \begin{array}{c} \text{mass flow rate owing} \\ \text{to evaporation} \end{array} \right\} = \rho_{H_2O} \, \beta \left(\begin{array}{c} \text{surface area} \\ \text{of the lake} \end{array} \right)$$

in which β is the convenience unit of *inches per year*. This parameter should be thought of as an average value for the entire lake, and for this problem β should be treated as a constant.

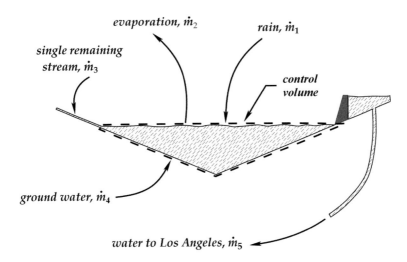

FIGURE 3-9b   Fixed control volume for the steady-state analysis of Mono Lake.

3.4.3   SECTION 3.3

3.10. A cylindrical tank having a diameter of 100 ft is used to store water for distribution to a suburban neighborhood. The average water consumption (stream 2 in Figure 3-10) during pre-dawn hours (midnight to 6 AM) is 100 m³/h. From 6 AM to 10 AM, the average water consumption increases to 500 m³/h, and then diminishes to 300 m³/h from 10 AM to 5 PM. During the night hours, from 5 PM to midnight, the average consumption falls even lower to 200 m³/h. The tank is replenished using a line (stream 1) that delivers water steadily into the tank at a rate of 1,120 gal/min. Assuming that the level of the tank at midnight is 3 m, plot the average level of the tank for a 24 hour period.

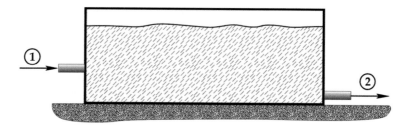

FIGURE 3-10   Water storage tank for distribution.

3.11. The seventh edition of Perry's Chemical Engineering Handbook[‡‡] gives the following formula for computing the volume of liquid inside a partially filled horizontal cylinder:

$$V = \pi L R^2 - \frac{L R^2}{2}(\alpha - \sin\alpha) \qquad (1)$$

Here, L is the length of the cylinder and R is the radius of the cylinder. The angle, α, illustrated in Figure 3-11a, is measured in radians. In this problem we wish to determine the depth of liquid in the tank, h, as a function of time when the *net flow* into the tank is Q. This flow rate is positive when the tank is being filled and negative when the tank is being emptied. The depth of the liquid in the tank is given in terms of the angle α by the trigonometric relation

$$h = R + R\cos(\alpha/2) \qquad (2)$$

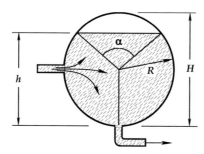

FIGURE 3-11a   Definition of geometric variables in horizontal cylindrical tank.

It follows that $h = 0$ when $\alpha = 2\pi$ and $h = 2R = H$ when $\alpha = 0$.

Part I. Choose an appropriate control volume and show that the macroscopic mass balance for a constant density fluid leads to

$$\frac{d\alpha}{dt} = \frac{2Q}{LR^2(\cos\alpha - 1)}, \qquad Q = Q_{in} - Q_{out} \qquad (3)$$

‡‡Perry, R. H., Green, D. W., and Maloney, J. O., 1997, *Perry's Chemical Engineers' Handbook*, 7th Edition, New York, McGraw-Hill Books.

Part II. Given an initial condition of the form

I.C. $\qquad\qquad\qquad\qquad \alpha = \alpha_o, \qquad t = 0 \qquad\qquad\qquad\qquad$ (4)

show that the implicit solution for $\alpha(t)$ is given by

$$\sin\alpha(t) - \alpha(t) = \sin\alpha_o - \alpha_o + \frac{2Qt}{LR^2} \qquad\qquad (5)$$

This equation can be solved using the methods described in Appendix B in order to determine $\alpha(t)$ which can then be used in Eq. 2 to determine the fluid depth, $h(t)$.

Part III. In Figure 3-11b, values of $\alpha(t)$ are shown as a function of the dimensionless time, $Q|t/\pi LR^2$ for $\alpha_o = 0$. The curve shown in Figure 3-11b represents values of $\alpha(t)$ when the tank is being drained, while the curve shown in Figure 3-11c represents the values of $\alpha(t)$ when the tank is being filled.

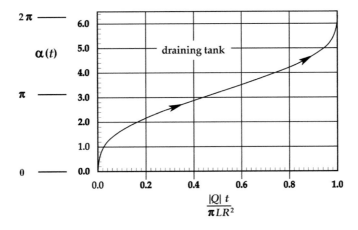

FIGURE 3-11b   Angle, $\alpha(t)$ for draining the system illustrated in Figure 3-11a.

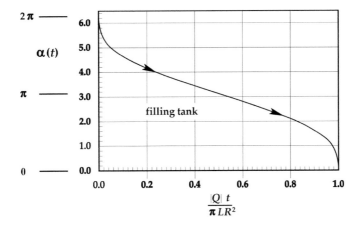

FIGURE 3-11c   Angle, $\alpha(t)$, for filling the system illustrated in Figure 3-11a.

The length and radius of the tank under consideration are given by

$$L = 8 \text{ ft}, \qquad\qquad R = 1.5 \text{ ft} \qquad\qquad (6)$$

Part IIIa. Given the conditions, $Q = -0.45$ gal/min and $\alpha = 0$ when $t = 0$, use Figure 3-11b to determine the time required to completely drain the tank.

Part IIIb. If the initial depth of the tank is $h = 0.6$ ft and the net flow into the tank is $Q = 0.55$ gal/min, use Figure 3-11c to determine the time required to fill the tank. While Figure 3-11c has been constructed on the basis that $\alpha = 2\pi$ when $t = 0$, a little thought will indicate that it can be used for other initial conditions.

3.12. A cylindrical tank of diameter D is filled to a depth h_o. As illustrated in Figure 3-12, at $t = 0$, a plug is pulled from the bottom of the tank and the volumetric flow rate through the orifice is given by what is sometimes known as Torricelli's law[§§]

$$Q = C_d A_o \sqrt{2\Delta p/\rho} \qquad\qquad (1)$$

Here, C_d is a discharge coefficient having a value of 0.6 and A_o is the area of the orifice. If the cross-sectional area of the tank is large compared to the area of the orifice, the pressure in the tank is essentially hydrostatic and Δp is given by

$$\Delta p = \rho g h \qquad\qquad (2)$$

where h is the depth of the fluid in the tank. This leads to Torricelli's law in the form

$$Q = C_d A_o \sqrt{2gh}, \quad \text{hydrostatic conditions} \qquad\qquad (3)$$

Use this information to derive an equation for the depth of the fluid as a function of time. For a tank filled with water to a depth of 1.6 m having a diameter of 20 cm, how long will it take to lower the depth to 1 cm if the diameter of the orifice is 3 mm?

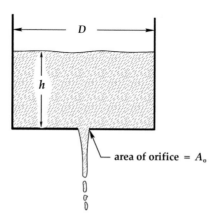

FIGURE 3-12   Draining tank.

[§§] Rouse, H. and Ince, S. 1957, *History of Hydraulics*, Dover Publications, Inc., New York.

3.13. The system illustrated in Figure 3-13 was analyzed in Example 3.5 and the depth at a single specified time was determined using the bisection method. In this problem, you are asked to repeat the type of calculation presented in Example 3.5 applying methods described in Appendix B. Determine a sufficient number of dimensionless times so that a curve of $x = \beta\sqrt{h}/\alpha$ versus $\theta = \beta^2 t/2\alpha$ can be constructed.

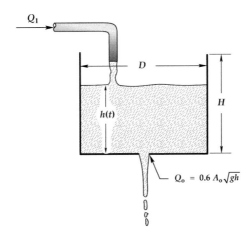

FIGURE 3-13 Tank filling process.

Part (a). The bi-section method
Part (b). The false position method
Part (c). Newton's method
Part (d). Picard's method
Part (e). Wegstein's method

3.14a. When full, a bathtub contains 25 gallons of water and the depth of the water is one foot. If the empty bathtub is filled with water from a faucet at a flow rate of 10 liters per minute, how long will it take to fill the bathtub?

3.14b. Suppose the bathtub has a leak and water drains out of the bathtub at a rate given by Torricelli's law (see Problem 3.12)

$$Q_{leak} = C_d A_o \sqrt{2gh} \tag{1}$$

Here, h is the depth of water in the bathtub, and C_d represents the discharge coefficient associated with the area of the leak in the bathtub, A_o. Since neither C_d nor A_o are known, we express Eq. 1 as

$$Q_{leak} = k\sqrt{h} \tag{2}$$

in which $k = C_d A_o \sqrt{2g}$. To find the value of k we have a single experimental condition given by

Experiment: $\qquad Q_{leak} = 3.16 \times 10^{-5}\,\text{m}^3/\text{s}, \quad h = 0.10\text{m} \tag{3}$

Given the experimental value of k, assume that the cross section of the bathtub is constant and determine how long it will take to fill the leaky bathtub.

3.15. The flow of blood in veins and arteries is a transient process in which the elastic conduits expand and contract. As a simplified example, consider the artery shown in Figure 3-15. At some instant in time, the inner radius has a radial velocity of 0.012 cm/s. The length of the artery is 13 cm and the volumetric flow rate at the *entrance* of the artery is 0.3 cm³/s. If the inner radius of the artery is 0.15 cm, at the particular instant of time, what is the volumetric flow rate at the exit of the artery?

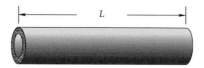

FIGURE 3-15   Expanding artery.

3.16. A variety of devices, such as ram pumps, hydraulic jacks, and shock absorbers, make use of moving solid cylinders to generate a desired fluid motion. In Figure 3-16, we have illustrated a cylindrical rod entering a cylindrical cavity in order to force the fluid out of that cavity. In order to determine the *force* acting on the cylindrical rod, we must know the velocity of the fluid in the annular region. If the density of the fluid can be treated as a constant, the velocity can be determine by application of the macroscopic mass balance and in this problem, you are asked to develop a general representation for the fluid velocity.

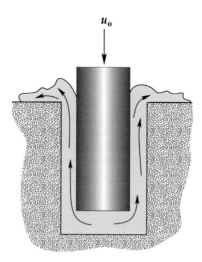

FIGURE 3-16   Flow in a hydraulic ram.

3.17. In Figure 3-17, we have illustrated a capillary tube that has just been immersed in a pool of water. The water is rising in the capillary so that the height of liquid in the tube is a function of time. Later, in a course on fluid mechanics, you will learn that the average velocity of the liquid, $\langle v_z \rangle$, can be represented by the equation

$$\underbrace{2\sigma/r_0}_{\substack{capillary \\ force}} - \underbrace{\rho g h}_{\substack{gravitational \\ force}} = \underbrace{\frac{8\mu \langle v_z \rangle h}{r_0^2}}_{\substack{viscous \\ force}} \tag{1}$$

in which $\langle v_z \rangle$ is the average velocity in the capillary tube. The surface tension σ, capillary radius r_0, and fluid viscosity μ can all be treated as constants in addition to the fluid density ρ and the gravitational constant g. From Eq. 1, it is easy to deduce that the final height (when $\langle v_z \rangle = 0$) of the liquid is given by

$$h_\infty = 2\sigma / \rho g r_0 \tag{2}$$

In this problem, you are asked to determine the height h as a function of time[***] for the initial condition given by

I.C. $$h = 0, \quad t = 0 \tag{3}$$

Part (a). Derive a governing differential equation for the height, $h(t)$, that is to be solved subject to the initial condition given by Eq. 3. Solve the initial value problem to obtain an implicit equation for $h(t)$.

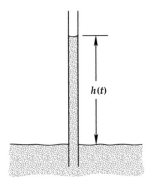

FIGURE 3-17 Transient capillary rise.

Part (b). By arranging the implicit equation for $h(t)$ in dimensionless form, demonstrate that this mathematical problem is identical in form to the problem described in Example 3.5 and Problem 3.13.

[***] Levich, V.G. 1962, *Physicochemical Hydrodynamics*, Prentice-Hall, Inc., Englewood Cliffs, N.J.

Part (c). Use the bisection method described in Example 3.5 and Sec. B1 of Appendix B to solve the governing equation in order to determine $h(t)$ for the following conditions:

$$\frac{\mu}{\rho} = 0.010 \text{ cm}^2/\text{s}, \quad g = 980 \text{ cm/s}^2, \quad r_o = 0.010 \text{ cm},$$

$$\lambda\sigma = 70 \text{ dyne/cm}, \quad \rho = 1 \text{ g/cm}^3$$

If a very fine capillary tube is available ($r_o = 0.010$ cm), you can test your analysis by doing a simple experiment in which the capillary rise is measured as a function of time.

3.18. In Figure 3-18, we have illustrated a cross-sectional view of a barge loaded with stones. The barge has sprung a lead as indicated, and the volumetric flow rate of the leak is given by

$$Q_{leak} = C_d \, A_o \sqrt{g(h - h_i)}$$

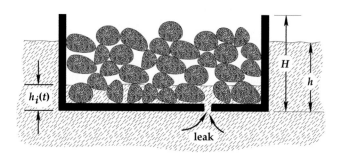

FIGURE 3-18   Leaking barge.

Here, C_d is a discharge coefficient equal to 0.6, A_o is the area of the hole through which the water is leaking, h is the height of the external water surface above the bottom of the barge, and h_i is the internal height of the water above the bottom of the barge. The initial conditions for this problem are

I.C. $h = h_o, \quad h_i = 0, \quad t = 0$

and you are asked to determine when the barge will sink. The length of the barge is L and the space available for water inside the barge is $\varepsilon\,HwL$. Here, ε is usually referred to as the void fraction and for this particular load of stones $\varepsilon = 0.35$.

In order to solve this problem, you will need to make use of the fact that the buoyancy force acting on the barge is

$$\text{buoyancy force} = (\rho g h)\,wL$$

where ρ is the density of water. This buoyancy force is equal and opposite to the gravitational force acting on the barge, and this is given by

$$\text{gravitational force} = mg$$

Here, m represents the mass of the barge, the stones, and the water that has leaked into the barge. The amount of water that has leaked into the barge is given by

$$\text{volume of water in the barge} = \varepsilon\, h_i\, (wL)$$

Given the following parameters:

$$w = 30 \text{ ft}, \quad L = 120 \text{ ft}, \quad A_o = 0.03 \text{ ft}^2, \quad h_o = 8 \text{ ft}, \quad H = 12 \text{ ft}$$

determine how long it will take before the barge sinks. You can compare your solution to this problem with an experiment done in your bathtub. Fill a coffee can with rocks and weigh it; then add water and weigh it again in order to determine the void fraction. Remove the water (but not the rocks) and drill a small hole in the bottom. Measure the diameter of the hole (it should be about 0.1 cm) so that you know the area of the leak, and place the can in a bathtub filled with water. Measure the time required for the can to sink.

3.19. The solution to Problem 3.9 indicates that the diversion of water from Mono Lake to Los Angeles would cause the level of the lake to drop 19 meters. A key parameter in this prediction is the evaporation rate of 36 in/year, and the steady-state analysis gave no indication of the time required for this reduction to occur. In this problem you are asked to develop the unsteady analysis of the Mono Lake water balance. Use available experimental data to predict the evaporation rate, and then use your solution and the new value of the evaporation rate to predict the final values of the radius and the depth of the lake. You are also asked to predict the number of years required for the maximum depth of the lake to come within 10 cm of its final value. The following information is available:

Year	Surface Area	Maximum Depth
prior to 1941	156,000 acres	181.5 ft.
1970	45,700 acres	164.0 ft.

During these years between 1941 and 1970, the diversion of water from Mono Lake Basin was 56,000 acre-ft. per year and in 1970 this was increased to 110,000 acre-ft. per year. The additional water was obtained from wells in the Mono Lake Basin, and as an approximation you can assume that this caused a decrease in $\dot{m}_4$ (see Figure 3-9b) by an amount equal to 54,000 acre-ft. per year. Your analysis will lead to an implicit equation for β and the methods described in Appendix B can be used to obtain a solution. Develop a solution based on the following methods:

Part (a). The bisection method
Part (b). The false position method
Part (c). Newton's method
Part (d). Picard's method
Part (e). Wegstein's method

3.20. During the winter months on many campuses across the country, students can be observed huddled in doorways contemplating an unexpected downpour. In order not to be accused of idle ways, engineering students will often devote this time to the problem of estimating the speed at which they should run to their next class in order to minimize the unavoidable soaking. This problem has such importance in the general scheme of things that in March of 1973 it became the subject of one of Ann Landers' syndicated columns entitled "What Way is Wetter?" Following typical Aristotelian logic, Ms. Landers sided with the common sense solution, "the faster you run, the quicker you get there, and the drier you will be." Clearly, a rational analysis is in order and this can be accomplished by means of the macroscopic mass balance for an arbitrary moving control volume.

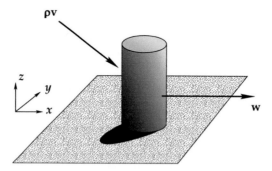

FIGURE 3-20   Student running in the rain.

In order to keep the analysis relatively simple, the running student should be modeled as a cylinder of height h and diameter D as illustrated in Figure 3-20. The rain should be treated as a continuum with the mass flux of water represented by $\rho\mathbf{v}$. Here, the density ρ will be equal to the density of water multiplied by the volume fraction of the raindrops and the velocity $\mathbf{v}$ will be equal to the velocity of the raindrops. The velocity of the student is given by $\mathbf{w}$, and both $\mathbf{v}$ and $\mathbf{w}$ should be treated as constants. You should consider the special case in which $\mathbf{v}$ and $\mathbf{w}$ have no component in the y-direction, and you should separate your analysis into two parts. In the first part, consider $\mathbf{i} \cdot \mathbf{v} > 0$ as indicated in Figure 3-20, and in the second part consider the case indicated by $\mathbf{i} \cdot \mathbf{v} < 0$. In the first part, one needs to consider both $\mathbf{i} \cdot \mathbf{v} > \mathbf{i} \cdot \mathbf{w}$ and $\mathbf{i} \cdot \mathbf{v} < \mathbf{i} \cdot \mathbf{w}$. When $\mathbf{i} \cdot \mathbf{v} > \mathbf{i} \cdot \mathbf{w}$, the runner gets wet on the top and back side. On the other hand, when the runner is moving faster than the horizontal component of the rain, the runner gets when on the top and the front side. When $\mathbf{i} \cdot \mathbf{v} < 0$ the runner only gets wet on the top and the front side and the analysis is somewhat easier. In your search for an extremum, it may be convenient to represent the accumulated mass in a dimensionless form according to

$$\frac{m(t) - m_{o}}{\rho L D h} = F(parameters)$$

in which t is equal to the distance run divided by $\mathbf{i} \cdot \mathbf{w} = u_{o}$, and m_{o} is the initial mass of the runner.

4 Multicomponent Systems

In the previous chapter, we considered single-component systems for which there was a single density, ρ, a single velocity, $\mathbf{v}$, and no chemical reactions. In multi-component systems, we must deal with the density of individual species and this leads to the characterization of systems in terms of *species mass densities* and species molar concentrations, in addition to mass fractions and mole fractions. Not only must we characterize the composition of multicomponent systems, but we must also consider the fact that different molecular species move at different velocities. This leads us to the concept of the *species velocity* which plays a dominant role in the detailed study of separation and purification processes and in the analysis of chemical reactors. In this chapter, we will discuss the concept of the species velocity and then illustrate how a certain class of macroscopic mass balance problems can be solved without dealing directly with this important aspect of multi-component systems. While chemical reactions represent an essential feature of multi-component systems, we will delay a thorough discussion of that matter until Chapter 6.

4.1 AXIOMS FOR THE MASS OF MULTICOMPONENT SYSTEMS

In Chapter 3, we studied the concept of *conservation of mass* for single-component systems, both for a *body* and for a *control volume*. The *words* associated with the control volume representation were

$$
\left\{ \begin{array}{c} \text{time rate of change} \\ \text{of mass in a} \\ \text{control volume} \end{array} \right\} = \left\{ \begin{array}{c} \text{rate at which} \\ \text{mass } enters \text{ the} \\ \text{control volume} \end{array} \right\} - \left\{ \begin{array}{c} \text{rate at which} \\ \text{mass } leaves \text{ the} \\ \text{control volume} \end{array} \right\} \qquad (4.1)
$$

and for a *fixed control volume*, the *mathematical representation* was given by

$$
\frac{d}{dt} \int_V \rho \, dV + \int_A \rho \, \mathbf{v} \cdot \mathbf{n} \, dA = 0 \qquad (4.2)
$$

One must keep in mind that the use of vectors allows us to represent both the mass *entering* the control volume and mass *leaving* the control volume in terms of $\rho \, \mathbf{v} \cdot \mathbf{n}$. This follows from the fact that $\mathbf{v} \cdot \mathbf{n}$ is negative over surfaces where mass is *entering* the control volume and $\mathbf{v} \cdot \mathbf{n}$ is positive over surfaces where mass is *leaving* the control volume. In addition, one should remember that the control volume, V, in Eq. 4.2 is arbitrary and this allows us to choose the control volume to suit our needs.

DOI: 10.1201/9781003283751-4

Now we are ready to consider N-component systems in which chemical reactions can take place and, in this case, we need to make use of the *two axioms* for the mass of multicomponent systems. The first axiom deals with the mass of species A, and when this species can undergo chemical reaction, we need to extend Eq. 4.1 to the form given by

$$\left\{\begin{array}{c} \text{time rate of change} \\ \text{of mass of species } A \\ \text{in a control volume} \end{array}\right\} = \left\{\begin{array}{c} \text{rate at which} \\ \text{mass of species } A \\ \textit{enters} \text{ the} \\ \text{control volume} \end{array}\right\}$$

$$-\left\{\begin{array}{c} \text{rate at which} \\ \text{mass of species } A \\ \textit{leaves} \text{ the} \\ \text{control volume} \end{array}\right\} + \left\{\begin{array}{c} \text{net rate of production} \\ \text{of the mass of species } A \\ \text{owing to} \\ \textit{chemical reactions} \end{array}\right\} \quad (4.3)$$

In order to develop a precise mathematical representation of this axiom, we require the following quantities:

$$\rho_A = \left\{\begin{array}{c} \text{mass density} \\ \text{of species } A \end{array}\right\} \quad (4.4)$$

$$\mathbf{v}_A = \left\{\begin{array}{c} \text{velocity of} \\ \text{species } A \end{array}\right\} \quad (4.5)$$

$$r_A = \left\{\begin{array}{c} \text{net mass rate of production} \\ \text{per unit volume of species } A \\ \text{owing to } \textit{chemical reactions} \end{array}\right\} \quad (4.6)$$

Here, it is important to understand that r_A represents both the *creation* of species A (when r_A is positive) and the *consumption* of species A (when r_A is negative). In terms of these primitive quantities, we can make use of an arbitrary fixed control volume to express Eq. 4.3 as

Axiom I: $\quad \dfrac{d}{dt}\displaystyle\int_V \rho_A \, dV + \int_A \rho_A \mathbf{v}_A \cdot \mathbf{n} \, dA = \int_V r_A \, dV, \quad A = 1, 2, \ldots, N \quad (4.7)$

Within the volume, V, the *total mass produced by chemical reactions* must be zero. This is our second axiom that we express in words as

Axiom II: $\qquad \left\{\begin{array}{c} \text{total rate of production} \\ \text{of mass owing to} \\ \textit{chemical reactions} \end{array}\right\} = 0 \quad (4.8)$

and in terms of the definition given by Eq. 4.6, this word statement takes the form

$$\sum_{A=1}^{A=N} \int_V r_A \, dV = 0 \tag{4.9}$$

The summation over all N molecular species can be interchanged with the volume integration in this representation of the second axiom, and this allows us to express Eq. 4.9 as

$$\int_V \sum_{A=1}^{A=N} r_A \, dV = 0 \tag{4.10}$$

Since the volume V is arbitrary, the integrand must be zero and we extract the preferred form of the second axiom given by

Axiom II:
$$\sum_{A=1}^{A=N} r_A = 0 \tag{4.11}$$

Here it should be clear that r_A represents both the *creation* of species A (when r_A is positive) and the *consumption* of species A (when r_A is negative). In Eqs. 4.7 and 4.11, we have used a *mixed mode nomenclature* to represent the chemical species, i.e., we have used both letters and numbers simultaneously. Traditionally, we use upper case Roman letters to designate various chemical species, thus the rates of production for species A, B, and C are designated by r_A, r_B, and r_C. When dealing with systems containing N different molecular species, we allow an indicator, such as A or D or G, to take on values from 1 to N in order to produce compact forms of the two axioms given by Eqs. 4.7 and 4.11. We could avoid this mixed mode nomenclature consisting of both letters and numbers by expressing Eq. 4.11 in the form;

Axiom II: $r_A + r_B + r_C + r_D + \cdots + r_N = 0$ \hfill (4.12)

however, this approach is rather cumbersome when dealing with N-component systems.

The concept that mass is neither created nor destroyed by chemical reactions (as indicated by Eq. 4.11) is based on the work of Lavoisier[*] who stated:

> We observe in the combustion of bodies generally four recurring phenomena which would appear to be invariable laws of nature; while these phenomena are implied in other memoirs which I have presented, I must recall them here in a few words.

Lavoisier went on to list four phenomena associated with combustion, the third of which was given by

> *Third Phenomenon.* In all combustion, pure air in which the combustion takes place is destroyed or decomposed and the burning body increases in weight exactly in proportion to the quantity of air destroyed or decomposed.

---

[*] Lavoisier, A. L. 1777, Memoir on combustion in general, Mem. Acad. r. Sci. Paris 592–600.

It is this *Third Phenomenon*, when extended to all reacting systems, that supports Axiom II in the form represented by Eq. 4.9. The experiments that led to the *Third Phenomenon* were difficult to perform in the 18th century and those difficulties have been recounted by Toulmin[†].

4.1.1   MOLAR CONCENTRATION AND MOLECULAR MASS

When chemical reactions occur, it is generally more convenient to work with the *molar form* of Eqs. 4.7 and 4.11. The appropriate measure of concentration is then the *molar concentration* defined by

$$c_A = \rho_A / MW_A = \left\{ \begin{array}{c} moles \text{ of species } A \\ \text{per unit volume} \end{array} \right\} \tag{4.13}$$

while the appropriate net rate of production for species A is given by

$$R_A = r_A / MW_A = \left\{ \begin{array}{c} \text{net } molar \text{ rate of production} \\ \text{per unit volume of species } A \\ \text{owing to chemical reactions} \end{array} \right\} \tag{4.14}$$

Here MW_A represents the *molecular mass* (see Sec. 2.1.1) of species A that is given explicitly by

$$MW_A = \frac{\text{kilograms of } A}{\text{moles of } A} \tag{4.15}$$

The numerical values of the molecular mass are obtained from the atomic masses associated with any particular molecular species, and values for both the atomic mass and the molecular mass are given in Tables A1 and A2 in Appendix A. In those tables, we have represented the atomic mass and the molecular mass in terms of grams per mole, thus the definition given by Eq. 4.15 for water leads to

$$MW_{H_2O} = \frac{0.018015 \text{ kg}}{\text{mol}} = \frac{18.015 \text{ g}}{\text{mol}} \tag{4.16}$$

In terms of c_A and R_A, the two axioms given by Eqs. 4.7 and 4.11 take the form

Axiom I: $$\frac{d}{dt} \int_V c_A \, dV + \int_A c_A \mathbf{v}_A \cdot \mathbf{n} \, dA = \int_V R_A \, dV, \quad A = 1, 2, \ldots, N \tag{4.17}$$

Axiom II: $$\sum_{A=1}^{A=N} MW_A R_A = 0 \tag{4.18}$$

[†] Toulmin, S.E. 1957, Crucial experiments: Priestley and Lavoisier, J. Hist. Ideas **18**, 205–220.

Here it is important to note that mass is conserved during chemical reactions while moles need not be conserved. For example, the decomposition of calcium carbonate (solid) to calcium oxide (solid) and carbon dioxide (gas) is described by

$$(CaCO_3)_{solid} \rightarrow (CaO)_{solid} + (CO_2)_{gas} \tag{4.19}$$

thus *one mole* is consumed and *two moles* are produced by this chemical reaction.

One must be very careful to understand that the *net molar rate of production per unit volume of species A owing to chemical reactions*, R_A, may be the result of many different chemical reactions. For example, in the chemical production system illustrated in Figure 4.1, carbon dioxide may be created by the oxidation of carbon monoxide, by the complete combustion of methane, or by other chemical reactions taking place within the control volume illustrated in Figure 4.1. The combination of all these individual chemical reactions is represented by R_{CO_2}.

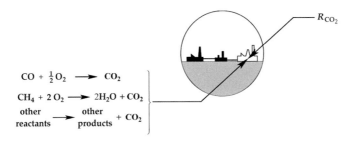

FIGURE 4.1   Chemical production system.

It is important to note that in Figure 4.1 we have suggested the *stoichiometry* of the reactions taking place while the actual *chemical kinetic processes* taking place may be much more complicated. The subject of *local* and *global* stoichiometry is discussed in detail in Chapter 6. There we make use of the concept that atomic species are conserved in order to develop constraints on the local and global net rates of production. In Chapter 9, we introduce the concept of reaction kinetics and *elementary* stoichiometry. There we begin to explore the actual chemical kinetic processes that are the origin of the net rate of production of carbon dioxide and other molecular species.

4.1.2   Moving Control Volumes

In Sec. 3.3, we developed the macroscopic mass balance for a single component system in terms of an *arbitrary* moving control volume, $V_a(t)$. The speed of displacement of the surface of a moving control volume is given by $\mathbf{w} \cdot \mathbf{n}$, and the flux of species A that crosses this moving surface is given by the normal component of the *relative velocity* for species A, i.e., $(\mathbf{v}_A - \mathbf{w}) \cdot \mathbf{n}$. On the basis of this concept, we can express the first axiom for the mass of species A in the form

Axiom I: $$\frac{d}{dt} \int_{V_a(t)} \rho_A \, dV + \int_{A_a(t)} \rho_A (\mathbf{v}_A - \mathbf{w}) \cdot \mathbf{n} \, dA = \int_{V_a(t)} r_A \, dV \tag{4.20}$$

When it is convenient to work with molar quantities, we divide this result by the molecular mass of species A in order to obtain the form for an *arbitrary* moving control volume given by

Axiom I: $$\frac{d}{dt} \int_{V_a(t)} c_A \, dV + \int_{A_a(t)} c_A (\mathbf{v}_A - \mathbf{w}) \cdot \mathbf{n} \, dA = \int_{V_a(t)} R_A \, dV \qquad (4.21)$$

The volume associated with a *specific* moving control volume will be designated by $V(t)$ while the volume associated with a fixed control volume will be designated by V. Throughout this chapter, we will restrict our studies to fixed control volumes in order to focus our attention on the new concepts associated with multicomponent systems. However, the world of chemical engineering is filled with moving, dynamic systems and the analysis of those systems will require the use of Eqs. 4.20 and 4.21.

4.2   SPECIES MASS DENSITY

The mass of species A per unit volume in a mixture of several components is known as the *species mass density*, and it is represented by ρ_A. The species mass density can range from zero, when no species A is present in the mixture, to the density of pure species A, when no other species are present. In order to understand what is meant by the species mass density, we consider a mixing process in which three *pure species* are combined to create a uniform mixture of species A, B, and C. This mixing process is illustrated in Figure 4.2 where we have indicated that three pure species are combined to create a uniform mixture having a measured volume of 45 cm³. The total volume of the three *pure species* is 50 cm³, thus there is a *change of volume* upon mixing as is usually the case with liquids. We denote this change of volume upon mixing by ΔV_{mix}, and for the process illustrated in Figure 4.2 we express this quantity as

$$\Delta V_{mix} = V - (V_A + V_B + V_C) \qquad (4.22)$$

The densities of the *pure species* have been denoted by a superscript zero, thus ρ_A^o represents the mass density of pure species A. The species mass density of species A is defined by

$$\left\{ \begin{array}{c} species\ mass\ density \\ of\ species\ A \end{array} \right\} = \frac{(mass\ of\ species\ A)}{\left(\begin{array}{c} volume\ in\ which\ species\ A \\ is\ contained \end{array} \right)} \qquad (4.23)$$

This definition applies to mixtures in which species A is present as well as to the case of pure species A.

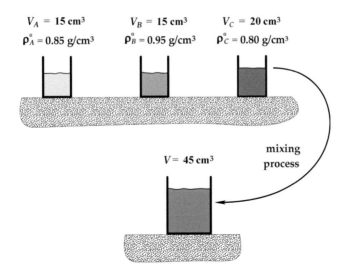

$V_A = 15 \text{ cm}^3$ $V_B = 15 \text{ cm}^3$ $V_C = 20 \text{ cm}^3$

$\rho_A^o = 0.85 \text{ g/cm}^3$ $\rho_B^o = 0.95 \text{ g/cm}^3$ $\rho_C^o = 0.80 \text{ g/cm}^3$

mixing
$V = 45 \text{ cm}^3$ process

FIGURE 4.2   Mixing process.

If we designate the mass of species A as m_A and the volume of the *uniform mixture* as V, the species mass density can be expressed as

$$\rho_A = m_A / V \qquad (4.24)$$

For the mixing process illustrated in Figure 4.2, we are given the density, ρ_A^o, and the volume, V_A, and this allows us to determine the mass of species A as

$$m_A = \rho_A^o V_A = (0.85 \text{ g/cm}^3)(15 \text{ cm}^3) = 12.75 \text{ g} \qquad (4.25)$$

From this we can determine the species mass density, ρ_A, in the mixture according to

$$\rho_A = m_A / V = \frac{12.75 \text{ g}}{45 \text{ cm}^3} = 0.283 \text{ g/cm}^3 \qquad (4.26)$$

This type of calculation can be carried out for species B and C in order to determine ρ_B and ρ_C.

The *total mass density* is simply the sum of all the species mass densities and it is defined by

$$\left\{ \begin{array}{c} \text{total mass} \\ \text{density} \end{array} \right\} = \rho = \sum_{A=1}^{A=N} \rho_A \qquad (4.27)$$

The total mass density can be determined experimentally by measuring the mass, m, and the volume, V, of a mixture. For any a particular mixture, it is difficult to measure directly the species mass density; however, one can prepare a mixture in which the species mass densities can be determined as we have suggested in Figure 4.2. When working with molar forms, we often need the total molar concentration and this is defined by

$$\left\{ \begin{array}{c} \text{total molar} \\ \text{concentration} \end{array} \right\} = c = \sum_{A=1}^{A=N} c_A \tag{4.28}$$

in which c_A is determined by Eq. 4.13.

4.2.1 MASS FRACTION AND MOLE FRACTION

For solid and liquid systems, it is sometimes convenient to use the *mass fraction* as a measure of concentration. The mass fraction of species A can be expressed in words as

$$\omega_A = \left\{ \begin{array}{c} \text{mass of species } A \text{ per} \\ \text{unit mass of the mixture} \end{array} \right\} \tag{4.29}$$

and in precise mathematical form we have

$$\omega_A = \frac{\rho_A}{\rho} = \rho_A \bigg/ \sum_{G=1}^{G=N} \rho_G \tag{4.30}$$

Note that the indicator, G, is often referred to as a *dummy indicator* since any letter would suffice to denote the summation over all species in the mixture. In this particular case, we would not want to use A as the dummy indicator since this could lead to confusion.

The *mole fraction* is analogous to the mass fraction and is defined by

$$x_A = \frac{c_A}{c} = c_A \bigg/ \sum_{G=1}^{G=N} c_G \tag{4.31}$$

If one wishes to avoid the *mixed-mode nomenclature* in Eqs. 4.30 and 4.31, one must express the mass fraction as

$$\omega_A = \frac{\rho_A}{\rho} = \frac{\rho_A}{\rho_A + \rho_B + \rho_C + \rho_D + \cdots + \rho_N} \tag{4.32}$$

while the mole fraction takes the form

$$y_A = \frac{c_A}{c} = \frac{c_A}{c_A + c_B + c_C + c_D + \cdots + c_N} \tag{4.33}$$

Very often x_A is used to represent mole fractions in *liquid mixtures* and y_A to represent mole fractions in *vapor mixtures*, thus Eq. 4.33 represents the mole fraction in a vapor mixture while Eq. 4.31 represents the mole fraction in a liquid mixture.

EXAMPLE 4.1   CONVERSION OF MOLE FRACTIONS TO MASS FRACTIONS

Sometimes we may be given the composition of a mixture in terms of the various mole fractions and we may require the mass fractions of the various constituents. To convert from x_A to ω_A, we proceed as follows:

$$x_A = c_A / c \tag{1}$$

$$c_A = x_A c \tag{2}$$

$$\rho_A = MW_A\, c_A = MW_A\, x_A\, c \tag{3}$$

$$\rho = \sum_{G=1}^{G=N} \rho_G = \left(\sum_{G=1}^{G=N} MW_G\, x_G \right) c \tag{4}$$

$$\omega_A = \frac{\rho_A}{\rho} = \frac{MW_A\, x_A\, c}{\left(\displaystyle\sum_{G=1}^{G=N} MW_G\, x_G \right) c} = \frac{MW_A\, x_A}{\displaystyle\sum_{G=1}^{G=N} MW_G\, x_G} \tag{5}$$

4.2.2   TOTAL MASS BALANCE

Given the total density defined by Eq. 4.27, we are ready to recover the *total mass balance* for multicomponent, reacting systems. For a fixed control volume, this is developed by summing Eq. 4.7 over all species to obtain

$$\sum_{A=1}^{A=N} \frac{d}{dt} \int_V \rho_A\, dV + \sum_{A=1}^{A=N} \int_A \rho_A \mathbf{v}_A \cdot \mathbf{n}\, dA = \sum_{A=1}^{A=N} \int_V r_A\, dV \tag{4.34}$$

The summation procedure can be interchanged with differentiation and integration so that this result takes the form

$$\frac{d}{dt} \int_V \sum_{A=1}^{A=N} \rho_A\, dV + \int_A \sum_{A=1}^{A=N} \rho_A \mathbf{v}_A \cdot \mathbf{n}\, dA = \int_V \sum_{A=1}^{A=N} r_A\, dV \tag{4.35}$$

On the basis of *definition* of the total mass density given by Eq. 4.27 and the *axiom* given by Eq. 4.11, this result simplifies to

$$\frac{d}{dt} \int_V \rho\, dV + \int_A \sum_{A=1}^{A=N} \rho_A \mathbf{v}_A \cdot \mathbf{n}\, dA = 0 \tag{4.36}$$

At this point, we identify the *total mass flux* according to

$$\left\{\begin{array}{c} \text{total} \\ \text{mass flux} \end{array}\right\} = \rho\mathbf{v} = \sum_{A=1}^{A=N} \rho_A \mathbf{v}_A \qquad (4.37)$$

Since ρ is defined by Eq. 4.27, this result represents a definition of the velocity $\mathbf{v}$ that can be expressed as

$$\mathbf{v} = \sum_{A=1}^{A=N} \omega_A \mathbf{v}_A \qquad (4.38)$$

This velocity is known as the *mass average velocity* and it plays a key role *both* in our studies of macroscopic mass balances and in subsequent studies of fluid mechanics, heat transfer, and mass transfer. Use of this definition for the mass average velocity allows us to express Eq. 4.36 as

$$\frac{d}{dt}\int_V \rho\,dV + \int_A \rho\mathbf{v}\cdot\mathbf{n}\,dA = 0 \qquad (4.39)$$

This is identical in form to the mass balance for a fixed control volume that was presented in Chapter 3; however, this result has *greater physical content* than our previous result for single-component systems. In this case, the density, ρ, is not the density of a single component, but it is the sum of all the species densities as indicated by Eq. 4.27. In addition, the velocity, $\mathbf{v}$, is not the velocity of a single component, but it is the mass average velocity defined by Eq. 4.38.

4.3 SPECIES VELOCITY

In our representation of the axioms for the mass of multicomponent systems, we have introduced the concept of a *species velocity*. This indicates that individual molecular species move at their own velocities designated by $\mathbf{v}_A$ where $A = 1, 2, \ldots, N$. In order to begin thinking about the species velocity, we consider a lump of sugar (species A) placed in the bottom of a teacup which is very carefully filled with water (species B). If we wait long enough, the solid sugar illustrated in Figure 4.3 will dissolve and become uniformly distributed throughout the cup.

FIGURE 4.3   Dissolution of sugar.

This is a clear indication that the velocity of the sugar molecules is *different* from the velocity of the water molecules, i.e.,

$$\mathbf{v}_{sugar} \neq \mathbf{v}_{water} \tag{4.40}$$

If the solution in the cup is not stirred, the velocity of the sugar molecules will be *very small* and the time required for the sugar to become uniformly distributed throughout the cup will be *very long*. We generally refer to this process as *diffusion* and diffusion velocities are generally *very small*. If we stir the liquid in the teacup, the sugar molecules will be transported away from the sugar cube by *convection* as we have illustrated in Figure 4.4. In this case, the sugar will become uniformly distributed throughout the cup in a relatively short time and we generally refer to this process as *mixing*. Mechanical mixing can accelerate the process by which the sugar becomes uniformly distributed throughout the teacup; however, a true mixture of sugar and water could never be achieved *unless the velocities of the two species were different*. The difference between species velocities is crucial. It is responsible for mixing, for separation and purification, and it is necessary for chemical reactions to occur. If all species velocities were equal, life on earth would cease immediately in terms of the continuum perspective.

In addition to mixing the sugar and water as indicated in Figures 4.3 and 4.4, we can also separate the sugar and water by allowing the water to evaporate. In that case, all the water in the teacup would appear in the surrounding air and the sugar would remain in the bottom of the cup. This separation would not be possible unless the velocity of the water were different than the velocity of the sugar. While the *difference between species velocities* is of crucial interest to chemical engineers, there is a class of problems for which we can ignore this difference and still obtain useful results. In the following paragraphs, we want to identify this class of problems.

To provide another example of the difference between species velocities and diffusion velocities, we consider the process of absorption of SO_2 in a falling film of water as illustrated in Figure 4.5. The gas mixture entering the column consists of air (nitrogen and oxygen), which is essentially *insoluble* in water, and SO_2, which is *soluble* in water. Because of the absorption of SO_2 in the water, the exit gas is less detrimental to the local environment. It should be *intuitively appealing* that the species velocities in the axial direction are constrained by

$$\mathbf{v}_{SO_2} \cdot \mathbf{k} \approx \mathbf{v}_{air} \cdot \mathbf{k} \tag{4.41}$$

FIGURE 4.4   Mixing of sugar.

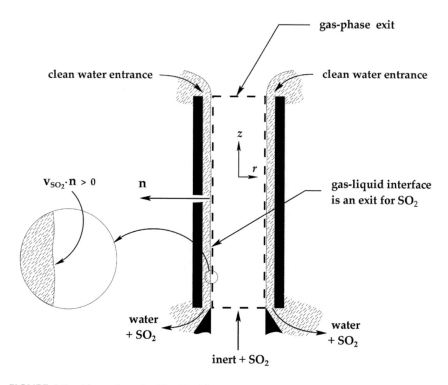

FIGURE 4.5 Absorption of sulfur dioxide.

in which $\mathbf{k}$ is the unit vector pointing in the z-direction. The situation for the radial components of $\mathbf{v}_{SO_2}$ and $\mathbf{v}_{air}$ is quite different because the radial components are *normal to the gas-liquid interface*. Since both nitrogen and oxygen are only *slightly soluble* in water, we neglect the transport of air at the gas-liquid interface and write

$$\mathbf{v}_{air} \cdot \mathbf{n} = 0, \quad \text{at the gas-liquid interface} \tag{4.42}$$

On the other hand, the sulfur dioxide is *very soluble* in water and the rate at which it leaves the gas stream and enters the liquid stream is significant. Because of this the radial component of $\mathbf{v}_{SO_2}$ must be positive and we express this idea as

$$\mathbf{v}_{SO_2} \cdot \mathbf{n} > 0, \quad \text{at the gas-liquid interface} \tag{4.43}$$

It should be clear that $\mathbf{v}_{SO_2} \cdot \mathbf{n}$ is a velocity associated with a *diffusion process* while $\mathbf{v}_{SO_2} \cdot \mathbf{k}$ is a velocity associated with a *convection process*. The latter process is generally much, much larger than the former, i.e.,

$$\underbrace{\mathbf{v}_{SO_2} \cdot \mathbf{k}}_{convection} \gg \underbrace{\mathbf{v}_{SO_2} \cdot \mathbf{n}}_{diffusion} \tag{4.44}$$

The motion of a chemical species can result from a *force applied to the fluid*, i.e., a fan might be used to move the gas mixture through the tube illustrated in Figure 4.5. The motion of a chemical species can also result from a *concentration gradient* such as the gradient that causes the sugar to diffuse throughout the teacup illustrated in

Figure 4.3. Because the motion of chemical species can be caused by both applied forces and concentration gradients, it is reasonable to decompose the species velocity into two parts: the *mass average velocity* and the *mass diffusion velocity*. We represent this decomposition as

$$\mathbf{v}_{SO_2} = \underbrace{\mathbf{v}}_{\substack{mass\ average \\ velocity}} + \underbrace{\mathbf{u}_{SO_2}}_{\substack{mass\ diffusion \\ velocity}} \qquad (4.45)$$

At *entrances and exits*, such as those illustrated in Figure 4.5, the *diffusion velocity* in the z-direction is usually small compared to the *mass average velocity* in the z-direction and Eq. 4.45 can be *approximated* by

$$\mathbf{v}_{SO_2} \cdot \mathbf{k} = \mathbf{v} \cdot \mathbf{k}, \quad \text{negligible diffusion velocity} \qquad (4.46)$$

In this text, we will repeatedly make use of this simplification in order to solve a variety of problems without the need to predict the diffusion velocity. However, in subsequent courses the subject of mass transfer across fluid-fluid interfaces will be studied in detail, and those studies will require a complete understanding of the role of the diffusion velocity.

In order to reinforce our thoughts about the species velocity, the mass average velocity, and the mass diffusion velocity, we return to the definition of the mass average velocity given by Eq. 4.38 and express the z-component of $\mathbf{v}$ according to

$$\mathbf{v} \cdot \mathbf{k} = \omega_{SO_2} (\mathbf{v}_{SO_2} \cdot \mathbf{k}) + \omega_{air} (\mathbf{v}_{air} \cdot \mathbf{k}) \qquad (4.47)$$

in which the mass fractions are constrained by

$$\omega_{SO_2} + \omega_{air} = 1 \qquad (4.48)$$

On the basis of the approximation given by Eq. 4.41, we conclude that the species velocities are constrained by

$$\mathbf{v} \cdot \mathbf{k} \approx \mathbf{v}_{SO_2} \cdot \mathbf{k} \approx \mathbf{v}_{air} \cdot \mathbf{k} \qquad (4.49)$$

However, the radial component of the species velocities is an entirely different matter. Once again, we can use Eq. 4.38 to obtain

$$\mathbf{v} \cdot \mathbf{n} = \omega_{SO_2} (\mathbf{v}_{SO_2} \cdot \mathbf{n}) + \omega_{air} (\mathbf{v}_{air} \cdot \mathbf{n}), \quad \text{at the gas-liquid interface} \qquad (4.50)$$

and on the basis of Eq. 4.42 this reduces to

$$\mathbf{v} \cdot \mathbf{n} = \omega_{SO_2} (\mathbf{v}_{SO_2} \cdot \mathbf{n}), \quad \text{at the gas-liquid interface} \qquad (4.51)$$

Under these circumstances, we see that

$$\mathbf{v} \cdot \mathbf{n} \neq \mathbf{v}_{SO_2} \cdot \mathbf{n} \neq \mathbf{v}_{air} \cdot \mathbf{n}, \quad \text{at the gas-liquid interface} \qquad (4.52)$$

and the type of approximation indicated by Eq. 4.46 is *not valid*. In this text, we will study a series of macroscopic mass balance problems for which Eq. 4.46 represents a reasonable approximation; however, one must always remember that neglect of the *diffusion velocity* is a very delicate matter, and it is considered further in Problem 4.10. As we have mentioned before, the *difference* between species velocities is responsible for separation and purification, and it is absolutely necessary in order for chemical reactions to occur.

4.4 MEASURES OF VELOCITY

In the previous section, we defined the *mass average velocity* according to

$$\mathbf{v} = \sum_{B=1}^{B=N} \omega_B\, \mathbf{v}_B, \quad \text{mass average velocity} \tag{4.53}$$

and we noted that the total mass flux vector was given by

$$\rho\, \mathbf{v} = \sum_{B=1}^{B=N} \rho_B\, \mathbf{v}_B = \left\{ \begin{array}{l} \text{total mass} \\ \text{flux vector} \end{array} \right\} \tag{4.54}$$

By analogy, we define the *molar average velocity by*

$$\mathbf{v}^* = \sum_{B=1}^{B=N} x_B\, \mathbf{v}_B, \quad \text{molar average velocity} \tag{4.55}$$

and it follows that the total molar flux vector is given by

$$c\, \mathbf{v}^* = \sum_{B=1}^{B=N} c_B\, \mathbf{v}_B = \left\{ \begin{array}{l} \text{total molar} \\ \text{flux vector} \end{array} \right\} \tag{4.56}$$

In the previous section, we used a decomposition of the species velocity of the form

$$\mathbf{v}_A = \mathbf{v} + \mathbf{u}_A \tag{4.57}$$

so that the species *mass flux vector* could be expressed as

$$\rho_A \mathbf{v}_A = \underbrace{\rho_A \mathbf{v}}_{\substack{\text{convective} \\ \text{flux}}} + \underbrace{\rho_A \mathbf{u}_A}_{\substack{\text{diffusive} \\ \text{flux}}} \tag{4.58}$$

In dealing with the molar flux vector, one finds it convenient to express the species velocity as

$$\mathbf{v}_A = \mathbf{v}^* + \mathbf{u}_A^* \tag{4.59}$$

so that the molar flux takes the form

$$c_A \mathbf{v}_A = \underbrace{c_A \mathbf{v}^*}_{\substack{convective \\ flux}} + \underbrace{c_A \mathbf{u}_A^*}_{\substack{diffusive \\ flux}} \qquad (4.60)$$

When *convective transport dominates*, the species velocity, the mass average velocity and the molar average velocity are all essentially equal, i.e.,

$$\mathbf{v}_A = \mathbf{v} = \mathbf{v}^* \qquad (4.61)$$

This is the situation that we encounter most often in our study of material balances, and we will make use of this result repeatedly to determine the flux of species A at *entrances and exits*. While Eq. 4.61 is widely used to describe velocities at entrances and exits, one must be very careful about the general use of this approximation. If Eq. 4.61 were true for all species under all conditions, there would be no separation, no purification, no mixing, and no chemical reactions!

Under certain circumstances, we may want to use a *total mole balance* and this is obtained by summing Eq. 4.17 over all N species in order to obtain

$$\frac{d}{dt} \int_V \sum_{A=1}^{A=N} c_A \, dV + \int_A \sum_{A=1}^{A=N} (c_A \, \mathbf{v}_A) \cdot \mathbf{n} \, dA = \int_V \sum_{A=1}^{A=N} R_A \, dV \qquad (4.62)$$

Use of the definitions given by Eqs. 4.28 and 4.56 allows us to write this result in the form

$$\frac{d}{dt} \int_V c \, dV + \int_A c \mathbf{v}^* \cdot \mathbf{n} \, dA = \int_V \sum_{A=1}^{A=N} R_A \, dV \qquad (4.63)$$

Here we should note that the overall net rate of production of moles need not be zero, thus the overall mole balance is more complex than the overall mass balance.

4.5   MOLAR FLOW RATES AT ENTRANCES AND EXITS

Here we direct our attention to the macroscopic mole balance for a fixed control volume

$$\frac{d}{dt} \int_V c_A \, dV + \int_A c_A \, \mathbf{v}_A \cdot \mathbf{n} \, dA = \int_V R_A \, dV, \quad A = 1, 2, \ldots, N \qquad (4.64)$$

with the intention of evaluating $c_A \, \mathbf{v}_A \cdot \mathbf{n}$ at entrances and exits. The area integral of the molar flux, $c_A \mathbf{v}_A \cdot \mathbf{n}$, can be represented as

$$\int_A c_A \, \mathbf{v}_A \cdot \mathbf{n} \, dA = \int_{A_e} c_A \, \mathbf{v}_A \cdot \mathbf{n} \, dA + \int_{A_i} c_A \, \mathbf{v}_A \cdot \mathbf{n} \, dA \qquad (4.65)$$

where A_e represents the entrances and exits at which *convection dominates* and A_i represents an *interfacial area* over which diffusive fluxes may dominate. In this text, our primary interest is the study of control volumes having entrances and exits at which *convective transport* is much more important than *diffusive transport*. We have illustrated this type of control volume in Figure 4.6 in which the entrances and exits for both the water and the air are at the top and bottom of the column. The surface of the control volume that coincides with the liquid-solid interface represents an impermeable boundary at which $c_A \mathbf{v}_A \cdot \mathbf{n} = 0$. For systems of this type, we express Eq. 4.65 as

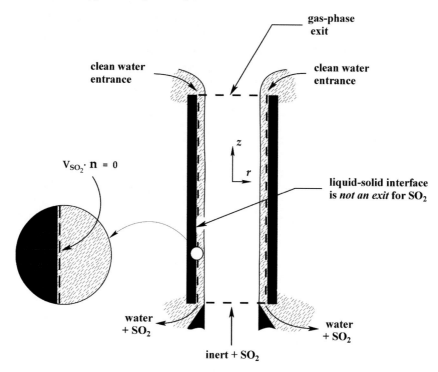

FIGURE 4.6 Entrances and exits at which convection dominates.

$$\int_A c_A \mathbf{v}_A \cdot \mathbf{n} \, dA = \int_{A_e} c_A \mathbf{v}_A \cdot \mathbf{n} \, dA$$

$$= \int_{A_{entrances}} c_A \mathbf{v}_A \cdot \mathbf{n} \, dA + \int_{A_{exits}} c_A \mathbf{v}_A \cdot \mathbf{n} \, dA \qquad (4.66)$$

In order to simplify our discussion about the flux at entrances and exits, we direct our attention to *an exit* and express the molar flow rate at that exit as

$$\dot{M}_A = \int_{A_{exit}} c_A \mathbf{v}_A \cdot \mathbf{n} \, dA \qquad (4.67)$$

On the basis of the discussion in Sec. 4.2, we assume that the *diffusive flux* is negligible ($\mathbf{v}_A \cdot \mathbf{n} = \mathbf{v} \cdot \mathbf{n}$) so that the above result takes the form

$$\dot{M}_A = \int_{A_{exit}} c_A \mathbf{v} \cdot \mathbf{n} \, dA \tag{4.68}$$

It is possible that both c_A and $\mathbf{v}$ vary across the exit and a detailed evaluation of the area integral is required in order to determine the molar flow rate of species A. In general, this is not the case; however, it is very important to be aware of this possibility.

4.5.1   AVERAGE CONCENTRATIONS

In Sec. 3.2.1, we defined a volume average density and we use the same definition here for the volume average concentration given by

$$\langle c_A \rangle = \frac{1}{V} \int_V c_A \, dV, \quad \text{volume average concentration} \tag{4.69}$$

At entrances and exits, we often work with the "bulk concentration" or "cup mixed concentration" that was defined earlier in Sec. 3.2.1. For the concentration, c_A, we repeat the definition according to

$$\langle c_A \rangle_b = \frac{1}{Q_{exit}} \int_{A_{exit}} c_A \mathbf{v} \cdot \mathbf{n} \, dA = \frac{\displaystyle\int_{A_{exit}} c_A \mathbf{v} \cdot \mathbf{n} \, dA}{\displaystyle\int_{A_{exit}} \mathbf{v} \cdot \mathbf{n} \, dA} \tag{4.70}$$

In terms of the *bulk concentration*, the molar flow rate given by Eq. 4.68 can be expressed as

$$\dot{M}_A = \langle c_A \rangle_b \, Q_{exit} \tag{4.71}$$

in which it is understood that $\dot{M}_A$ and $\langle c_A \rangle_b$ represent the molar flow rate and bulk concentration at the exit. In addition to the bulk or cup-mixed concentration, one may encounter the area average concentration denoted by $\langle c_A \rangle$ and defined at an exit according to

$$\langle c_A \rangle = \frac{1}{A_{exit}} \int_{A_{exit}} c_A \, dA \tag{4.72}$$

If the concentration is constant over A_{exit}, the area average concentration is equal to this constant value, i.e.

$$\langle c_A \rangle = c_A, \quad \text{when } c_A \text{ is constant} \tag{4.73}$$

We often refer to this condition as a "flat" concentration profile, and for this case we have

$$\langle c_A \rangle_b = \langle c_A \rangle = c_A, \quad \text{flat concentration profile} \tag{4.74}$$

Under these circumstances, the molar flow rate takes the form

$$\dot{M}_A = c_A Q_{exit}, \quad \textit{flat concentration profile} \tag{4.75}$$

The conditions for which c_A can be treated as a constant over an exit or an entrance are likely to occur in many practical applications.

When the flow is turbulent, there are rapid velocity fluctuations about the mean or time-averaged velocity. The velocity fluctuations tend to create uniform velocity profiles and they play a crucial role in the transport of mass *orthogonal* to the direction of the mean flow. The contribution of turbulent fluctuations to mass transport *parallel* to the direction of the mean flow can normally be neglected and we will do so in our treatment of macroscopic mass balances. In subsequent courses on fluid mechanics and mass transfer, the influence of turbulence will be examined more carefully. In our treatment, we will make use of the reasonable approximation that the turbulent velocity profile is *flat* and this means that $\mathbf{v} \cdot \mathbf{n}$ is constant over A_{exit}. Both turbulent and laminar velocity profiles are illustrated in Figure 4.7 and there we see that the velocity for turbulent flow is nearly constant over a major portion of the flow field. If we make the "flat velocity profile" assumption, we can express Eq. 4.70 as

$$\langle c_A \rangle_b = \frac{\displaystyle\int_{A_{exit}} c_A \mathbf{v} \cdot \mathbf{n}\, dA}{\displaystyle\int_{A_{exit}} \mathbf{v} \cdot \mathbf{n}\, dA} = \frac{\displaystyle\int_{A_{exit}} c_A\, dA}{\displaystyle\int_{A_{exit}} dA} \frac{\mathbf{v} \cdot \mathbf{n}}{\mathbf{v} \cdot \mathbf{n}}$$

$$= \frac{1}{A_{exit}} \int_{A_{exit}} c_A\, dA = \langle c_A \rangle \tag{4.76}$$

For this case, the molar flow rate at the exit takes the form

$$\dot{M}_A = \langle c_A \rangle Q_{exit}, \quad \textit{flat velocity profile} \tag{4.77}$$

To summarize, we note that Eq. 4.71 is an *exact* representation of the molar flow rate in terms of the *bulk concentration* and the volumetric flow rate.

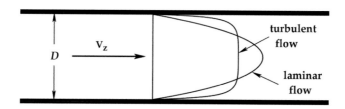

FIGURE 4.7   Laminar and turbulent velocity profiles for flow in a tube.

When the concentration profile can be *approximated* as flat, the molar flow rate can be represented in terms of the *constant concentration* and the volumetric flow rate as indicated by Eq. 4.75. When the velocity profile can be *approximated* as flat, the molar flow rate can be represented in terms of the *area average concentration* and the volumetric flow rate as indicated by Eq. 4.77. If one is working with the species mass balance given by Eq. 4.7, the development represented by Eqs. 4.64–4.77 can be applied simply by replacing c_A with ρ_A.

4.6   ALTERNATE FLOW RATES

There are a number of relations between species flow rates and total flow rates that are routinely used in solving macroscopic mass or mole balance problems *provided* that either the velocity profile is flat or the concentration profile is flat. For example, we can always write Eq. 4.68 in the form

$$\dot{M}_A = \int_{A_{exit}} c_A \mathbf{v} \cdot \mathbf{n} \, dA = \int_{A_{exit}} \frac{c_A}{c} c \mathbf{v} \cdot \mathbf{n} \, dA = \int_{A_{exit}} x_A c \mathbf{v} \cdot \mathbf{n} \, dA \qquad (4.78)$$

If either $c\mathbf{v} \cdot \mathbf{n}$ or x_A is constant over the area of the exit, we can express this result as

$$\dot{M}_A = \langle x_A \rangle \dot{M} \qquad (4.79)$$

where $\dot{M}$ is the total molar flow rate defined by

$$\dot{M} = \sum_{A=1}^{A=N} \dot{M}_A \qquad (4.80)$$

If the individual molar flow rates are known and one desires to determine the area averaged mole fraction at an entrance or an exit, it is given by

$$\langle x_A \rangle = \dot{M}_A \bigg/ \sum_{B=1}^{B=N} \dot{M}_B \qquad (4.81)$$

provided that either $c\mathbf{v} \cdot \mathbf{n}$ or x_A is constant over the area of the entrance or the exit. It will be left as an exercise for the student to show that similar relations exist between mass fractions and mass flow rates. For example, a form analogous to Eq. 4.79 is given by

$$\dot{m}_A = \langle \omega_A \rangle \dot{m} \qquad (4.82)$$

and the mass fraction at an entrance or an exit can be expressed as

$$\langle \omega_A \rangle = \dot{m}_A \Big/ \sum_{G=1}^{G=N} \dot{m}_G \tag{4.83}$$

One *must* keep in mind that the results given by Eqs. 4.79–4.83 are only valid when either the concentration (density) is constant or the molar (mass) flux is constant over the entrance or exit.

When neither of these simplifications is valid, we express Eq. 4.78 as

$$\dot{M}_A = \int_{A_{exit}} x_A c \mathbf{v} \cdot \mathbf{n} \, dA = \langle x_A \rangle_b \dot{M} \tag{4.84}$$

where $\langle x_A \rangle_b$ is the cup mixed mole fraction of species A. The definition of the mole fraction requires that

$$\sum_{A=1}^{A=N} x_A = 1 \tag{4.85}$$

and it will be left as an exercise for the student to show that

$$\sum_{A=1}^{A=N} \langle x_A \rangle_b = 1 \tag{4.86}$$

This type of constraint on the mole fractions (and mass fractions) applies at every entrance and exit and it often represents an important equation in the *set of equations* that are used to solve macroscopic mass balance problems.

4.7   SPECIES MOLE/MASS BALANCE

In this section, we examine the problem of solving the N equations represented by either Eq. 4.7 or Eq. 4.17 under steady-state conditions in the *absence of chemical reactions*. The distillation process illustrated in Figure 4.8 represents a very simple example. Most distillation processes are more complex and most are integrated into a chemical plant as discussed in Chapter 1. It was also pointed out in Chapter 1 that complex chemical plants can be understood by first understanding the individual units that make up the plants (see Figures 1.5 and 1.6), thus understanding the simple process illustrated in Figure 4.8 is an important step in our studies. For this particular ternary distillation process, we are given the information listed in Table 4.1. Often the input conditions for a process are completely specified, and this means that the input flow rate and all the compositions would be specified. However, in Table 4.1, we have not specified x_C since this mole fraction will be determined by Eq. 4.85. If we list this mole fraction as $x_C = 0.5$, we would be *over-specifying* the problem and this would lead to difficulties with our degree of freedom analysis.

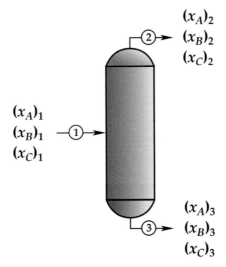

$$(x_A)_2$$
$$(x_B)_2$$
$$(x_C)_2$$

$$(x_A)_1$$
$$(x_B)_1$$
$$(x_C)_1$$

$$(x_A)_3$$
$$(x_B)_3$$
$$(x_C)_3$$

FIGURE 4.8   Distillation column.

In our application of macroscopic balances to single component systems in Chapter 3, we began each problem by identifying a control volume and we listed rules that should be followed for the construction of control volumes. For multi-component systems, we change those rules only slightly to obtain

Rule I. Construct a *primary cut* where information is *required*.
Rule II. Construct a *primary cut* where information is *given*.
Rule III. Join these cuts with a surface located where $\mathbf{v}_A \cdot \mathbf{n}$ is *known*.
Rule IV. When joining the primary cuts to form control volumes, minimize the number of new or *secondary cuts* since these introduce information that is neither given nor required.
Rule V. Be sure that the surface specified by Rule III encloses regions in which volumetric information is either *given* or *required*.

In Rule III, it is understood that $\mathbf{v}_A$ represents the species velocity for all N species, and in Rule II, it is assumed that the *given* information is necessary for the solution of the problem. For the system illustrated in Figure 4.8, it should be obvious that we need to cut the *entrance* and *exit* streams and then join the cuts as illustrated in Figure 4.9. There we have shown the details of the cut at stream #2, and we have illustrated that the cuts at the entrance and exit streams are joined by a surface that is coincident with the solid-air interface where $\mathbf{v}_A \cdot \mathbf{n} = 0$.

4.7.1   DEGREES-OF-FREEDOM ANALYSIS

In order to solve the macroscopic balance equations for this distillation process, we require that the number of constraining equations be equal to the number of unknowns. To be certain that this is the case, we perform a *degrees-of-freedom analysis* which

TABLE 4.1

Specified Conditions

	Stream #1	Stream #2	Stream #3
	$M_1 = 1,200$ mol/h		
	$x_A = 0.3$	$x_A = 0.6$	$x_A = 0.1$
	$x_B = 0.2$	$x_B = 0.3$	

consists of three parts. We begin this analysis with a *generic part* in which we identify the process variables that apply to a *single control volume* in which there are N molecular species and M streams. We assume that every molecular species is present in every stream, and this leads to the *generic degrees of freedom*. Having determined the generic degrees of freedom, we direct our attention to the *generic specifications and constraints* which also apply to the control volume in which there are N molecular species and M streams. Finally, we consider the *particular specifications and constraints* that reduce the generic degrees of freedom to zero if we have a well-posed problem in which all process variables can be determined. If the last part of our analysis does not reduce the degrees of freedom to zero, we need *more information* in order to solve the problem. The inclusion of chemical reactions in the degree of freedom analysis will be delayed until we study stoichiometry in Chapter 6.

The *first step* in our analysis is to prepare a list of the process variables, and this leads to

Mole fractions: $(x_A)_i, (x_B)_i, (x_C)_i$ $i = 1, 2, 3$ (4.87)

Molar flow rates: $\dot{M}_i$, $i = 1, 2, 3$ (4.88)

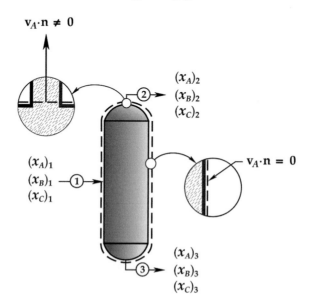

FIGURE 4.9   Control volume for distillation column.

For a system containing three molecular species and having three streams, we determine that there are twelve generic process variables as indicated below.

I. Three mole fractions in each of three streams   9
II. Three molar flow rates   3

For this process, the generic degrees of freedom are given by

Generic Degrees of Freedom (A)   12

In this first step, it is important to recognize that we have assumed that all species are present in all streams, and it is for this reason that we obtain the *generic* degrees of freedom.

The *second step* in this process is to determine the *generic specifications and constraints* associated with a system containing three molecular species and three streams. In order to solve this ternary distillation problem, we will make use of the three molecular species balances given by

$$\text{Species balances:} \quad \int_A x_A\, c\mathbf{v}\cdot\mathbf{n}\,dA = 0 \quad \int_A x_B\, c\mathbf{v}\cdot\mathbf{n}\,dA = 0 \quad \int_A x_C\, c\mathbf{v}\cdot\mathbf{n}\,dA = 0 \quad (4.89)$$

along with the three mole fraction constraints that apply at the streams that are cut by the control volume illustrated in Figure 4.9.

$$\text{Mole fraction constraints:} \qquad (x_A)_i + (x_B)_i + (x_C)_i = 1, \quad i = 1,2,3 \qquad (4.90)$$

We list these specifications and constraints as

I. Balance equations for *three* molecular species   3
II. Mole fraction constraints for the *three* streams   3

and this leads to the *generic* specifications and constraints given by

Generic Specifications and Constraints (B)   6

Moving on to the *third step* in our degree of freedom analysis, we list the *particular* specifications and constraints according to

I. Conditions for Stream #1: $\dot{M}_1 = 1200$ mol/h, $x_A = 0.3$, $x_B = 0.2$ 3
II. Conditions for Stream #2: $x_A = 0.6$, $x_B = 0.3$ 2
III. Conditions for Stream #3: $x_A = 0.1$ 1

This leads us to the *particular* specifications and constraints indicated by

Particular Specifications and Constraints (C)   6

TABLE 4.2

Degrees-of-Freedom

Stream Variables	
Compositions	$N \times M = 9$
flow rates	$M = 3$
Generic Degrees of Freedom (**A**)	$(N \times M) \pm M = \mathbf{12}$
Number of Independent Balance Equations	
mass/mole balance equations	$N = 3$
Number of Constraints for Compositions	$M = 3$
Generic Constraints (**B**)	$N \pm M = \mathbf{6}$
Specified Stream Variables	
Compositions	5
flow rates	1
Constraints for Compositions	0
Auxiliary Constraints	0
Particular Specifications and Constraints (**C**)	6
Degrees of Freedom (**A** – **B** – **C**)	**0**

and we can see that there are *zero degrees of freedom* for this problem. We sum-marize our degree of freedom analysis in Table 4.2 that provides a template for subsequent problems in which we have N molecular species and M streams.

When developing the particular specifications and constraints, it is extremely important to understand that the three mole fractions can be specified *only* in the following manner:

    I. *None* of the mole fractions are specified in a particular stream.
    II. *One* of the mole fractions is specified in a particular stream.
    III. *Two* of the mole fractions are specified in a particular stream.

The point here is that one cannot specify all three mole fractions in a particular stream because of the constraint on the mole fractions given by Eq. 4.90. If one specifies all three mole fractions in a particular stream, Eq. 4.90 for that stream must be deleted and the generic specifications and constraints are *no longer generic*.

There are two important results associated with this degree of freedom analysis. First, we are certain that a solution exits, and this provides motivation for persevering when we encounter difficulties. Second, we are now familiar with the nature of this problem and this should help us to organize a procedure for the development of a solution.

4.7.2 SOLUTION OF MACROSCOPIC BALANCE EQUATIONS

Before beginning the solution procedure, we should clearly identify what is known and what is unknown, and we do this with an extended version of Table 4.1 (see Table 4.3).

When the spaces identified by question marks have been filled with results, our solution will be complete. We begin our analysis with the simplest calculations

TABLE 4.3

Specified and Unknown Conditions

Stream #1	Stream #2	Stream #3
$M_1 = 1{,}200$ mol/h	?	?
$x_A = 0.3$	$x_A = 0.6$	$x_A = 0.1$
$x_B = 0.2$	$x_B = 0.3$	?
?	?	?

TABLE 4.4

Unknowns to Be Determined

Stream #1	Stream #2	Stream #3
$M_1 = 1{,}200$ mol/h	?	?
$x_A = 0.3$	$x_A = 0.6$	$x_A = 0.1$
$x_B = 0.2$	$x_B = 0.3$	$x_B = 0.9$
		$- x_C$
$x_C = 0.5$	$x_C = 0.1$	?

and make use of the constraints given by Eqs. 4.90. These can be used to express Table 4.3, as follows (see Table 4.4).

This table indicates that we have *three unknowns* to be determined on the basis of the *three species balance equations*. Use of the results given in Table 4.4 allows us to express the balance equations given by Eqs. 4.89 as

Species A:
$$0.6\,\dot{M}_2 + 0.1\,\dot{M}_3 + 0 = 360 \text{ mol/h} \tag{4.91a}$$

Species B:
$$0.3\,\dot{M}_2 + 0.9\,\dot{M}_3 - \underbrace{\dot{M}_3\,\langle x_C \rangle_3}_{bi-linear} = 240 \text{ mol/h} \tag{4.91b}$$

Species C:
$$0.1\,\dot{M}_2 + 0 + \underbrace{\dot{M}_3\,\langle x_C \rangle_3}_{bi-linear} = 600 \text{ mol/h} \tag{4.91c}$$

in which the *product* of unknowns, $\dot{M}_3$ and $\langle x_C \rangle_3$, has been identified as a *bi-linear form*. This is different from a *linear form* in which $\dot{M}_3$ and $\langle x_C \rangle_3$ would appear separately, or a *non-linear form* such as $\sqrt{\dot{M}_3}$ or $\langle x_C \rangle_3^2$. In this problem, we are confronted with three unknowns, $\dot{M}_2$, $\dot{M}_3$ and $\langle x_C \rangle_3$, and three equations that can easily be solved to yield

$$\dot{M}_2 = 480 \text{ mol/h}, \quad \dot{M}_3 = 720 \text{ mol/h}, \quad \langle x_C \rangle_3 = 0.767 \tag{4.92}$$

This information can be summarized as shown in Table 4.5.

TABLE 4.5

Solution for Molar Flows and Mole Fractions

Stream #1	Stream #2	Stream #3
$M_1 = 1{,}200$ mol/h	$M_2 = 480$ mol/h	$M_3 = 720$ mol/h
$x_A = 0.3$	$x_A = 0.6$	$x_A = 0.100$
$x_B = 0.2$	$x_B = 0.3$	$x_B = 0.133$
$x_C = 0.5$	$x_C = 0.1$	$x_C = 0.767$

The structure of this ternary distillation process is typical of macroscopic mass balance problems for multicomponent systems. These problems become increasing complex (in the algebraic sense) as the number of components increases and as chemical reactions are included, thus it is important to understand the *general structure*. Macroscopic mass balance problems are always linear in terms of the compositions and flow rates even though these quantities may appear in bi-linear forms. This is the case in Eqs. 4.91 where an unknown flow rate is multiplied by an unknown composition; however, these equations are still linear in $\dot{M}_3$ and $\langle x_C \rangle_3$, thus a unique solution is possible. When chemical reactions occur, and the reaction rate expressions (see Chapter 9) are non-linear, numerical methods are generally necessary and one must be aware that *nonlinear problems may have more than one solution or no solution.*

4.7.3   SOLUTION OF SETS OF EQUATIONS

To illustrate a classic procedure for solving sets of algebraic equations, we direct our attention to Eqs. 4.91. We begin by *eliminating* the term, $\dot{M}_3 \langle x_C \rangle_3$, from Eq. 4.91b to obtain the following pair of linear equations:

Species A: $0.6 \dot{M}_2 + 0.1 \dot{M}_3 = 360$ mol/h   (4.93a)

Species B: $0.4 \dot{M}_2 + 0.9 \dot{M}_3 = 840$ mol/h   (4.93b)

To solve this set of linear equations, we make use of a simple scheme known as Gaussian elimination. We begin by dividing Eq. 4.93a by the coefficient 0.6 in order to obtain the following form of Eqs. 4.93a and 4.93b:

$$\dot{M}_2 + 0.1667 \dot{M}_3 = 600 \text{ mol/h} \qquad (4.94a)$$

$$0.4 \dot{M}_2 + 0.9 \dot{M}_3 = 840 \text{ mol/h} \qquad (4.94b)$$

Next, we multiply the first equation by 0.4 and subtract that result from the second equation to provide

$$\dot{M}_2 + 0.1667 \dot{M}_3 = 600 \text{ mol/h} \qquad (4.95a)$$

$$0.8333 \dot{M}_3 = 600 \text{ mol/h} \qquad (4.95b)$$

We now divide the last equation by 0.8333 to obtain the solution for the unknown molar flow rate, $\dot{M}_3$.

$$\dot{M}_2 + 0.1667\,\dot{M}_3 = 600 \text{ mol/h} \tag{4.96a}$$

$$\dot{M}_3 = 720 \text{ mol/h} \tag{4.96b}$$

At this point, we begin the procedure of "back substitution" which requires that Eq. 4.96b be substituted into Eq. 4.96a in order to obtain the final solution for the two molar flow rates.

$$\dot{M}_2 = 480 \text{ mol/h} \tag{4.97a}$$

$$\dot{M}_3 = 720 \text{ mol/h} \tag{4.97b}$$

The procedure leading from Eqs. 4.93 to the solution given by Eqs. 4.97 is trivial for a pair of equations; however, if we were working with a five-component system the algebra would be overwhelmingly difficult and a computer solution would be required.

4.8 MULTIPLE UNITS

When more than a single unit is under consideration, some care is required in the choice of control volumes, and the two-column distillation unit illustrated in Figure 4.10 provides an example. In that figure, we have indicated that all the mass fractions in the streams entering and leaving the two-column unit are specified, i.e., the problem is *over-specified* and we will need to be careful in our degree of freedom analysis. In addition to the mass fractions, we are also given that the mass flow rate to the first column is $\dot{m}_1 = 1{,}000$ kg/h. On the basis of this information, we want to predict the following:

1. The mass flow rate of both overhead (or *distillate*) streams (streams #2 and #3).
2. The mass flow rate of the bottoms from the second column (stream #4).
3. The mass flow rate of the feed to the second column (stream #5).

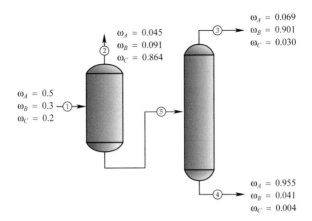

FIGURE 4.10 Two-column distillation unit.

We begin the process of constructing control volumes by making the *primary cuts* shown in Figure 4.11. Those cuts have been made where information is *given* (stream #1) and information is *required* (streams #2, #3, #4, and #5). In order to join the primary cuts to form control volumes, we are forced to construct two control volumes. We first form a control volume that connects three *primary cuts* (streams #1, #2, and #5) and encloses the first column of the two-column distillation unit.

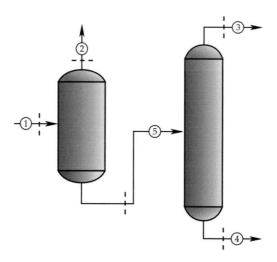

FIGURE 4.11   Primary cuts for the two-column distillation unit.

In order to construct a control volume that joins the primary cuts of streams #3 and #4, we have two choices. One choice is to *enclose the second column* by joining the primary cuts of streams #3, #4, and #5, while the second choice is illustrated in Figure 4.12 where we have shown Control Volumes I and II. If Control Volume II were constructed so that it joined streams #3, #4, and #5, it would not be connected to the *single source* of the necessary information, i.e., the mass flow rate of stream #1. In that case, the information about stream #5 would cancel in the balance equations and we would not be able to determine the mass flow rate in stream #5.

Since the data are given in terms of mass fractions and the mass flow rate of stream #1, the appropriate macroscopic balance is given by Eq. 4.7. For steady-state conditions in the absence of chemical reactions, the three species mass balances are given by

$$\int_A \rho_A \mathbf{v}_A \cdot \mathbf{n} \, dA = 0, \quad \int_A \rho_B \mathbf{v}_B \cdot \mathbf{n} \, dA = 0, \quad \int_A \rho_C \mathbf{v}_C \cdot \mathbf{n} \, dA = 0 \qquad (4.98)$$

Since convective effects will dominate at the entrances and exits of the two control volumes, we can neglect diffusive effects and express the three mass balances in the form

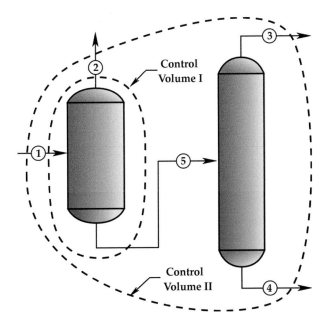

FIGURE 4.12 Control volumes for two-column distillation unit.

Species balances: $\displaystyle\int_A \omega_A \rho \mathbf{v}\cdot\mathbf{n}\,dA = 0 \quad \int_A \omega_B \rho \mathbf{v}\cdot\mathbf{n}\,dA = 0 \quad \int_A \omega_C \rho \mathbf{v}\cdot\mathbf{n}\,dA = 0$ (4.99)

Here we have represented the fluxes in terms of the mass fractions since the stream compositions are given in terms of mass fractions that are constrained by

Constraints: $\quad (\omega_A)_i + (\omega_B)_i + (\omega_C)_i = 1, \quad i = 1,2,3,4,5$ (4.100)

Before attempting to determine the flow rates in streams 2, 3, 4, and 5, we need to perform a degree-of-freedom analysis to be certain that the problem is well-posed.

In the degree-of-freedom analysis, we begin with Control Volume II and list the process variables as

<u>Control Volume II</u>

Mass fractions: $\quad (\omega_A)_i, (\omega_B)_i, (\omega_C)_i, \quad i = 1, 2, 3, 4$ (4.101)

Mass flow rates: $\quad \dot{m}_i, \quad i = 1, 2, 3, 4$ (4.102)

Next, we indicate the number of process variables explicitly as

1. Three mole fractions in each of four streams   12
2. Four mass molar flow rates   4

For this process the generic degrees of freedom are given by

Generic Degrees of Freedom (A)   <u>16</u>

Moving on to the generic specifications and constraints, we list the *three* molecular species balances given by

Species balances: $\int_A \omega_A \rho \mathbf{v} \cdot \mathbf{n}\, dA = 0 \quad \int_A \omega_B \rho \mathbf{v} \cdot \mathbf{n}\, dA = 0 \quad \int_A \omega_C \rho \mathbf{v} \cdot \mathbf{n}\, dA = 0$

$$(4.103)$$

There are *four* mass-fraction constraints that apply to the streams cut by Control Volume II and these are given by

Constraints: $\qquad (\omega_A)_i + (\omega_B)_i + (\omega_C)_i = 1, \quad i = 1, 2, 3, 4 \qquad (4.104)$

We now move to the *second step* in our degree of freedom analysis that we express as

    I. Balance equations for three molecular species   <u>3</u>
    II. Mole fraction constraints for the four streams   <u>4</u>

Generic Specifications and Constraints (B)   <u>7</u>

Our *third step* in the degree of freedom analysis requires that we list the *particular specifications and constraints* according to

    I. Conditions for Stream #1:
          $\dot{m}_1 = 1{,}000$ kg/h, $\omega_A = 0.5,$ $\omega_B = 0.3$ <u>3</u>
    II. Conditions for Stream #2: $\omega_A = 0.045,$ $\omega_B = 0.091$ <u>2</u>
    III. Conditions for Stream #3: $\omega_A = 0.069,$ $\omega_B = 0.901$ <u>2</u>
    IV. Conditions for Stream #4: $\omega_A = 0.955,$ $\omega_B = 0.041$ <u>2</u>

This leads us to the particular specifications and constraints indicated by

Particular Specifications and Constraints (C)   <u>9</u>

We summarize these results in Table 4.6 which indicates that use of Control Volume II will lead to a well-posed problem. Thus, we can use Eqs. 4.103 and 4.104 to determine the mass flow rates in streams 2, 3, 4 and this calculation is carried out in the following paragraphs.

    Often in problems of this type, it is convenient to work with $N-1$ species mass balances and the total mass balance that is given by Eq. 4.39. In terms of the use of Eqs. 4.103 with Control Volume II, this approach leads to

<u>Control Volume II</u>:

Species A: $\qquad (\omega_A)_2\, \dot{m}_2 + (\omega_A)_3\, \dot{m}_3 + (\omega_A)_4\, \dot{m}_4 = (\omega_A)_1\, \dot{m}_1 \qquad (4.105\text{a})$

Species B: $\qquad (\omega_B)_2\, \dot{m}_2 + (\omega_B)_3\, \dot{m}_3 + (\omega_B)_4\, \dot{m}_4 = (\omega_B)_1\, \dot{m}_1 \qquad (4.105\text{b})$

Total: $\qquad\qquad\qquad \dot{m}_2 + \dot{m}_3 + \dot{m}_4 = \dot{m}_1 \qquad\qquad\qquad (4.105\text{c})$

TABLE 4.6

Degrees-of-Freedom

Stream Variables

Compositions	$N \times M = 12$
flow rates	$M = 4$
Generic Degrees of Freedom (**A**)	$(N \times M) \pm M = \mathbf{16}$
Number of Independent Balance Equations	
mass/mole balance equations	$N = 3$
Number of Constraints for Compositions	$M = 4$
Generic Specifications and Constraints (**B**)	$N + M = \mathbf{7}$
Specified Stream Variables	
Compositions	8
flow rates	1
Constraints for Compositions	0
Auxiliary Constraints	0
Particular Specifications and Constraints (**C**)	9
Degrees of Freedom (**A − B − C**)	**0**

Here we are confronted with *three equations* and *three unknowns*, and our problem is quite similar to that encountered in Sec. 4.7 where our study of a single distillation column led to a set of *two equations* and *two unknowns*. That problem was solved by Gaussian elimination as indicated by Eqs. 4.93–4.97. The same procedure can be used with Eqs. 4.105, and one begins by dividing Eq. 4.105a by $(\omega_A)_2$ to obtain the following results

Control Volume II:

A: $\dot{m}_2 + [(\omega_A)_3/(\omega_A)_2]\,\dot{m}_3 + [(\omega_A)_4/(\omega_A)_2]\,\dot{m}_4 = [(\omega_A)_1/(\omega_A)_2]\,\dot{m}_1$ (4.106a)

B: $(\omega_B)_2\,\dot{m}_2 + (\omega_B)_3\,\dot{m}_3 + (\omega_B)_4\,\dot{m}_4 = (\omega_B)_1\,\dot{m}_1$ (4.106b)

Total: $\dot{m}_2 + \dot{m}_3 + \dot{m}_4 = \dot{m}_1$ (4.106c)

In order to eliminate $\dot{m}_2$ from Eq. 4.106b, one multiplies Eq. 4.106c by $(\omega_B)_2$ and then subtracts the result from Eq. 4.106b. To eliminate $\dot{m}_2$ from Eq. 4.106c, one need only subtract Eq. 4.106a from Eq. 4.106c. These two operations lead to the following balance equations:

Control Volume II:

A: $\dot{m}_2 + [(\omega_A)_3/(\omega_A)_2]\,\dot{m}_3 + [(\omega_A)_4/(\omega_A)_2]\,\dot{m}_4 = [(\omega_A)_1/(\omega_A)_2]\,\dot{m}_1$ (4.107a)

B: $\{\ldots\}\,\dot{m}_3 + \{\ldots\}\,\dot{m}_4 = \{\ldots\}\,\dot{m}_1$ (4.107b)

Total: $[\ldots]\,\dot{m}_3 + [\ldots]\,\dot{m}_4 = [\ldots]\,\dot{m}_1$ (4.107c)

Clearly, the algebra is becoming quite complex, and it will become worse when we use Eq. 4.107b to eliminate $\dot{m}_3$ from Eq. 4.107c. Without providing the details, we continue the elimination process to obtain the solution to Eq. 4.107c and this leads to the following expression for $\dot{m}_4$:

$$\dot{m}_4 = \dot{m}_1 \frac{\left[1 - \dfrac{(\omega_A)_1}{(\omega_A)_2}\right] - \dfrac{\dfrac{(\omega_B)_1}{(\omega_B)_3}\left[1 - \dfrac{(\omega_B)_2}{(\omega_B)_1}\dfrac{(\omega_A)_1}{(\omega_A)_2}\right]}{\left[1 - \dfrac{(\omega_B)_2}{(\omega_B)_3}\dfrac{(\omega_A)_3}{(\omega_A)_2}\right]}\left[1 - \dfrac{(\omega_A)_3}{(\omega_A)_2}\right]}{\left[1 - \dfrac{(\omega_A)_4}{(\omega_A)_2}\right] - \dfrac{\dfrac{(\omega_B)_4}{(\omega_B)_3}\left[1 - \dfrac{(\omega_B)_2}{(\omega_B)_4}\dfrac{(\omega_A)_4}{(\omega_A)_2}\right]}{\left[1 - \dfrac{(\omega_B)_2}{(\omega_B)_3}\dfrac{(\omega_A)_3}{(\omega_A)_2}\right]}\left[1 - \dfrac{(\omega_A)_3}{(\omega_A)_2}\right]} \tag{4.108}$$

Equally complex expressions can be obtained for $\dot{m}_2$ and $\dot{m}_3$, and the numerical values for the three mass flow rates are given by

$$\dot{m}_2 = 219 \text{ kg/h}, \quad \dot{m}_3 = 288 \text{ kg/h}, \quad \dot{m}_4 = 492 \text{ kg/h} \tag{4.109}$$

In order to determine $\dot{m}_5$, we must make use of the balance equations for Control Volume I that are given by Eqs. 4.103. These can be expressed in terms of two species balances and one total mass balance leading to

Control Volume I:

Species A: $-(\omega_A)_1 \dot{m}_1 + (\omega_A)_2 \dot{m}_2 + (\omega_A)_5 \dot{m}_5 = 0$ $\qquad$ (4.110a)

Species B: $-(\omega_B)_1 \dot{m}_1 + (\omega_B)_2 \dot{m}_2 + (\omega_B)_5 \dot{m}_5 = 0$ $\qquad$ (4.110b)

Total: $-\dot{m}_1 + \dot{m}_2 + \dot{m}_5 = 0$ $\qquad$ (4.110c)

and the last of these quickly leads us to the result for $\dot{m}_5$.

$$\dot{m}_5 = \dot{m}_1 - \dot{m}_2 = 780 \text{ kg/h} \tag{4.111}$$

The algebraic complexity associated with the simple process represented in Figure 4.10 encourages the use of matrix methods. We begin a study of those methods in the following section and we continue to study and to apply matrix methods throughout the remainder of the text. Our studies of stoichiometry in Chapter 6 and reaction kinetics in Chapter 9 rely heavily on matrix methods that are presented as needed. In addition, a general discussion of matrix methods is available in Appendix C.

4.9   MATRIX ALGEBRA

In Sec. 4.7, we examined a distillation process with the objective of determining molar flow rates, and our analysis led to a set of *two equations* and *two unknowns* given by

Species A: $\qquad 0.6\,\dot{M}_2 + 0.1\,\dot{M}_3 = 360$ mol/h $\qquad$ (4.112a)

Species B: $\qquad 0.4\,\dot{M}_2 + 0.9\,\dot{M}_3 = 840$ mol/h $\qquad$ (4.112b)

These mole balances are easily obtained from Eqs. 4.91 and the algebraic effort required to solve them is trivial. In Sec. 4.8, we considered the system illustrated in Figure 4.10 and the analysis led to the following set of *three equations* and *three unknowns*:

Species A: $\qquad (\omega_A)_2\,\dot{m}_2 + (\omega_A)_3\,\dot{m}_3 + (\omega_A)_4\,\dot{m}_4 = (\omega_A)_1\,\dot{m}_1$ $\qquad$ (4.113a)

Species B: $\qquad (\omega_B)_2\,\dot{m}_2 + (\omega_B)_3\,\dot{m}_3 + (\omega_B)_4\,\dot{m}_4 = (\omega_B)_1\,\dot{m}_1$ $\qquad$ (4.113b)

Total: $\qquad \dot{m}_2 + \dot{m}_3 + \dot{m}_4 = \dot{m}_1$ $\qquad$ (4.113c)

The algebraic effort required to solve these three equations for $\dot{m}_2$, $\dot{m}_3$, and $\dot{m}_4$ was considerable as one can see from the solution given by Eq. 4.108. It should not be difficult to imagine that solving sets of four or five equations can become exceedingly difficult to do by hand; however, computer routines are available that can be used to solve virtually any set of equations that have the form given by Eqs. 4.113.

In dealing with sets of many equations, it is convenient to use the language of matrix algebra. For example, in matrix notation we would express Eqs. 4.113 according to

$$
\begin{bmatrix}
(\omega_A)_2 & (\omega_A)_3 & (\omega_A)_4 \\
(\omega_B)_2 & (\omega_B)_3 & (\omega_B)_4 \\
1 & 1 & 1
\end{bmatrix}
\begin{bmatrix}
\dot{m}_2 \\
\dot{m}_3 \\
\dot{m}_4
\end{bmatrix}
=
\begin{bmatrix}
(\omega_A)_1\,\dot{m}_1 \\
(\omega_B)_1\,\dot{m}_1 \\
\dot{m}_1
\end{bmatrix}
\qquad (4.114)
$$

In this representation of Eq. 4.113, the 3×3 matrix of mass fractions multiplies the 3×1 column matrix of mass flow rates to produce a 3×1 column matrix that is equal to the right-hand side of Eq. 4.114. In working with matrices, it is generally convenient to make use of a nomenclature in which subscripts are used to identify the row and column in which an element is located. We used this type of nomenclature in Chapter 2 where the $m \times n$ matrix A was represented by

$$
A =
\begin{bmatrix}
a_{11} & a_{12} & \cdots & a_{1n} \\
a_{21} & a_{22} & \cdots & a_{2n} \\
\cdots & \cdots & \cdots & \cdots \\
a_{m1} & a_{m2} & \cdots & a_{mn}
\end{bmatrix}
\qquad (4.115)
$$

Here the first subscript identifies the *row* in which an element is located while the second subscript identifies the *column*. In Chapter 2, we discussed matrix addition and subtraction, and here we wish to discuss matrix multiplication and the matrix operation that is analogous to division. Matrix multiplication between A and B is defined only if the number of columns of A (in this case n) is equal to the number of rows of B (in this case m). Given an $m \times n$ matrix A and an $n \times p$ matrix B, the product between A and B is illustrated by the following equation:

$$
\begin{bmatrix}
a_{11} & \cdots & a_{1n} \\
\cdot & \cdots & \cdot \\
a_{i1} & \cdots & a_{in} \\
\cdot & \cdots & \cdot \\
a_{m1} & \cdots & a_{mn}
\end{bmatrix}
\begin{bmatrix}
b_{11} & \cdots & b_{1j} & \cdots & b_{1p} \\
\cdot & \cdots & \cdot & \cdots & \cdot \\
\cdot & \cdots & \cdot & \cdots & \cdot \\
\cdot & \cdots & \cdot & \cdots & \cdot \\
b_{n1} & \cdots & b_{nj} & \cdots & b_{np}
\end{bmatrix}
=
\begin{bmatrix}
c_{11} & \cdots & c_{1p} \\
\cdot & \cdots & \cdot \\
\cdot & c_{ij} & \cdot \\
\cdot & \cdots & \cdot \\
c_{m1} & \cdots & c_{mp}
\end{bmatrix}
\tag{4.116}
$$

Here we see that the elements of the *ith* row in matrix A multiply the elements of the *jth* column in matrix B to produce the element in the *ith* row and the *jth* column of the matrix C. For example, the specific element c_{11} is given by

$$
c_{11} = a_{11} b_{11} + a_{12} b_{21} +, \cdots, + a_{1n} b_{n1}
\tag{4.117}
$$

while the general element c_{ij} is given by

$$
c_{ij} = a_{i1} b_{1j} + a_{i2} b_{2j} +, \cdots, + a_{in} b_{nj}
\tag{4.118}
$$

In Eq. 4.116, we see that an $m \times n$ matrix can multiply an $n \times p$ matrix to produce an $m \times p$ matrix, and we see that the matrix multiplication represented by AB is *only defined* when the number of columns in A is equal to the number of rows in B. The matrix multiplication illustrated in Eq. 4.114 conforms to this rule since there are *three columns* in the matrix of mass fractions and *three rows* in the column matrix of mass flow rates. The configuration illustrated in Eq. 4.114 is extremely common since it is associated with the solution of n equations for n unknowns. Our generic representation for this type of matrix equation is given by

$$
Au = b
\tag{4.119}
$$

in which A is a *square matrix*, u is a column matrix of *unknowns*, and b is a column matrix of *knowns*. These matrices are represented explicitly by the following forms:

$$
A =
\begin{bmatrix}
a_{11} & a_{12} & \cdots & a_{1n} \\
a_{21} & a_{22} & \cdots & a_{2n} \\
\cdots & \cdots & \cdots & \cdots \\
a_{n1} & a_{n2} & \cdots & a_{nn}
\end{bmatrix},
\quad
u =
\begin{bmatrix}
u_1 \\
\cdot \\
\cdot \\
u_n
\end{bmatrix},
\quad
b =
\begin{bmatrix}
b_1 \\
\cdot \\
\cdot \\
b_n
\end{bmatrix}
\tag{4.120}
$$

Sometimes, the coefficients in A depend on the unknowns, u, and the matrix equation may be *bi-linear* as indicated in Eqs. 4.91.

The transpose of the matrix A is constructed by interchanging the rows and columns to obtain the new matrix A^{T}:

$$
A =
\begin{bmatrix}
a_{11} & a_{12} & \cdots & a_{1n} \\
a_{21} & a_{22} & \cdots & a_{2n} \\
\cdot & \cdot & \cdots & \cdot \\
a_{m1} & a_{m2} & \cdots & a_{mn}
\end{bmatrix},
\quad
A^{\mathrm{T}} =
\begin{bmatrix}
a_{11} & a_{21} & \cdots & a_{m1} \\
a_{12} & a_{22} & \cdots & a_{m2} \\
\cdot & \cdot & \cdots & \cdot \\
\cdot & \cdot & \cdots & \cdot \\
a_{1n} & a_{2n} & \cdots & a_{mn}
\end{bmatrix}
\tag{4.121}
$$

Here it is important to note that A is an $m \times n$ matrix while A^{T} is an $n \times m$ matrix.

4.9.1 INVERSE OF A SQUARE MATRIX

In order to solve Eq. 4.119, one cannot "divide" by A to determine the unknown, u, since *matrix division is not defined*. There is, however, a related operation involving the *inverse* of a matrix. The inverse of a matrix, A, is another matrix, A^{-1}, such that the product of A and A^{-1} is given by

$$A A^{-1} = I \tag{4.122}$$

in which I is the *identity matrix*. Identity matrices have ones in the diagonal elements and zeros in the off-diagonal elements as illustrated by the following 4×4 matrix:

$$I = \begin{bmatrix} 1 & 0 & 0 & 0 \\ 0 & 1 & 0 & 0 \\ 0 & 0 & 1 & 0 \\ 0 & 0 & 0 & 1 \end{bmatrix} \tag{4.123}$$

For the inverse of a matrix to exist, the matrix must be a *square matrix*, i.e., the number of rows must be equal to the number of columns. In addition, the *determinant* of the matrix must be different from zero. Thus for Eq. 4.122 to be valid, we require that the determinant of A be different from zero, i.e.,

$$|A| \neq 0 \tag{4.124}$$

This type of requirement plays an important role in the derivation of the *pivot theorem* (see Sec. 6.4) that forms the basis for one of the key developments in Chapter 6.

As an example of the use of the inverse of a matrix, we consider the following set of four equations containing four unknowns:

$$\begin{aligned} a_{11}u_1 + a_{12}u_2 + a_{13}u_3 + a_{14}u_4 &= b_1 \\ a_{21}u_1 + a_{22}u_2 + a_{23}u_3 + a_{24}u_4 &= b_2 \\ a_{31}u_1 + a_{32}u_2 + a_{33}u_3 + a_{34}u_4 &= b_3 \\ a_{41}u_1 + a_{42}u_2 + a_{43}u_3 + a_{44}u_4 &= b_4 \end{aligned} \tag{4.125}$$

In compact notation, we would represent these equations according to

$$A u = b \tag{4.126}$$

We suppose that A is invertible and that the inverse, along with u and b, are given by

$$A^{-1} = \begin{bmatrix} \bar{a}_{11} & \bar{a}_{12} & \bar{a}_{13} & \bar{a}_{14} \\ \bar{a}_{21} & \bar{a}_{22} & \bar{a}_{23} & \bar{a}_{24} \\ \bar{a}_{31} & \bar{a}_{32} & \bar{a}_{33} & \bar{a}_{34} \\ \bar{a}_{41} & \bar{a}_{42} & \bar{a}_{43} & \bar{a}_{44} \end{bmatrix}, \quad u = \begin{bmatrix} u_1 \\ u_2 \\ u_3 \\ u_n \end{bmatrix}, \quad b = \begin{bmatrix} b_1 \\ b_2 \\ b_3 \\ b_n \end{bmatrix} \tag{4.127}$$

We can multiply Eq. 4.126 by A^{-1} to obtain

$$A^{-1} A u = I u = u = A^{-1} b \tag{4.128}$$

and the details are given by

$$
\begin{bmatrix} 1 & 0 & 0 & 0 \\ 0 & 1 & 0 & 0 \\ 0 & 0 & 1 & 0 \\ 0 & 0 & 0 & 1 \end{bmatrix} \begin{bmatrix} u_1 \\ u_2 \\ u_3 \\ u_n \end{bmatrix} = \begin{bmatrix} \bar{a}_{11} & \bar{a}_{12} & \bar{a}_{13} & \bar{a}_{14} \\ \bar{a}_{21} & \bar{a}_{22} & \bar{a}_{23} & \bar{a}_{24} \\ \bar{a}_{31} & \bar{a}_{32} & \bar{a}_{33} & \bar{a}_{34} \\ \bar{a}_{41} & \bar{a}_{42} & \bar{a}_{43} & \bar{a}_{44} \end{bmatrix} \begin{bmatrix} b_1 \\ b_2 \\ b_3 \\ b_n \end{bmatrix}
$$

(4.129)

Carrying out the matrix multiplication on the left-hand side leads to

$$
\begin{bmatrix} u_1 \\ u_2 \\ u_3 \\ u_n \end{bmatrix} = \begin{bmatrix} \bar{a}_{11} & \bar{a}_{12} & \bar{a}_{13} & \bar{a}_{14} \\ \bar{a}_{21} & \bar{a}_{22} & \bar{a}_{23} & \bar{a}_{24} \\ \bar{a}_{31} & \bar{a}_{32} & \bar{a}_{33} & \bar{a}_{34} \\ \bar{a}_{41} & \bar{a}_{42} & \bar{a}_{43} & \bar{a}_{44} \end{bmatrix} \begin{bmatrix} b_1 \\ b_2 \\ b_3 \\ b_n \end{bmatrix}
$$

(4.130)

while the more complex multiplication on the right-hand side provides

$$
\begin{bmatrix} u_1 \\ u_2 \\ u_3 \\ u_n \end{bmatrix} = \begin{bmatrix} \bar{a}_{11}\,b_1 & \bar{a}_{12}\,b_2 & \bar{a}_{13}\,b_3 & \bar{a}_{14}\,b_4 \\ \bar{a}_{21}\,b_1 & \bar{a}_{22}\,b_2 & \bar{a}_{23}\,b_3 & \bar{a}_{24}\,b_4 \\ \bar{a}_{31}\,b_1 & \bar{a}_{32}\,b_2 & \bar{a}_{33}\,b_3 & \bar{a}_{34}\,b_4 \\ \bar{a}_{41}\,b_1 & \bar{a}_{42}\,b_2 & \bar{a}_{43}\,b_3 & \bar{a}_{44}\,b_4 \end{bmatrix}
$$

(4.131)

Each element of the column matrix on the left-hand side is equal to the corresponding element on the right-hand side and this leads to the solution for the unknowns given by

$$
\begin{aligned}
u_1 &= \bar{a}_{11}b_1 + \bar{a}_{12}b_2 + \bar{a}_{13}b_3 + \bar{a}_{14}b_4 \\
u_2 &= \bar{a}_{21}b_1 + \bar{a}_{22}b_2 + \bar{a}_{23}b_3 + \bar{a}_{24}b_4 \\
u_3 &= \bar{a}_{31}b_1 + \bar{a}_{32}b_2 + \bar{a}_{33}b_3 + \bar{a}_{34}b_4 \\
u_4 &= \bar{a}_{41}b_1 + \bar{a}_{42}b_2 + \bar{a}_{43}b_3 + \bar{a}_{44}b_4
\end{aligned}
$$

(4.132)

It should be clear that there are two main problems associated with the use of Eqs. 4.125 to obtain the solution for the unknowns given by Eq. 4.132. The first problem is the correct interpretation of a physical process to arrive at the original set of equations, and the second problem is the determination of the inverse of the matrix A.

There are a variety of methods for developing the inverse for a matrix; however, sets of equations are usually solved numerically without calculating the inverse of a matrix. One of the classic methods is known as Gaussian elimination and we can illustrate this technique with the problem that was studied in Sec. 4.8. We can make use of the data provided in Figure 4.10 to express Eqs. 4.105 in the form

Species A: $0.045\,\dot{m}_2 + 0.069\,\dot{m}_3 + 0.955\,\dot{m}_4 = 500$ kg/h   (4.133a)

Species B: $0.091\,\dot{m}_2 + 0.901\,\dot{m}_3 + 0.041\,\dot{m}_4 = 300$ kg/h   (4.133b)

Total: $\dot{m}_2 + \dot{m}_3 + \dot{m}_4 = 1{,}000$ kg/h   (4.133c)

and our objective is to determine the three mass flow rates, $\dot{m}_2$, $\dot{m}_3$, and $\dot{m}_4$. The solution is obtained by making use of the following three rules which are referred to as *elementary row operations*:

I. Any equation in the set can be modified by multiplying or dividing by a non-zero scalar without affecting the solution.
II. Any equation can be added or subtracted from the set without affecting the solution.
III. Any two equations can be interchanged without affecting the solution.

For this particular problem, it is convenient to arrange the three equations in the form

$$
\begin{aligned}
\dot{m}_2 + \dot{m}_3 + \dot{m}_4 &= 1{,}000 \text{ kg/h} \\
0.045\,\dot{m}_2 + 0.069\,\dot{m}_3 + 0.955\,\dot{m}_4 &= 500 \text{ kg/h} \\
0.091\,\dot{m}_2 + 0.901\,\dot{m}_3 + 0.041\,\dot{m}_4 &= 300 \text{ kg/h}
\end{aligned}
\tag{4.134}
$$

We begin by eliminating the first term in the second equation. This is accomplished by multiplying the first equation by 0.045 and subtracting the result from the second equation to obtain

$$
\begin{aligned}
\dot{m}_2 + \dot{m}_3 + \dot{m}_4 &= 1{,}000 \text{ kg/h} \\
0 + 0.024\,\dot{m}_3 + 0.910\,\dot{m}_4 &= 455 \text{ kg/h} \\
0.091\,\dot{m}_2 + 0.901\,\dot{m}_3 + 0.041\,\dot{m}_4 &= 300 \text{ kg/h}
\end{aligned}
\tag{4.135}
$$

Directing our attention to the third equation, we multiply the first equation by 0.091 and subtract the result from the third equation to obtain

$$
\begin{aligned}
\dot{m}_2 + \dot{m}_3 + \dot{m}_4 &= 1{,}000 \text{ kg/h} \\
0 + 0.024\,\dot{m}_3 + 0.910\,\dot{m}_4 &= 455 \text{ kg/h} \\
0 + 0.810\,\dot{m}_3 - 0.050\,\dot{m}_4 &= 209 \text{ kg/h}
\end{aligned}
\tag{4.136}
$$

The second equation can be conditioned by dividing by 0.024 so that the equation set takes the form

$$
\begin{aligned}
\dot{m}_2 + \dot{m}_3 + \dot{m}_4 &= 1{,}000 \text{ kg/h} \\
0 + \dot{m}_3 + 37.917\,\dot{m}_4 &= 18{,}958 \text{ kg/h} \\
0 + 0.810\,\dot{m}_3 - 0.050\,\dot{m}_4 &= 209 \text{ kg/h}
\end{aligned}
\tag{4.137}
$$

We now multiply the second equation by 0.810 and subtract the result from the third equation to obtain

$$
\begin{aligned}
\dot{m}_2 + \dot{m}_3 + \dot{m}_4 &= 1{,}000 \text{ kg/h} \\
0 + \dot{m}_3 + 37.917\,\dot{m}_4 &= 18{,}958 \text{ kg/h} \\
0 + 0 - 30.763\,\dot{m}_4 &= -15{,}147 \text{ kg/h}
\end{aligned}
\tag{4.138}
$$

and division of the third equation by −30.763 leads to

$$\begin{aligned}
\dot{m}_2 + \dot{m}_3 + \quad\;\; \dot{m}_4 \;&= \;1{,}000 \text{ kg/h} \\
0 + \dot{m}_3 + 37.917\,\dot{m}_4 &= 18{,}958.3 \text{ kg/h} \\
0 + 0 + \quad\;\; \dot{m}_4 \;&= \;492.39 \text{ kg/h}
\end{aligned} \tag{4.139}$$

Having worked our way *forward* through this set of three equations in order to determine $\dot{m}_4$, we can work our way *backward* through the set to determine $\dot{m}_3$ and $\dot{m}_2$. The results are the same as we obtained earlier in Sec. 4.8 and we list the result again as

$$\dot{m}_2 = 219 \text{ kg/h}, \quad \dot{m}_3 = 288 \text{ kg/h}, \quad \dot{m}_4 = 492 \text{ kg/h} \tag{4.140}$$

The procedure represented by Eqs. 4.133–4.139 is extremely convenient for automated computation and large systems of equations can be quickly solved using a variety of software. Using systems such as MATLAB or *Mathematica*, problems of this type become quite simple.

4.9.2 DETERMINATION OF THE INVERSE OF A SQUARE MATRIX

The Gaussian elimination procedure described in the previous section is closely related to the determination of the inverse of a square matrix. In order to illustrate how the inverse matrix can be calculated, we consider the coefficient matrix associated with Eq. 4.134 which we express as

$$A = \begin{bmatrix} 1 & 1 & 1 \\ 0.045 & 0.069 & 0.955 \\ 0.091 & 0.901 & 0.041 \end{bmatrix} \tag{4.141}$$

We can express Eq. 4.134 in the compact form represented by Eq. 4.119 to obtain

$$Au = b \tag{4.142}$$

in which u is the column matrix of *unknown* mass flow rates and b is the column matrix of known mass flow rates.

$$u = \begin{bmatrix} \dot{m}_2 \\ \dot{m}_3 \\ \dot{m}_4 \end{bmatrix}, \quad b = \begin{bmatrix} 1{,}000 \text{ kg/h} \\ 500 \text{ kg/h} \\ 300 \text{ kg/h} \end{bmatrix} = \begin{bmatrix} b_2 \\ b_3 \\ b_4 \end{bmatrix} \tag{4.143}$$

We can also make use of the unit matrix given by

$$I = \begin{bmatrix} 1 & 0 & 0 \\ 0 & 1 & 0 \\ 0 & 0 & 1 \end{bmatrix} \tag{4.144}$$

to express Eq. 4.142 in the form

$$Au = Ib \qquad (4.145)$$

Now we wish to repeat the Gaussian elimination used in Sec. 4.9.1, but in this case, we will make use of Eq. 4.145 rather than Eq. 4.142 in order to retain the terms associated with Ib. Multiplying the first of Eqs. 4.134 by -0.045 and adding the result to the second equation leads to

$$
\begin{aligned}
\dot{m}_2 \;+\; \dot{m}_3 \;+\; \dot{m}_4 &= b_2 + 0 + 0 = 1{,}000 \text{ kg/h} \\
0 + 0.024\,\dot{m}_3 + 0.910\,\dot{m}_4 &= -0.045\,b_2 + b_3 + 0 = 455 \text{ kg/h} \qquad (4.146)\\
0.091\,\dot{m}_2 + 0.901\,\dot{m}_3 + 0.041\,\dot{m}_4 &= 0 + 0 + b_4 = 300 \text{ kg/h}
\end{aligned}
$$

Note that this set of equations is identical to Eqs. 4.135 except that we have included the terms associated with Ib. We now proceed with the Gaussian elimination represented by Eqs. 4.136–4.139 to arrive at

$$
\begin{aligned}
\dot{m}_2 + \dot{m}_3 + \dot{m}_4 &= b_2 + 0 + 0 = 1{,}000 \text{ kg/h} \\
0 + \dot{m}_3 + 37.917\,\dot{m}_4 &= -1.875\,b_2 + 41.667\,b_3 + 0 = 18{,}958.3 \text{ kg/h} \\
0 + 0 + \dot{m}_4 &= -0.0464\,b_2 + 1.0971\,b_3 - 0.03251\,b_4 = 492.39 \text{ kg/h}
\end{aligned}
$$

$$(4.147)$$

This result is identical to Eq. 4.139 except for the fact that we have retained the terms in the second column matrix that are associated with Ib in Eq. 4.145. In Sec. 4.9.1, we used Eq. 4.139 to carry out a *backward elimination* in order to solve for the mass flow rates given by Eq. 4.140, and here we want to present that backward elimination explicitly in order to demonstrate that it leads to the inverse of the coefficient matrix, A. This procedure is known as the Gauss-Jordan algorithm and it consists of elementary row operations that reduce the left-hand side of Eq. 4.147 to a diagonal form.

We begin to construct a diagonal form by multiplying the third equation by 37.917 and subtracting it from the second equation so that our set of equations takes the form

$$
\begin{aligned}
\dot{m}_2 + \dot{m}_3 + \dot{m}_4 &= b_2 + 0 + 0 = 1{,}000 \text{ kg/h} \\
0 + \dot{m}_3 + 0 &= -0.1152\,b_2 + 0.0677\,b_3 + 1.2326\,b_4 = 288.42 \text{ kg/h} \qquad (4.148)\\
0 + 0 + \dot{m}_4 &= -0.0464\,b_2 + 1.0971\,b_3 - 0.03251\,b_4 = 492.39 \text{ kg/h}
\end{aligned}
$$

We now multiply the third equation by -1 and add the result to the first equation to obtain

$$
\begin{aligned}
\dot{m}_2 + \dot{m}_3 + 0 &= 1.0464\,b_2 - 1.09711\,b_3 + 0.03251\,b_4 = 507.61 \text{ kg/h} \\
0 + \dot{m}_3 + 0 &= -0.1152\,b_2 + 0.0677\,b_3 + 1.2326\,b_4 = 288.42 \text{ kg/h} \qquad (4.149)\\
0 + 0 + \dot{m}_4 &= -0.0464\,b_2 + 1.0971\,b_3 - 0.03251\,b_4 = 492.39 \text{ kg/h}
\end{aligned}
$$

Finally, we multiply the second equation by -1 and add the result to the first equation to obtain the desired diagonal form

$$\dot{m}_2 + 0 + 0 = 1.1616 b_2 - 1.16484 b_3 - 1.20005 b_4 = 219.187 \text{ kg/h}$$
$$0 + \dot{m}_3 + 0 = -0.1152 b_2 + 0.0677 b_3 + 1.2326 b_4 = 288.42 \text{ kg/h} \quad (4.150)$$
$$0 + 0 + \dot{m}_4 = -0.0464 b_2 + 1.0971 b_3 - 0.03251 b_4 = 492.39 \text{ kg/h}$$

Here we see that the unknown mass flow rates are determined by

$$\dot{m}_2 = 219 \text{ kg/h}, \quad \dot{m}_3 = 288 \text{ kg/h}, \quad \dot{m}_4 = 492 \text{ kg/h} \quad (4.151)$$

where only three significant figures have been listed since it is unlikely that the input data are accurate to more than 1%. The coefficient matrix contained in the central term of Eqs. 4.150 is the inverse of A and we express this result as

$$A^{-1} = \begin{bmatrix} 1.1616 & -1.16484 & -1.20005 \\ -0.1152 & 0.0677 & 1.2326 \\ -0.0464 & 1.0971 & -0.0325 \end{bmatrix} \quad (4.152)$$

The solution procedure that led from Eq. 4.142 to the final answer can be expressed in compact form according to,

$$Au = b, \Rightarrow A^{-1}Au = A^{-1}b, \Rightarrow Iu = A^{-1}b, \Rightarrow u = A^{-1}b \quad (4.153)$$

and in terms of the details of the inverse matrix the last of these four equations can be expressed as

$$\begin{bmatrix} \dot{m}_2 \\ \dot{m}_3 \\ \dot{m}_4 \end{bmatrix} = \begin{bmatrix} 1.1616 & -1.16484 & -1.20005 \\ -0.1152 & 0.0677 & 1.2326 \\ -0.0464 & 1.0971 & -0.0325 \end{bmatrix} \begin{bmatrix} 1,000 \text{ kg/h} \\ 500 \text{ kg/h} \\ 300 \text{ kg/h} \end{bmatrix} \quad (4.154)$$

Given that computer routines are available to carry out the Gauss-Jordan algorithm, it should be clear that the solution of a set of linear mass balance equations is a routine matter.

4.10 PROBLEMS

SECTION 4.1

4.1. Use Eq. 4.20 to obtain a macroscopic *mole balance* for species A in terms of a moving control volume. Indicate how your result can be used to obtain Eq. 4.17.

SECTION 4.2

4.2. Determine the mass density, ρ, for the mixing process illustrated in Figure 4.2.

4.3. A liquid hydrocarbon mixture was made of 295 kg of benzene, 289 kg of toluene, and 287 kg of p-xylene. Assume there is no change of volume upon mixing, i.e., $\Delta V_{mix} = 0$, in order to determine:

1. The species density of each species in the mixture.
2. The total mass density.
3. The mass fraction of each species.

4.4. A gas mixture contains the following quantities: (1) carbon monoxide, 0.5 mol/m^3; (2) carbon dioxide, 0.5 mol/m^3; and (3) hydrogen, 0.6 mol/m^3. Determine the species mass density and mass fraction of each of the components in the mixture.

4.5. The species mass densities of a three-component (A, B, and C) liquid mixture are: acetone, $\rho_A = 326.4 \text{ kg/m}^3$, acetic acid, $\rho_B = 326.4 \text{ kg/m}^3$, and ethanol, $\rho_C = 217.6 \text{ kg/m}^3$. Determine the following for this mixture:

1. The mass fraction of each species in the mixture.
2. The mole fraction of each species in the mixture.
3. The mass of each component required to make one cubic meter of mixture.

4.6. A mixture of gases contains one kilogram of each of the following species: methane (A), ethane (B), propane (C), carbon dioxide (D), and nitrogen (E). Calculate the following:

1. The mole fraction of each species in the mixture
2. The average molecular mass of the mixture

4.7. Two gas streams, having the flow rates and properties indicated in Table 4-7, are mixed in a pipeline. Assume *perfect mixing*, i.e., no change of volume upon mixing, and determine the composition of the mixed stream in mol/m^3.

TABLE 4-7

Composition of Gas Streams

	Stream #1	Stream #2
Mass flow rate	0.226 kg/s	0.296 kg/s
methane	0.48 kg/m^3	0.16 kg/m^3
ethane	0.90 kg/m^3	0.60 kg/m^3
propane	0.88 kg/m^3	0.220 kg/m^3

4.8. Develop a representation for the *mole fraction* of species A in an N-component system in terms of the *mass fractions* and molecular masses of the species. Use the result to prove that the mass fractions and mole fractions in a binary system are equal when the two molecular masses are equal.

4.9. Derive the total mass balance for an arbitrary moving control volume beginning with the species mass balance given by Eq. 4.20.

SECTION **4.3**

4.10. The species velocities, in a *binary system*, can be decomposed according to

$$\mathbf{v}_A = \mathbf{v} + \mathbf{u}_A, \quad \mathbf{v}_B = \mathbf{v} + \mathbf{u}_B \tag{1}$$

in which $\mathbf{v}$ represents the mass average velocity defined by Eq. 4.38. One can use this result, along with the definition of the mass average velocity, *to prove* that

$$\mathbf{v}_A = \mathbf{v}_B + \left(\frac{1}{1 - \omega_A} \right) \mathbf{u}_A \tag{2}$$

This means that the *approximation*, $\mathbf{v}_A \approx \mathbf{v}_B$ requires the restriction

$$\left| \mathbf{u}_A \right| << (1 - \omega_A) \left| \mathbf{v}_A \right| \tag{3}$$

Since $1 - \omega_A$ is always less than one, we can always satisfy this inequality whenever the mass diffusion velocity is small compared to the species velocity, i.e.,

$$\left| \mathbf{u}_A \right| << \left| \mathbf{v}_A \right| \tag{4}$$

For the sulfur dioxide mass transfer process illustrated in Figure 4.4, this means that the approximation

$$\mathbf{v}_{SO_2} \cdot \mathbf{k} \approx \mathbf{v}_{air} \cdot \mathbf{k} \tag{5}$$

is valid whenever the mass diffusion velocity is restricted by

$$\left| \mathbf{u}_{SO_2} \cdot \mathbf{k} \right| << \left| \mathbf{v}_{SO_2} \cdot \mathbf{k} \right| \tag{6}$$

In many practical cases, this restriction is satisfied and all species velocities can be approximated by the mass average velocity.

Next, we direct our attention to the mass transfer process at the gas-liquid interface illustrated in Figure 4.5. If we assume that there is no mass transfer of air into or out of the liquid phase, we can *prove* that

$$\mathbf{u}_{SO_2} \cdot \mathbf{n} = (1 - \omega_{SO_2}) \, \mathbf{v}_{SO_2} \cdot \mathbf{n}, \quad \textit{at the gas} - \textit{liquid interface} \tag{7}$$

Under these circumstances, the mass diffusion velocity is *never small* compared to the species velocity for practical conditions. Thus, the type of approximation indicated by Eq. 4.46 is *never valid* for the component of the velocity *normal* to the gas-liquid interface. As a simplification, we can treat the sulfur dioxide-air system as a binary system with species A representing the sulfur dioxide and species B representing the air.

   a. Use the definition of the mass average velocity given by Eq. 4.38 to prove
      Eq. 2.
   b. Use $\mathbf{v}_{air} \cdot \mathbf{n} = \mathbf{v}_B \cdot \mathbf{n} = 0$ in order to prove Eq. 7.

SECTION 4.4

4.11. A three-component liquid mixture flows in a pipe with a mass averaged velocity of $v = 0.9$ m/s. The density of the mixture is $\rho = 850$ kg/m³. The components of the mixture and their mole fractions are: n-pentane, $x_P = 0.2$, benzene, $x_B = 0.3$, and naphthalene, $x_N = 0.5$. The diffusion fluxes of each component in the streamwise direction are:

$$\text{pentane, } \rho_P u_P = 1.564 \times 10^{-6} \text{ kg/m}^2\text{s}$$

$$\text{benzene, } \rho_B u_B = 1.563 \times 10^{-6} \text{ kg/m}^2\text{s}$$

$$\text{naphthalene, } \rho_N u_N = -3.127 \times 10^{-6} \text{ kg/m}^2\text{s}.$$

Determine the diffusion velocities and the species velocities of the three components. Use this result to determine the molar averaged velocity, v^*. Note that you must use eight significant figures in your computation.

SECTION 4.5

4.12. Sometimes, heterogeneous chemical reactions take place at the walls of tubes in which reactive mixtures are flowing. If species A is being consumed at a tube wall because of a chemical reaction, the concentration profile may be of the form

$$c_A(r) = c_A^o \left[1 - \phi (r/r_o)^2 \right] \tag{1}$$

Here r is the radial position and r_o is the tube radius. The parameter ϕ depends on the net rate of production of chemical species at the wall and the molecular diffusivity, and it is bounded by $0 \le \phi \le 1$. If ϕ is zero, the concentration across the tube is uniform at the value c_A^o. If the flow in the tube is laminar, the velocity profile is given by

$$v_z(r) = 2\langle v_z \rangle \left[1 - (r/r_o)^2 \right] \tag{2}$$

and the volumetric flow rate is

$$Q = \langle v_z \rangle \pi r_o^2 \tag{3}$$

For this process, determine the molar flow rate of species A in terms of c_A^o, ϕ, and $\langle v_z \rangle$. When $\phi = 0.5$, determine the bulk concentration, $\langle c_A \rangle_b$, and the area-averaged concentration, $\langle c_A \rangle$. Use these results to determine the difference between $\dot{M}_A$ and $\langle c_A \rangle Q$.

4.13. A flash unit is used to separate vapor and liquid streams from a liquid stream by lowering its pressure before it enters the flash unit. The feed stream is pure liquid water and the mass flow rate is 1,000 kg/h. Twenty percent (by mass) of the feed stream leaves the flash unit with a density $\rho = 10$ kg/m³. The remainder of the feed stream leaves the flash unit as liquid water with a density $\rho = 1,000$ kg/m³. Determine the following:

1. The mass flow rates of the exit streams in kg/s.
2. volumetric flow rates of exit streams in m³/s.

Section 4.6

4.14. Show that Eq. 4.79 results from Eq. 4.78 when either $c\mathbf{v} \cdot \mathbf{n}$ or x_A is constant over the area of the exit.

4.15. Use Eq. 4.79 to prove Eq. 4.80.

4.16. Derive Eq. 4.82 given that either $\rho\mathbf{v} \cdot \mathbf{n}$ or ω_A is constant over the area of the exit.

4.17. Prove Eq. 4.86.

Section 4.7

4.18. Determine $\dot{M}_3$ and the unknown mole fractions for the distillation process described in Sec. 4.7 subject to the following conditions:

Stream #1	Stream #2	Stream #3
$M_1 = 1{,}200$ mol/h	$M_2 = 250$ mol/h	$M_3 = ?$
$x_A = 0.3$	$x_A = 0.8$	$x_A = ?$
$x_B = 0.2$	$x_B = ?$	$x_B = 0.25$
$x_C = ?$	$x_C = ?$	$x_C = ?$

4.19. A continuous filter is used to separate a clear filtrate from alumina particles in a slurry. The slurry is 30% by weight of alumina that has a specific gravity of 4.5. The cake retains 5% by weight of water. For a feed stream of 1,000 kg/h, determine the following:

1. The mass flow rate of particles and water in the input stream
2. The volumetric flow rate of the inlet stream in m³/s.
3. The mass flow rate of filtrate and cake in kg/s.

4.20. A BTX unit, shown in Figure 4-20, is associated with a refinery that produces benzene, toluene, and xylenes. Stream #1 leaving the reactor-reforming unit has a volumetric flow rate of 10 m³/h and is a mixture of benzene (A), toluene (B), and xylenes (C) with the following composition:

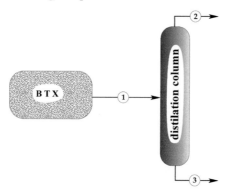

FIGURE 4-20   Reactor and distillation unit.

$$\langle c_A \rangle_1 = 6{,}000 \text{ mol/m}^3, \quad \langle c_B \rangle_1 = 2{,}000 \text{ mol/m}^3, \quad \langle c_C \rangle_1 = 2{,}000 \text{ mol/m}^3$$

Stream (1) is the feed to a distillation unit where the separation takes place according to the following specifications:

1. 98% of the benzene leaves with the distillate stream (stream #2).
2. 99% of the toluene in the feed leaves with the bottoms stream (stream #3)
3. 100% of the xylenes in the feed leaves with the bottoms stream (stream #3).

Assuming that the volumes of components are additive, and using the densities of pure components from Table A2 in the Appendix A, compute the concentration and volumetric flow rate of the distillate (stream #2) and bottoms (stream #3) stream leaving the distillation unit.

4.21. A standard practice in refineries is to use a holding tank in order to mix the light naphtha output of the refinery for quality control. During the first six hours of operation of the refinery, the stream feeding the holding tank at 200 kg/min had 30% by weight of n-pentane, 40% by weight of n-hexane, 30% by weight of n-heptane. During the next 12 hours of operation, the mass flow rate of the feed stream was 210 kg/min and the composition changed to 40% by weight of n-pentane, 40% by weight of n-hexane, and 20% by weight of n-heptane. Determine the following:

The average density of the feed streams
The concentration of the feed streams in mol/m³.

After 12 hours of operation, and assuming the tank was empty at the beginning, determine:

The volume of liquid in the tank in m³.
The concentration of the liquid in the tank, in mol/m³.
The partial density of the species in the tank.

4.22. A distillation column is used to separate a mixture of methanol, ethanol, and isopropyl alcohol. The feed stream, with a mass flow rate of 300 kg/h, has the following composition:

Component	Species Mass Density
methanol	395.5 kg/m³
ethanol	197.3 kg/m³
isopropyl alcohol	196.5 kg/m³

Separation of this mixture of alcohol takes place according to the following specifications:

a. 90% of the methanol in the feed leaves with the distillate stream
b. 5% of the ethanol in the feed leaves with the distillate stream
c. 3% of the isopropyl alcohol in the feed leaves with the distillate stream.

Assuming that the volumes of the components are additive, compute the concentration and volumetric flow rates of the distillate and bottom streams.

4.23. A mixture of ethanol (A) and water (B) is separated in a distillation column. The volumetric flow rate of the feed stream is 5 m³/h. The concentration of ethanol in the feed is $c_A = 2,800$ mol/m³. The distillate leaves the column with a concentration of ethanol $c_A = 13,000$ mol/m³. The volumetric flow rate of distillate is 1 m³/h. How much ethanol is lost through the bottoms of the column, in kilograms of ethanol per hour?

4.24. A ternary mixture of benzene, ethylbenzene, and toluene is fed to a distillation column at a rate of 10^5 mol/h. The composition of the mixture in % moles is: 74% benzene, 20% toluene, and 6% ethylbenzene. The distillate flows at a rate of 75×10^3 mol/h. The composition of the distillate in % moles is 97.33% benzene, 2% toluene, and the remainder is ethylbenzene. Find the molar flow rate of the bottoms stream and the mass fractions of the three components in the distillate and bottoms stream.

4.25. A complex mixture of aromatic compounds leaves a chemical reactor and is fed to a distillation column. The mass fractions and flow rates of distillate and bottoms streams are given in Table 4-25. Compute the molar flow rate and composition, in molar fractions, of the feed stream.

TABLE 4-25
Flow Rate and Composition of Distillate and Bottoms Streams

	(kg/h)	$\omega_{Benzene}$	$\omega_{Toluene}$	$\omega_{Benzaldehide}$	$\omega_{BenzoicAcid}$	$\omega_{MethylBenzoate}$
Distillate	125	0.1	0.85	0.03	0.0	0.02
Bottoms	76	0.0	0.05	0.12	0.8	0.03

4.26. A hydrocarbon feedstock is available at a rate of 10^6 mol/h, and consists of propane ($x_A = 0.2$), n-butane ($x_B = 0.3$), n-pentane ($x_C = 0.2$), and n-hexane ($x_D = 0.3$). The distillate contains all of the propane in the feed to the unit and 80% of the pentane fed to the unit. The mole fraction of butane in the distillate is $y_B = 0.4$. The bottom stream contains all of the hexane fed to the unit. Calculate the distillate and bottoms streams flow rate and composition in terms of mole fractions.

Section 4.8

Note: Problems marked with the symbol 🖳 will be difficult to solve without the use of computer software.

4.27. It is possible that the process illustrated in Figure 4.12 could be analyzed beginning with Control Volume I rather than beginning with Control Volume II. Begin the problem with Control Volume I and carry out a degree-of-freedom analysis to see what difficulties might be encountered.

4.28🖳. In a glycerol plant, a 10% (mass basis) aqueous glycerin solution containing 3% NaCl is treated with butyl alcohol as illustrated in Figure 4-28.

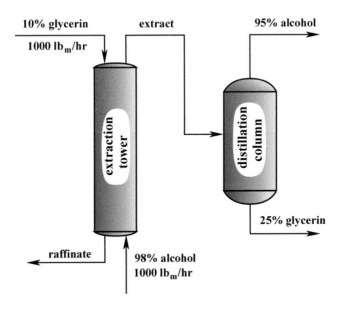

FIGURE 4-28   Solvent extraction process.

The alcohol fed to the extraction tower contains 2% water on a mass basis. The raffinate leaving the extraction tower contains all the original salt, 1.0% glycerin and 1.0% alcohol. The extract from the tower is sent to the distillation column. The distillate from this column is the alcohol containing 5% water. The bottoms from the distillation column are 25% glycerin and 75% water. The two feed streams to the extraction tower have equal mass flow rates of 1,000 lb_m/h. Determine the output of glycerin in pounds per hour from the distillation column.

Section 4.9

4.29🖳. In Sec. 4.8, the solution to the distillation problem was shown to reduce to solving the matrix equation, $Au = b$, in which

$$A = \begin{bmatrix} 1 & 1 & 1 \\ 0.045 & 0.069 & 0.955 \\ 0.091 & 0.901 & 0.041 \end{bmatrix}, \quad u = \begin{bmatrix} \dot{m}_2 \\ \dot{m}_3 \\ \dot{m}_4 \end{bmatrix}, \quad b = \begin{bmatrix} 1,000 \\ 500 \\ 300 \end{bmatrix}$$

Here, it is understood that the mass flow rates have been made dimensionless by dividing by lb_m/h. In addition to the matrix A, one can form what is known as an *augmented matrix*. This is designated by $A\!:\!b$ and it is constructed by adding the column of numbers in b to the matrix A in order to obtain

$$A\!:\!b = \begin{bmatrix} 1 & 1 & 1 & . & 1{,}000 \\ 0.045 & 0.069 & 0.955 & . & 500 \\ 0.091 & 0.901 & 0.041 & . & 300 \end{bmatrix}$$

Define the following *lists* in *Mathematica* corresponding to the rows of the augmented matrix, $A\!:\!b$.

$$R1 = \{1,\ 1,\ 1,\ 1{,}000\}$$

$$R2 = \{0.045,\ 0.069,\ 0.955,\ 500\}$$

$$R3 = \{0.091,\ 0.901,\ 0.041,\ 300\}$$

Write a sequence of *Mathematica* expressions that correspond to the elementary row operations for solving this system. The first elementary row operation that given Eq. 4.135 is

$$R2 = (-0.045)\,R2 + R1$$

Show that you obtain an augmented matrix that defines Eq. 4.139.

4.30⌨. In this problem you are asked to continue exploring the use of *Mathematica* in the analysis of the set of linear equations studied in Sec. 4.9.2, i.e., $Au = b$ where the matrices are defined by

$$A = \begin{bmatrix} 1 & 1 & 1 \\ 0.045 & 0.069 & 0.955 \\ 0.091 & 0.901 & 0.041 \end{bmatrix}, \quad u = \begin{bmatrix} \dot{m}_2 \\ \dot{m}_3 \\ \dot{m}_4 \end{bmatrix}, \quad b = \begin{bmatrix} 1{,}000 \\ 500 \\ 300 \end{bmatrix}$$

1. Construct the augmented matrix $A\!:\!I$ according to

$$A\!:\!I = \begin{bmatrix} 1 & 1 & 1 & . & 1 & 0 & 0 \\ 0.045 & 0.069 & 0.955 & . & 0 & 1 & 0 \\ 0.091 & 0.901 & 0.041 & . & 0 & 0 & 1 \end{bmatrix}$$

and use elementary row operations to transform this augmented matrix to the form

$$A\!:\!I = \begin{bmatrix} 1 & 0 & 0 & . & \\ 0 & 1 & 0 & . & B \\ 0 & 0 & 1 & . & \end{bmatrix}$$

Show that the elements represented by B make up the matrix B having the property that $B = A^{-1}$. Use your result to calculate $u = A^{-1}b$.

2. Show that the inverse found in Part 1 satisfies $AA^{-1} = I$.
3. Use *Mathematica*'s built-in function **Inverse** to find the inverse of A.
4. Use *Mathematica*'s **Row Reduce** function on the augmented matrix $A \vdots b$ and show that from the row echelon form you can obtain the same results as in (1).
5. Use *Mathematica*'s **Solve** function to solve $Au = b$.

5 Two-Phase Systems and Equilibrium Stages

In the previous chapter, we began our study of macroscopic mass and mole balances for multicomponent systems. There we encountered a variety of *measures of concentration* and here we summarize these measures as

$$\rho_A = \left(\begin{array}{c} \text{mass of species } A \\ \text{per unit volume} \end{array} \right) \tag{5.1a}$$

$$\rho = \sum_{A=1}^{A=N} \rho_A, \quad \text{total mass density} \tag{5.1b}$$

$$\omega_A = \rho_A / \rho, \quad \text{mass fraction} \tag{5.1c}$$

$$c_A = \rho_A / MW_A, \quad \text{molar concentration} \tag{5.1d}$$

$$c = \sum_{A=1}^{A=N} c_A, \quad \text{total molar concentration} \tag{5.1e}$$

$$y_A \text{ or } x_A = c_A / c, \quad \text{mole fraction} \tag{5.1f}$$

In the analysis of gas-phase systems, it is often important to relate the concentration to the pressure and temperature. This is done by means of an equation of state, often known as a *p-V-T* relation. In this chapter, we will make use of the *ideal gas* relations; however, many processes operate under conditions such that the ideal gas laws do not apply and one must make use of more general *p-V-T* relations. Non-ideal gas behavior is studied in a course on thermodynamics.

5.1 IDEAL GAS BEHAVIOR

For an *N*-component ideal gas mixture, we have the following relations

$$p_A V = n_A RT, \quad A = 1, 2, \dots, N \tag{5.2}$$

Here p_A is the *partial pressure* of species A and R is the gas constant. Values of the gas constant in different units are given in Table 5.1. If we sum Eq. 5.2, over all N species we obtain

$$pV = nRT \tag{5.3}$$

DOI: 10.1201/9781003283751-5

where

$$p = \sum_{A=1}^{A=N} p_A \tag{5.4}$$

$$n = \sum_{A=1}^{A=N} n_A \tag{5.5}$$

Eqs. 5.2–5.5 are sometimes referred to as *Dalton's Laws.*

In Figure 5.1, we have illustrated a *constant pressure, isothermal* mixing process. In the compartment containing species A illustrated in part a of Figure 5.1, we can use Eq. 5.3 to obtain

$$pV_A = n_A RT \tag{5.6}$$

TABLE 5.1

Numerical Values of the Gas Constant, *R*

Numerical Value	Units
8.314	m³ Pa/mol K
8.314	J/mol K
0.08314	L bar/mol K
82.06	atm-cm³/mol K
1.986	cal/mol K

---

while in the compartment containing species B we have

$$pV_B = n_B RT \tag{5.7}$$

Here we have made use of the fact that $p_A = p_B = p$ for this particular process. Upon removal of the partition and mixing, we have the situation illustrated in part b of Figure 5.1. For that condition, we can use Eq. 5.3 to obtain

$$pV = (n_A + n_B)RT \tag{5.8}$$

where the volume is given by

$$V = V_A + V_B + \Delta V_{mix} \tag{5.9}$$

By definition, an ideal gas mixture obeys what is known as *Amagat's Law,* i.e.

$$\Delta V_{mix} = 0, \quad \text{Amagat's Law} \tag{5.10}$$

Amagat's law can also be expressed as

$$\sum_{A=1}^{A=N} V_A = V, \quad \text{Amagat's Law} \tag{5.11}$$

In the gas phase, the mole fraction is generally denoted by y_A while x_A is reserved for mole fractions in the liquid phase. For ideal gas mixtures, it is easy to show, using Eqs. 5.1f, 5.2, and 5.3, that the mole fraction is given by

$$y_A = p_A/p \tag{5.12}$$

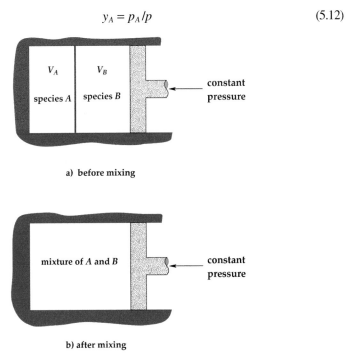

FIGURE 5.1 Constant pressure, isothermal mixing process.

This is an extremely convenient representation of the mole fraction in terms of the partial pressure; however, one must always remember that it is strictly valid only for an ideal gas mixture.

EXAMPLE 5.1 FLOW OF AN IDEAL GAS IN A PIPELINE

A large pipeline is used to transport natural gas from Oklahoma to Nebraska as illustrated in Figure 5.1a.

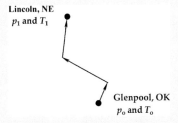

FIGURE 5.1a Transport of natural gas from Oklahoma to Nebraska.

Natural gas, consisting of methane with small amounts of ethane, propane, and other low molecular mass hydrocarbons, can be assumed to behave as an ideal gas at ambient temperature. At the pumping station in Glenpool, OK, the pressure in the 20-inch pipeline is 2,900 psia and the temperature is $T_o = 90$ F. At the receiving point in Lincoln, NE, the pressure is 2,100 psia and the temperature is $T_1 = 45$ F. The mass average velocity of the gas at Glenpool is 50 ft/s. Assuming ideal gas behavior and treating the gas as 100% methane, we want to compute the following:

  a. Mass average velocity at the end of the pipeline.
  b. Mass and molar flow rates at both ends of the pipeline.

This problem has been presented in terms of a variety of units, and it is sometimes convenient to express all variables in terms of SI units as follows:

$$\left(2,900 \text{ psia}\right) \times \frac{6,895 \text{ Pa}}{\text{psia}} \times \frac{\text{MPa}}{10^6 \text{ Pa}} = 20.00 \text{ MPa}$$

$$\left(2,100 \text{ psia}\right) \times \frac{6,895 \text{ Pa}}{\text{psia}} \times \frac{\text{MPa}}{10^6 \text{ Pa}} = 14.48 \text{ MPa}$$

$$\left(90 \text{ F} - 32 \text{ F}\right) \times \frac{5 \text{ C}}{9 \text{ F}} \times \frac{\text{K}}{\text{C}} + 273.15 \text{ K} = 305.4 \text{ K}$$

$$\left(45 \text{ F} - 32 \text{ F}\right) \times \frac{5 \text{ C}}{9 \text{ F}} \times \frac{\text{K}}{\text{C}} + 273.15 \text{ K} = 280.4 \text{ K}$$

$$\left(50 \text{ ft/s}\right) \times \frac{0.3048 \text{ m}}{\text{ft}} = 15.24 \text{ m/s}$$

$$D_o = D_1 = \left(20 \text{ in}\right) \times \frac{0.0254 \text{ m}}{\text{in}} = 0.508 \text{ m}$$

$$A_o = A_1 = \frac{\pi D^2}{4} = \frac{\pi \left(0.508 \text{ m}\right)^2}{4} = 0.2027 \text{ m}^2$$

Since the natural gas is assumed to be pure methane, we can obtain the molecular mass from Table A2 in Appendix A where we find $MW_{CH_4} = 16.043$ g/mol. The volume per mole of methane can be determined using the ideal gas law given by Eq. 5.3 along with the value of the gas constant found in Table 5.1. The volume per mole at the entrance is determined as

$$V_o / n = \frac{RT}{p_o} = \frac{8.314 \frac{\text{m}^3 \text{ Pa}}{\text{mol K}} \left(305.4 \text{ K}\right)}{19.98 \times 10^6 \text{ Pa}} = 1.2 \times 10^{-4} \text{ m}^3/\text{mol} \qquad (1)$$

while the volume per mole at the exit is given by

$$V_1/n = \frac{RT}{p_1} = 1.53 \times 10^{-4} \, \text{m}^3/\text{mol} \tag{2}$$

The gas densities at the entrance and exit of the pipeline are given by

$$(\rho_{\text{CH}_4})_o = \frac{MW_{\text{CH}_4}}{V_o/n} = \frac{16.043 \, \text{g/mol}}{1.2 \times 10^{-4} \, \text{m}^3/\text{mol}}$$

$$= 133{,}700 \, \text{g/m}^3 = 133.7 \, \text{kg/m}^3 \tag{3}$$

$$(\rho_{\text{CH}_4})_1 = \frac{MW_{\text{CH}_4}}{V_1/n} = 104.8 \, \text{kg/m}^3 \tag{4}$$

To perform a mass balance for the pipeline, we begin with the species mass balance given by

$$\frac{d}{dt} \int_V \rho_A \, dV + \int_A \rho_A \mathbf{v}_A \cdot \mathbf{n} \, dA = \int_V r_A \, dV \tag{5}$$

Since there are no chemical reactions and there is no accumulation, the first and last terms in this result are zero and Eq. 5 simplifies to

$$\int_A \rho_A \mathbf{v}_A \cdot \mathbf{n} \, dA = 0 \tag{6}$$

For a single component system, the species velocity is equal to the mass average velocity (see Chapter 4) and Eq. 6 takes the form

$$\int_A \rho_A \mathbf{v} \cdot \mathbf{n} \, dA = 0 \tag{7}$$

The control volume is constructed in the obvious manner, thus there is an entrance in Glenpool, OK and an exit in Lincoln, NE. Since the mass average velocity and the diameter of the pipeline are given, we express the mass balance as

$$(\rho_{\text{CH}_4})_o v_o A_o = (\rho_{\text{CH}_4})_1 v_1 A_1 \tag{8}$$

The only unknown in this result is the velocity of the gas at the exit of the pipeline. Solving for v_1 we obtain

$$v_1 = \left[\frac{A_o}{A_1} \frac{(\rho_{CH_4})_o}{(\rho_{CH_4})_1} \right] v_o = \frac{133.7 \, kg/m^3}{104.8 \, kg/m^3} (15.24 \, m/s) = 19.44 \, m/s \qquad (9)$$

The mass flow rate is a constant given by

$$\dot{m}_o = \dot{m}_1 = (\rho_{CH_4})_o v_o A_o$$

$$= \left(133.7 \, kg/m^3 \right) (15.24 \, m/s) \left(0.2027 \, m^2 \right) = 413 \, kg/s \qquad (10)$$

from which we determine the constant molar flow rate to be

$$\dot{M}_o = \dot{M}_1 = \frac{(\dot{m}_{CH_4})_o}{MW_{CH_4}} = \frac{413 \, kg/s}{16.043 \, g/mol} = 25.74 \times 10^3 \, mol/s \qquad (11)$$

In order to use Eq. 5.3 to estimate the density of a pure gas, we divide both sides by VRT and multiply both sides by MW to obtain

$$\rho = \frac{n \, MW}{V} = \frac{p \, MW}{R \, T} \qquad (5.13)$$

For an ideal gas mixture, one uses the definitions of the total mass density of a mixture (Eq. 5.1b) and total pressure (Eq. 5.4) along with Eq. 5.2 to obtain

$$\rho = \sum_{A=1}^{A=N} \rho_A = \sum_{A=1}^{A=N} \frac{n_A \, MW_A}{V} = \sum_{A=1}^{A=N} \frac{p_A \, MW_A}{R \, T} = \frac{p \, \overline{MW}}{R \, T} \qquad (5.14)$$

Here, we have used $\overline{MW}$ to represent the molar average molecular mass defined by

$$\overline{MW} = \frac{1}{p} \sum_{A=1}^{A=N} p_A \, MW_A = \sum_{A=1}^{A=N} y_A \, MW_A \qquad (5.15)$$

These results are applicable when molecule-molecule interactions are negligible, and this occurs for many gases under ambient conditions. At low temperatures and high pressures, gases depart from ideal gas behavior. Under those conditions, one should use more accurate equations of state such as those presented in a standard course on thermodynamics[*].

---

[*] Sandler, S.I. 2006, *Chemical, Biochemical, and Engineering Thermodynamics*, 4th edition, John Wiley and Sons, New York.

EXAMPLE 5.2   MOLECULAR MASS OF AIR

The air we breathe has a composition that depends on position. Air pollution sources abound and these sources add minute amounts of chemicals to the atmosphere. Combustion of fuels in cars and power plants are a source for sulfur dioxide, oxides of nitrogen, and carbon monoxide. Chemical industries add pollutants such as ammonia, chlorine, and even hydrogen cyanide to the atmosphere. In some locations, there are minute amounts of other gases such as argon, helium, and radon. *Standard dry air*, for the purpose of combustion computations (see Sec. 7.2), is assumed to be a mixture of 79% by volume of nitrogen and 21% by volume of oxygen. In this example, we want to determine the molar average molecular mass of standard dry air.

For an ideal gas, the volume percentage is also the molar percentage. Thus, the volume percentages of nitrogen and oxygen can be simply translated to mole fractions according to

$$79\% \text{ by volume of nitrogen} \rightarrow y_{N_2} = 0.79 \tag{1}$$

$$21\% \text{ by volume of oxygen} \rightarrow y_{O_2} = 0.21 \tag{2}$$

We can use Eq. 5.15 to compute the molar average molecular mass of standard dry air according to

$$\overline{MW} = MW_{air} = y_{N_2} MW_{N_2} + y_{O_2} MW_{O_2} = 28.85 \text{ g/mol}$$

Here we have used the subscript "air" to represent standard dry air as described by Eqs. 1 and 2.

5.2   LIQUID PROPERTIES AND LIQUID MIXTURES

When performing material balances for liquid systems, one must have access to reliable liquid properties. Unlike gases, the densities of liquids are weak functions of pressure and temperature, i.e., *large changes* in pressure and temperature result in *small changes* in the density. The changes in liquid density due to changes in pressure are determined by the coefficient of compressibility which is defined by

$$\kappa = \left\{ \begin{array}{c} \text{coefficient of} \\ \text{compressibility} \end{array} \right\} = \frac{1}{\rho}\left(\frac{\partial \rho}{\partial p}\right)_T \tag{5.16}$$

Changes in liquid density due to changes in temperature are determined by the coefficient of thermal expansion which is defined by

$$\beta = \left\{ \begin{array}{c} \text{coefficient of thermal} \\ \text{expansion} \end{array} \right\} = -\frac{1}{\rho}\left(\frac{\partial \rho}{\partial T}\right)_p \tag{5.17}$$

The coefficient of thermal expansion, β, is defined with a negative sign since the density of most liquids decreases with increasing temperature. Using a negative sign in the definition of the thermal expansion coefficient makes β positive for most liquids. There is an interesting counter example, however, and it is the density of liquid water at low temperatures. For liquid water between 4°C and the freezing point, 0°C, the coefficient of expansion is negative, i.e., $\beta < 0$. If it were not for this characteristic, water in lakes and rivers would freeze from the bottom during the winter, and this would destroy most aquatic life.

The density of *ideal liquid mixtures* is computed using Amagat's law that leads to

$$V = \sum_{A=1}^{A=N} V_A = \sum_{A=1}^{A=N} m_A/\rho_A^o \tag{5.18}$$

Here V is the volume of the mixture, while m_A and ρ_A^o represent the masses and densities of the pure components. Eq. 5.18 can be used to compute the density of the mixture according to

$$\rho = \frac{m}{V} = \frac{m}{\displaystyle\sum_{A=1}^{A=N} m_A/\rho_A^o} = \frac{1}{\displaystyle\sum_{A=1}^{A=N} \omega_A/\rho_A^o} \tag{5.19}$$

At low-to-moderate pressures, this version of Amagat's law is a satisfactory approximation. However, non-ideal behavior of liquid mixtures is a very complex topic. When some of the components of a liquid mixture are above their boiling points, or if the components are polar, the use of Eq. 5.19 may lead to significant errors[†].

5.3   VAPOR PRESSURE OF LIQUIDS

If we study the p-V-T characteristics of a real gas using the experimental system shown in Figure 5.2, we find the type of results illustrated in Figure 5.3. In the system illustrated in Figure 5.2, a single component is contained in a cylinder immersed in a constant temperature bath. We can increase or decrease the pressure inside the cylinder by simply moving the piston. When the molar volume (volume per mole) is sufficiently large, the distance between molecules is large enough (on the average) so that molecular interaction becomes unimportant. For example, at the temperature T_3, and a large value of V/n, we observe ideal gas behavior in Figure 5.3. This is illustrated by the fact that at a fixed temperature we have

$$pV/n = \text{constant} \tag{5.20}$$

[†] Reid, R.C., Prausnitz, J.M., and Sherwood, T.K., 1977, *The Properties of Gases and Liquids*, 6th Edition, New York, McGraw-Hill Books.

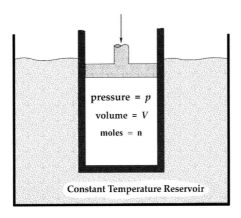

FIGURE 5.2 Experimental study of *p-V-T* behavior.

However, as the pressure is increased (and the volume decreased) for the system illustrated in Figure 5.3, a point is reached where liquid appears and the pressure remains constant as the volume continues to decrease. This pressure is referred to as the *vapor pressure* and we will identify it as p_{vap}. Obviously, the vapor pressure is a function of the temperature and knowledge of this temperature dependence is crucial for the solution of many engineering problems.

In a course on thermodynamics, students will learn that the *Clausius-Clapeyron equation* provides a reasonable approximation for the vapor pressure as a function of temperature. The Clausius-Clapeyron equation can be expressed as

$$p_{A,vap} = p_{A,vap}(T_o) \exp\left[-\frac{\Delta H_{vap}}{R}\left(\frac{1}{T} - \frac{1}{T_o}\right)\right] \qquad (5.21)$$

in which $p_{A,vap}$ represents the vapor pressure at the temperature *T*. We have used $p_{A,vap}(T_o)$ to represent the vapor pressure at the reference temperature T_o, while ΔH_{vap} represents the molar heat of vaporization. A more accurate empirical expression for the vapor pressure is given by Antoine's equation[‡].

$$\log(p_{A,vap}) = A - \frac{B}{(\theta + T)} \qquad (5.22)$$

in which $p_{A,vap}$ is determined in mmHg and *T* is specified in °C. The coefficients *A*, *B*, and θ are given in Table A3 of Appendix A for a variety of compounds. Note that Eq. 5.22 is *dimensionally incorrect* and must be used with great care as we indicated in our discussion of units in Sec. 2.3.

[‡] Wisniak, J. 2001, Historical development of the vapor pressure equation from Dalton to Antoine, J. Phase Equilibrium **22**, 622–630.

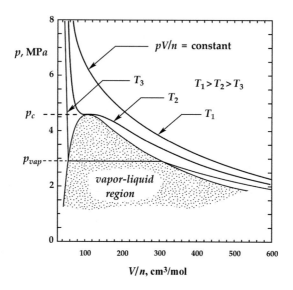

FIGURE 5.3 p-V-T behavior of methane.

EXAMPLE 5.3   VAPOR PRESSURE OF A SINGLE COMPONENT

In this example, we wish to estimate the vapor pressure of methanol at 25°C using the Clausius-Clapeyron equation. The heat of vaporization is $\Delta H_{vap} = 8,426$ cal/mol at the normal boiling point of methanol, 337.8 K.

The heat of vaporization is a function of temperature and pressure. The data given for the heat of vaporization is for the temperature $T = 337.8$ K $= 64.6$°C. At this temperature, the vapor pressure of methanol is equal to atmospheric pressure. In order to estimate the vapor pressure at 25°C, we use the normal boiling temperature as the reference temperature. Normally, we would compute the value of the heat of vaporization at 25°C using a thermodynamic relationship and then use an average value for ΔH_{vap} in Eq. 5.21. However, in this example, we will estimate the vapor pressure at 25°C using the heat of vaporization at 64.6°C. All variables can be converted into SI units as follows:

$$25°C + 273.16 \text{ K} = 298.16 \text{ K} \tag{1}$$

$$T_o = 337.8 \text{ K} \tag{2}$$

$$p_{M,vap}(T_o) = 1 \text{ atm} = 101,300 \text{ Pa} \tag{3}$$

$$\Delta H_{vap} = (8,426 \text{ cal/mol})(4.186 \text{ J/cal}) = 35,271 \text{ J/mol} \tag{4}$$

Substitution of these results into Eq. 5.21 gives

$$p_{M,vap} = (101,300 \text{ Pa}) \exp\left[-\frac{35,271 \text{ J/mol}}{8.314 \text{ m}^3 \text{ Pa/mol K}}\left(\frac{1}{298.2 \text{ K}} - \frac{1}{337.8 \text{ K}}\right)\right]$$

$$= 19,112 \text{ Pa} \tag{5}$$

Vapor pressures estimated using the Clausius-Clapeyron equation can exhibit substantial errors with respect to experimental values of vapor pressure. This is caused by the fact that the assumptions made in the development of this equation are not always valid. The semi-empirical equation known as Antoine's equation has the advantage that it is based on the correlation of experimental values of the vapor pressure.

EXAMPLE 5.4   VAPOR PRESSURE OF SINGLE COMPONENTS USING ANTOINE'S EQUATION

In this example, we determine the vapor pressure of methanol at 25°C using Antoine's equation, Eq. 5.22, and compare the result with the vapor pressure computed in Example 5.3. The numerical values of the coefficients in Antoine's equation are obtained from Table A3 of Appendix A and they are given by

$$A = 8.07246, \quad B = 1,574.99, \quad \theta = 238.86$$

Once again, we note that Antoine's equation is a dimensionally incorrect empiricism, and one must follow the rules of application that are given above and in Table A3 of Appendix A. Substitution of the values for A, B, and θ into Eq. 5.22 gives

$$\log p_{M,vap} = 8.07246 - \frac{1,574.99}{238.86 + 25} = 2.10342$$

and the vapor pressure of methanol at 25°C is $p_{M,vap} = 126.9$ mmHg $= 16,912$ Pa. The result computed using the Clausius-Clapeyron equation was $p_{M,vap} = 19,112$ Pa, thus the two results differ by 11%. The results of this example clearly indicate that it is misleading to represent calculated values of the vapor pressure to five significant figures.

5.3.1 MIXTURES

The behavior of vapor-liquid systems having more than one component can be quite complex; however, some mixtures can be treated as *ideal*. In an ideal vapor-liquid multi-component system, the partial pressure of species A in the gas phase is given by

Equilibrium relation: $p_A = p_{A,vap}\, x_A$ (5.23)

Here p_A is the partial pressure of species A in the *gas phase*, x_A is the mole fraction of species A in the *liquid phase*, and $p_{A,vap}$ is the vapor pressure of species A at the temperature under consideration. It is important to remember that Eq. 5.23 is an *equilibrium relation*; however, when the condition of *local thermodynamic equilibrium* is valid Eq. 5.23 can be used to calculate values of p_A for dynamic processes.

The equilibrium relation given by Eq. 5.23 represents a special case of a more general relation that is described in many texts[§,**] and will be studied in a course on thermodynamics. The general equilibrium relation is based on the *partial molar Gibbs free energy*, or the chemical potential, and it takes the form

Equilibrium relation: $(\mu_A)_{gas} = (\mu_A)_{liquid}$,   at the gas-liquid interface   (5.24)

Here, we have used μ_A to represent the *chemical potential* of species A that depends on the temperature (strongly), the pressure (weakly), and the composition of the phase under consideration. The matter of extracting Eq. 5.23 from Eq. 5.24 will be taken up in a subsequent course on thermodynamics.

For an ideal gas, use of Eq. 5.2 along with Eq. 5.3 indicates that the gas-phase mole fraction can be expressed as

$$y_A = p_A/p \tag{5.25}$$

This result can be used with Eq. 5.23 to obtain a relation between the gas and liquid-phase mole fractions that is given by

Equilibrium relation: $$y_A = x_A \left(p_{A,vap}/p \right) \tag{5.26}$$

This equilibrium relation is sometimes referred to as *Raoult's law*. For a two-component system, we can use Eq. 5.26 along with the constraint on the mole fractions

$$x_A + x_B = 1, \quad y_A + y_B = 1 \tag{5.27}$$

to obtain the following expression for the mole fraction of species A in the gas phase:

Equilibrium relation: $$y_A = \frac{\alpha_{AB}\, x_A}{1 + x_A(\alpha_{AB} - 1)} \tag{5.28}$$

Here α_{AB} is the *relative volatility* defined by

$$\alpha_{AB} = \frac{p_{A,vap}}{p_{B,vap}} \tag{5.29}$$

For a dilute binary solution of species A, one can express Eq. 5.28 as

$$y_A = \alpha_{AB}\, x_A, \quad \text{for } x_A(\alpha_{AB} - 1) \ll 1 \tag{5.30}$$

[§] Gibbs, J.W. 1928, *The Collected Works of J. Willard Gibbs*, Longmans, Green and Company, London.

[**] Prigogine, I. and Defay, R. 1954, *Chemical Thermodynamics*, Longmans Green and Company, London.

and this special form of Raoult's law is often referred to as *Henry's law*. For an N-component system, one can express Henry's law as

Henry's Law: $$y_A = K_{eq,A}\, x_A \qquad (5.31)$$

Here $K_{eq,A}$ is referred to as the Henry's law *constant* even though it is not a constant since it depends on the temperature and composition of the liquid, i.e.

$$K_{eq,A} = F\left(T, x_A, x_B, x_C, \ldots x_{N-1}\right) \qquad (5.32)$$

This treatment of gas-liquid systems is extremely brief and devoid of the rigor that will be encountered in a comprehensive discussion of phase equilibrium. However, we now have sufficient information to solve a few simple mass balance problems that involve two-phase systems.

5.4 SATURATION, DEW POINT AND BUBBLE POINT OF LIQUID MIXTURES

When a *pure liquid* phase (species A) is in equilibrium with the *pure gas* phase, the partial pressure of the component in the gas phase is equal to the vapor pressure of the pure component.

$$p_A = p_{A,vap} \qquad (5.33)$$

This result is consistent with setting $x_A = 1$ in Eq. 5.23 so that the liquid is pure component A. In general, when air is the gas phase, we will assume that the gas phase is saturated with the liquid component and that the concentration of the air in the liquid is negligible. When the liquid phase is a mixture, Raoult's law (Eq. 5.26) must be used to compute the composition of the gas phase.

If a liquid mixture is in equilibrium with its own vapors, the total pressure is equal to the sum of the partial pressures of the individual components.

$$p = \sum_{A=1}^{A=N} p_A = \sum_{A=1}^{A=N} p_{A,vap}\, x_A \qquad (5.34)$$

When a liquid mixture is heated, the vapor pressure of the components in the mixture increases and the sum of the partial pressures, given by Eq. 5.34, increases accordingly. When the sum of the partial pressures of the components of the mixture is equal to the atmospheric pressure, the liquid mixture boils. The difference between a liquid mixture and a pure liquid is that the boiling temperature of a mixture is not constant. For a mixture in equilibrium with its own vapors, the *bubble point* of a mixture is the pressure at which the liquid starts to vaporize. Similarly, for a vapor mixture, the *dew point* of the mixture is the pressure at which the vapors start to condense. These terms are also used when the liquid is in contact with air, i.e. it is customary to refer to the *bubble point* as the temperature at which the liquid mixture starts to boil and *dew point* as the temperature at which the first condensed liquid appears.

EXAMPLE 5.5   BUBBLE POINT OF A WATER-ALCOHOL MIXTURE

A mixture of ethanol (C_2H_5OH) and water (H_2O) with the mole fractions given by $x_{Et} = x_{H_2O} = 0.5$ is slowly heated under well-stirred conditions in an open beaker. Using Antoine's equation to determine the vapor pressure of the components and Raoult's law to estimate the partial pressures, we can estimate the bubble point of this mixture, i.e., the temperature at which the first bubbles will start forming at the bottom of the beaker as well as the composition of the first bubbles.

The vapor pressure of pure ethanol and water can be computed using Antoine's equation. The partial pressure of the components in the gas phase in equilibrium with the liquid mixture is computed using Raoult's law given by Eq. 5.26. The bubble point will be determined as the temperature at which the sum of the partial pressures of ethanol and water is equal to atmospheric pressure, i.e.,

$$p_{H_2O} + p_{Et} = p_{atm}$$

$$\text{or} \hspace{4cm} (1)$$

$$x_{H_2O} \, p_{H_2O,vap} + x_{Et} \, p_{Et,vap} = 760 \text{ mmHg}$$

This problem can be solved by substitution of Antoine's equation for the vapor pressures of the pure components in Eq. 1, and then solving for the temperature. A much simpler route consists of guessing values of the temperature until we satisfy Eq. 1. This procedure is easily done using a spreadsheet as illustrated in Table 5.5a. The values of Antoine's coefficients for water and ethanol are available in Table A3 of Appendix A and are given by

---

TABLE 5.5a

Computational Determination of the Boiling Point of a Water-Ethanol Mixture

Computation of Dew Point of Ethanol and Water Mixture			
Temp °C	p_{H_2O} mmHg	p_{Et} mmHg	Residue mmHg
60	74.7483588	175.7127	−509.539
70	116.9527	270.8179	−372.229
80	177.72766	405.8736	−176.399
86	225.542371	511.0501	−23.4076
86.8	232.662087	526.6464	−0.6915
86.9	233.565109	528.6235	2.188576
87	234.471058	530.6067	5.077746
90	263.047594	593.0441	96.09172

Water: $A = 7.94915$, $B = 1{,}657.46$, $\theta = 227.03$ (2)

Ethanol: $A = 8.1629$, $B = 1{,}623.22$, $\theta = 228.98$ (3)

The computed values of the vapor pressure are listed in Table 5.5a. We could continue the computation by inserting additional rows between the temperatures $T = 86.8°C$ and $T = 86.9°C$. However, for the purpose of this example, we will accept the boiling point of the mixture as $T = 86.85 \pm 0.05°C$.

5.4.1 Humidity

In air-water mixtures, the *humidity* is often used as a *measure of concentration* that is vaguely described according to

$$\text{humidity} = \frac{\text{mass of water}}{\text{mass of dry air}} \qquad (5.35)$$

In Sec. 4.5.1, we have been more precise in terms of *measures of concentration* and there we have identified point concentrations, area-average concentrations, and volume-average concentrations. An analogous set of definitions exists for the humidity. For example, the point version of the humidity is given explicitly by

$$\text{point humidity} = \frac{\rho_{H_2O}}{\rho_{air}} = \frac{\rho_{H_2O}}{\rho_{O_2} + \rho_{N_2}} \qquad (5.36)$$

Here, we note that "air" or "dry air" in terms of humidity calculations means oxygen and nitrogen and this is not to be confused with "standard dry air" that is used in combustion computations (see Example 5.2). The two are quite similar; however, the density of "air" for humidity calculations is given explicitly by $\rho_{O_2} + \rho_{N_2}$ and thus does not include the water that is present in humid air. It will be left as an exercise for the student to show that

$$\text{point humidity} = \frac{MW_{H_2O} \, p_{H_2O}}{MW_{air} \left(p - p_{H_2O} \right)} \qquad (5.37)$$

in which p is the total pressure and p_{H_2O} is the *partial pressure* of the water vapor. This result can be derived from the definition given by Eq. 5.36 only if the air-water mixture is treated as an ideal gas. The percent *relative humidity* is often used as a measure of concentration since our personal comfort may be closely connected to this quantity. It is defined by

$$\% \text{ relative humidity} = \frac{p_{H_2O}}{p_{H_2O, \, vap}} \times 100 \qquad (5.38)$$

where $p_{H_2O, vap}$ is the vapor pressure of water at the temperature of the system. When the percent relative humidity is 100%, the air is completely saturated and the addition of further water will result in condensation. Values of the vapor pressure of water are listed in Table 5.2 as a function of the temperature. The data are given in terms of *millimeters of mercury* and *inches of mercury*.

TABLE 5.2

Vapor Pressure of Water as a Function of Temperature

T, °C	Vapor Pressure, mmHg	T, °F	Vapor Pressure, inHg
0	4.579	32	0.180
5	6.543	40	0.248
10	9.209	50	0.363
15	12.788	60	0.522
20	17.535	70	0.739
25	23.756	80	1.032
30	31.824	90	1.422
35	42.175	100	1.932
40	55.324	110	2.596
45	71.88	120	3.446
50	92.51	130	4.525
55	118.04	140	5.881
60	149.38	150	7.569
65	187.54	160	9.652
70	233.7	170	12.199
75	289.1	180	15.291
80	355.1	190	19.014
85	433.6	200	23.467
90	525.76	212	29.922
95	633.90	220	34.992
100	760.00	230	42.308
105	906.07	240	50.837
110	1,074.56	250	60.725
115	1,267.98	260	72.134
120	1,489.14	270	85.225
125	1,740.93	280	100.18
130	2,026.16	290	117.19
135	2,347.26	300	136.44

EXAMPLE 5.6   HUMID AIR FLOW

Humid air exits a dryer at atmospheric pressure, 75°C, 25% relative humidity, and at a volumetric flow rate of 100 m³/min. In this example, we wish to determine:

a. Absolute humidity of the air in kg water/kg air.
b. Molar flow rates of water and dry air.

The vapor pressure of water at 75°C is found in Table 5.2 to be $p_{H_2O, vap} = 289.1$ mmHg. The density of mercury is found in Table A2 of Appendix A. We convert all parameters into SI units according to

$$p_{H_2O} = (0.25)\frac{289.1 \text{ mmHg}}{1,000 \text{ mm/m}} \left(9.81 \text{ m/s}^2\right)\left(13,546 \text{ kg/m}^3\right) = 9,605 \text{ Pa} \quad (1)$$

and we use Eq. 5.25 to compute the mole fraction of water in the air as

$$y_{H_2O} = \frac{p_{H_2O}}{p} = \frac{9,605 \text{ Pa}}{101,300 \text{ Pa}} = 0.095 \quad (2)$$

In order to determine the absolute humidity, we use Eq. 5.35 in the more precise form given by Eq. 5.36

$$\text{humidity} = \frac{\text{mass of water}}{\text{mass of dry air}} = \frac{\text{mass of water/volume}}{\text{mass of dry air/volume}} \quad (3)$$

and this leads to an expression for the humidity given by

$$\text{humidity} = \frac{\rho_{H_2O}}{\rho_{air}} = \frac{MW_{H_2O} \; y_{H_2O}}{MW_{air} \; y_{air}} = \frac{MW_{H_2O} \; y_{H_2O}}{MW_{air} \left(1 - y_{H_2O}\right)}$$

$$= \frac{(18.05 \text{ g water/mol})(0.095)}{(28.85 \text{ g dry air/mol})(1 - 0.095)} = 0.066 \text{ g water/g dry air} \quad (4)$$

Once again, we have used the subscript "air" as a convenient substitute for dry air and we will continue to make use of this simplification throughout our study of humidification processes. Assuming that water vapor and air behave as ideal gases at atmospheric pressure, we use the ideal gas law given by Eq. 5.3 to compute the total concentration of the mixture. The concentration of the gas mixture is the total number of moles of air and water per unit volume of the mixture. This can be expressed as

$$c = \frac{n}{V} = \frac{p}{RT} = \frac{101,300 \text{ Pa}}{8.314 \dfrac{\text{m}^3 \text{ Pa}}{\text{mol K}} \times 348.16 \text{K}} = 35 \text{ mol/m}^3 \quad (5)$$

This result gives the total number of moles of gas per unit volume of mixture. In order to determine the molar flow rates of water and dry air, we carry out the following calculations to obtain

$$\dot{M}_{H_2O} = c_{H_2O} \, Q = y_{H_2O} \, c \, Q$$

$$= 0.095 \times 35 \text{ mol/m}^3 \times 100 \text{ m}^3/\text{min} = 332.5 \text{ mol water/min} \quad (6a)$$

$$\dot{M}_{air} = c_{air} \, Q = y_{air} \, c \, Q$$

$$= (1 - y_{H_2O}) \, c \, Q = 3,167.5 \text{ mol dry air/min} \quad (6b)$$

5.4.2 MODIFIED MOLE FRACTION

In general, the most useful measures of concentration are the molar concentration c_A and the species mass density ρ_A. Associated with these concentrations are the mole fraction defined by Eq. 5.1f and the mass fraction defined by Eq. 5.1c. Sometimes, it is convenient to use a modified mole fraction or *mole ratio* which is based on all the species *except one*. If we identify that one species as species N, we express the modified mole fraction as

$$X_A = c_A \Big/ \sum_{B=1}^{B=N-1} c_B \qquad (5.39)$$

When it is convenient to work in terms of this modified mole fraction, one usually needs to be able to convert from x_A to X_A and it will be left as an exercise for the student to show that this relation is given by

$$X_A = x_A/(1-x_N), \quad A = 1, 2, ..., N \qquad (5.40)$$

In some types of analysis, it is convenient to choose species N to be species A. Under those circumstances we need to express Eq. 5.39 as

$$X_D = c_D \Big/ \sum_{\substack{B=1 \\ B \neq 1}}^{B=N} c_B \qquad (5.41)$$

and Eq. 5.41 takes the form

$$X_D = x_D/(1-x_A), \quad D = 1, 2, ..., N \qquad (5.42)$$

Similar relations can be developed for the modified mass fraction Ω_A and they will be left as exercises for the student. It is important to note that the sum over *all species* of the modified mass or mole fraction is not one.

5.5 EQUILIBRIUM STAGES

Mass transfer of a chemical species from one phase to another phase is an essential feature of the mixing and purification processes that are ubiquitous in the chemical and biological process industries. A comprehensive analysis of mass transfer requires an understanding of the prerequisite subjects of fluid mechanics, thermodynamics, and heat transfer; however, there are some mass transfer processes that can be approximated as *equilibrium stages* and these processes can be analyzed using the techniques presented in this text. Most students are familiar with an equilibrium stage when it is carried out in a *batch-wise manner*, since this is a common purification technique used in organic chemistry laboratories.

If an organic reaction produces a *desirable* product that is soluble in an organic phase and an *undesirable* product that is soluble in an aqueous phase, the desirable product can be purified by liquid-liquid extraction as illustrated in Figure 5.4. In the first step of this process, the mixture from a reactor is placed in a separatory funnel. Water is added, the system is agitated, and the phases are allowed to *equilibrate and*

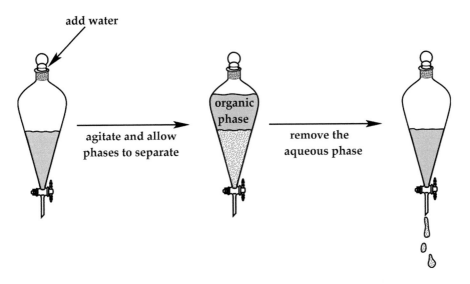

FIGURE 5.4 Batch liquid-liquid extraction.

separate. The amount of the undesirable product in the organic phase is reduced by an amount related to the volumes of the organic and aqueous phases and the equilibrium relation that determines how species A is distributed between the two phases. If the equilibrium relation is linear, the concentrations in the aqueous phase (β-phase) and organic phase (γ-phase) can be related by

Equilibrium relation: $$c_{A\gamma} = \kappa_{eq,A}\left(c_{A\beta}\right) \qquad (5.43)$$

This relation is based on the general concept of thermodynamic equilibrium illustrated by Eq. 5.24. In this case, we have used $\kappa_{eq,A}$ to represent the *equilibrium coefficient*; however, this information may be given in terms of a *distribution coefficient* defined as

Distribution coefficient: $$\kappa_{dist,A} = \frac{c_{A\beta}}{c_{A\gamma}} \qquad (5.44)$$

When working with equilibrium coefficients or distribution coefficients, one must be very careful to note the definition, since it is quite easy to invert these relations and create an enormous error. In addition, equilibrium relations are usually given in terms of mole fractions, but they are also given in terms of molar concentrations (as in Eq. 5.43), and they are sometime given in terms of species densities. A clear and unambiguous definition of any and all equilibrium relations is essential to avoid errors.

The analysis of the process illustrated in Figure 5.4 is relatively simple provided that the following conditions are valid: (1) There are negligible changes in the volumes of the organic and aqueous phases caused by the mass transfer process. (2) There is no species A in the aqueous phase used in the extraction process. (3) The linear equilibrium relation given by Eq. 5.43 is valid. If the batch process illustrated in Figure 5.4 is repeated N times, the concentration of species A in the organic phase is given by

$$(c_{A\gamma})_N = \frac{(c_{A\gamma})_o}{\left(1 + \dfrac{V_\beta}{\kappa_{eq,A} V_\gamma}\right)^N} \tag{5.45}$$

Here we have used V_β to represent the volume of the aqueous phase, V_γ to represent the volume of the organic phase, and $(c_{A\gamma})_N$ to represent the concentration of the *undesirable product* of the chemical reaction in the organic phase after N extractions. Eq. 5.45 indicates that repeated batch-wise extractions can be used to reduce the concentration of species A in the organic phase to arbitrarily small values.

To characterize the behavior of the repeated batch extraction process, it is convenient to define an *absorption factor* according to

$$A = \left(V_\beta / \kappa_{eq,A} V_\gamma \right) \tag{5.46}$$

so that Eq. 5.45 takes the form

$$(c_{A\gamma})_N = \frac{(c_{A\gamma})_o}{(1+A)^N} \tag{5.47}$$

This allows us to express two important limiting cases as

$$A \to 0, \quad (c_{A\gamma})_N \to (c_{A\gamma})_o, \quad \text{no change occurs} \tag{5.48a}$$

$$A \to \infty, \quad (c_{A\gamma})_N \to 0, \quad \text{maximum change occurs} \tag{5.48b}$$

Here, it is clear that one would like the absorption factor to be as large as possible; however, the definition given by Eq. 5.46 indicates that the value of A is limited by the *process* illustrated in Figure 5.4 and the *equilibrium coefficient* defined by Eq. 5.43. The derivation of Eq. 5.45 is left as an exercise for the student as is the case in which the concentration of species A in the original aqueous phase, $(c_{A\beta})_o$ is not zero.

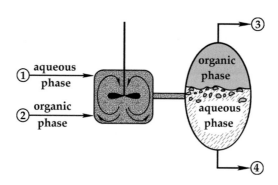

FIGURE 5.5   Liquid-liquid extraction.

While the batch extraction process illustrated in Figure 5.4 is convenient for use in an organic chemistry laboratory, a continuous process is preferred for a large-scale commercial purification process such as we have illustrated in Figure 5.5. There we have shown a system consisting of a *mixer* that provides a large surface area for mass transfer followed by a *settler* that separates the aqueous and organic phases. If the mixer-settler system is efficient, the two phases will be in equilibrium as they leave the settler. In this example, we are given an equilibrium relation in the form of Henry's law

Equilibrium relation: $y_A = K_{eq,A}\, x_A$,   at the fluid-fluid interface   (5.49)

which is the *mole fraction* version of the equilibrium relation given earlier by Eq. 5.43. If the flow rates of the aqueous and organic phases are *slow enough* and the mass transfer of species A between the two phases is *fast enough*, we can use Eq. 5.49 to construct a process equilibrium relation as

Process equilibrium relation: $(y_A)_3 = K_{eq,A}\,(x_A)_4$ (5.50)

We refer to this as a *process equilibrium relation* because it expresses the organic phase mole fraction in Stream #3 in terms of the aqueous phase mole fraction in Stream #4. Knowing when Eq. 5.50 is a valid approximation requires a detailed study of the mass transfer process of species A.

In a problem of this type, we wish to know how the concentrations in the *exit streams* are related to the concentrations in the *inlet streams*, and this leads to the use of the control volume illustrated in Figure 5.6.

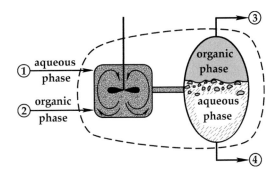

FIGURE 5.6   Control volume for mixer-settler.

The macroscopic mole balance for species A takes the form

Species A: $\displaystyle\int_{A_e} c_A \mathbf{v}\cdot\mathbf{n}\,dA = 0$ (5.51)

in which $\mathbf{v}\cdot\mathbf{n}$ is used in place of $\mathbf{v}_A\cdot\mathbf{n}$ with the idea that diffusive effects are negligible at the *entrances and exits* of the system. Since the process equilibrium relation given by Eq. 5.50 is expressed in terms of mole fractions, it is convenient to express Eq. 5.51 in the form (see Sec. 4.6)

Species A:
$$\int_{A_e} x_A \, c\mathbf{v} \cdot \mathbf{n} \, dA = 0 \tag{5.52}$$

When this result is applied to the control volume illustrated in Figure 5.6, we obtain

Species A:
$$-(x_A)_1 \dot{M}_1 - (y_A)_2 \dot{M}_2 + (y_A)_3 \dot{M}_3 + (x_A)_4 \dot{M}_4 = 0 \tag{5.53}$$

Here, we have used x_A to represent the mole fraction of species A in the aqueous phase and y_A to represent the mole fraction of species A in the organic phase. In addition, we have used $\dot{M}$ to represent the total molar flow rate in each of the four streams. At the two exits of the settler, we assume that the organic and aqueous streams are in equilibrium, thus the mixer-settler is considered to be an *equilibrium stage* and Eq. 5.50 is applicable. When changes in the mole fraction of species A are sufficiently small, we can use the approximations given by

$$\dot{M}_1 = \dot{M}_4 = \dot{M}_\beta, \quad \dot{M}_2 = \dot{M}_3 = \dot{M}_\gamma \tag{5.54}$$

Here we note that the change in the molar flow rates can be estimated as

$$\Delta\dot{M}_\beta \approx \Delta(x_A) \, \dot{M}_\beta, \quad \Delta\dot{M}_\gamma \approx \Delta(y_A) \, \dot{M}_\gamma \tag{5.55}$$

in which $\Delta(x_A)$ and $\Delta(y_A)$ represent the changes in the mole fractions that occur between the inlet and outlet streams. From these estimates, we conclude that

$$\Delta\dot{M}_\beta \ll \dot{M}_\beta, \quad \Delta\dot{M}_\gamma \ll \dot{M}_\gamma \tag{5.56}$$

whenever the change in the mole fractions are constrained by

$$\Delta(x_A) \ll 1, \quad \Delta(y_A) \ll 1 \tag{5.57}$$

If we impose these constraints on the process illustrated in Figure 5.6, we can use the approximation given by Eq. 5.54 in order to express Eq. 5.53 as

$$(y_A)_3 = (y_A)_2 + [(x_A)_1 - (x_A)_4] \left(\dot{M}_\beta / \dot{M}_\gamma \right) \tag{5.58}$$

We now make use of the process equilibrium relation given by Eq. 5.50 to eliminate $(x_A)_4$ leading to

$$(y_A)_3 = (y_A)_2 + (x_A)_1 \left(\dot{M}_\beta / \dot{M}_\gamma \right) - \frac{(y_A)_3}{K_{eq,A}} \left(\dot{M}_\beta / \dot{M}_\gamma \right) \tag{5.59}$$

Here, it is convenient to arrange the macroscopic mole balance for species A in the form

$$(y_A)_3 \left[1 + \frac{\left(\dot{M}_\beta / \dot{M}_\gamma \right)}{K_{eq,A}} \right] = (y_A)_2 + K_{eq,A} \, (x_A)_1 \left[\frac{\left(\dot{M}_\beta / \dot{M}_\gamma \right)}{K_{eq,A}} \right] \tag{5.60}$$

which suggests that we define an *absorption factor* according to

$$A = \frac{\left(\dot{M}_\beta / \dot{M}_\gamma \right)}{K_{eq,A}} \qquad (5.61)$$

Use of this expression in Eq. 5.60 and solving for $(y_A)_3$ leads to the following expression for the mole fraction of species A in the organic stream (γ-phase) leaving the settler illustrated in Figure 5.6.

$$(y_A)_3 = \frac{(y_A)_2}{1+A} + \frac{A}{1+A}\left[K_{eq,A}(x_A)_1 \right] \qquad (5.62)$$

This allows us to express two important limiting cases given by

$$A \rightarrow 0, \qquad (y_A)_3 \rightarrow (y_A)_2, \qquad \text{no change occurs} \qquad (5.63a)$$

$$A \rightarrow \infty, \qquad (y_A)_3 \rightarrow K_{eq,A}(x_A)_1, \qquad \text{maximum change occurs} \qquad (5.63b)$$

In order to design a mixer-settler to achieve a specified mole fraction of species A in Stream #3, one needs only to specify the absorption factor.

In the previous paragraphs, we examined a purification process from the point of view of an *equilibrium stage*, i.e., we assumed that the organic and aqueous streams leaving the settler are in equilibrium. The true state of equilibrium will never be achieved in a dynamic system such as the mixer-settler illustrated in Figure 5.6. However, the assumption of equilibrium is often a reasonable approximation and when that is the case, various mass transfer systems can be successfully designed and analyzed using the concept of an equilibrium stage.

If we are confronted with the problem of species A being transferred between a gas stream and a liquid stream, we require a contacting device that is quite different from that shown in Figure 5.6. In this case, we employ a gas-liquid contacting device of the type illustrated in Figure 5.7. Here a gas is forced through a perforated plate and then up through a liquid stream that flows across the plate. If the gas bubbles are *small enough* and if the liquid is *deep enough* and *completely mixed*, the gas will be in equilibrium with the liquid as it leaves the control volume. When this is the case,

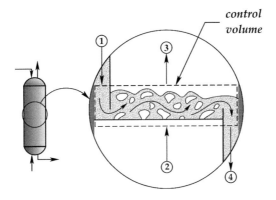

FIGURE 5.7   Gas-liquid contacting device.

we can treat the system illustrated in Figure 5.7 as an *equilibrium stage*. When the analysis is restricted to dilute solutions, one can usually employ a linear equilibrium relation, thus the mole fractions of species A in the exiting gas and liquid streams are related by

Process equilibrium relation: $(y_A)_3 = K_{eq,A} (x_A)_4$ (5.64)

Under these circumstances, the analysis of this system is identical to the analysis of the liquid-liquid extraction process illustrated in Figure 5.6.

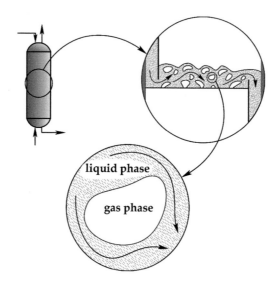

FIGURE 5.8   Gas-liquid mass transfer.

When the process equilibrium relation given by Eq. 5.64 is *not valid*, the simplification of an *equilibrium stage* can no longer be applied and one must move to a smaller length scale to analyze the mass transfer process. This situation is illustrated in Figure 5.8 where we have indicated that the mass transfer process must be studied at a smaller scale. The analysis of mass transfer at this scale will occur in a chemical engineering course where the equilibrium relation given by Eq. 5.64 will be applied at the gas-liquid interface. We express this type of equilibrium condition as

Equilibrium relation: $y_A = K_{eq,A} x_A$,   at the gas-liquid interface   (5.65)

When the form given by Eq. 5.64 represents a valid approximation, the techniques studied in this section can be applied with confidence.

In the preceding paragraphs, we have illustrated liquid-liquid and gas-liquid systems that can be treated as *equilibrium stages*. Such systems are ubiquitous in the world of chemical engineering where purification of liquid and gas streams is a major activity. In addition to the contacting devices illustrated in Figures 5.6 and 5.7, there are many other processes that can often be approximated as equilibrium stages and some of these are examined in the following paragraphs.

EXAMPLE 5.7   CONDENSATION OF WATER IN HUMID AIR

On a warm spring day in Baton Rouge, LA, the atmospheric pressure is 755 mmHg, the temperature is 80°F, and the relative humidity is 80%. A large industrial air conditioner operating at atmospheric pressure is illustrated in Figure 5.7a. It treats 1,000 kg/h of air (dry air basis) and lowers the air temperature from 80°F to 15°C. The cool air leaving the unit is assumed to be in equilibrium with the water that leaves the unit at 15°C. Thus the air conditioner is treated as an *equilibrium stage*.

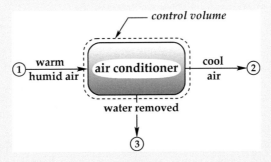

FIGURE 5.7a   Air conditioning as a separation process.

Here we want to determine how much liquid water, in kg/h, is removed from the warm, humid air by the air conditioning system. In this case, the obvious control volume cuts all three streams and encloses the air conditioner. If we assume that the system operates at steady state and we conclude that there are no chemical reactions, the appropriate form of the species mass balance is given by

$$\int_A \rho_A \mathbf{v}_A \cdot \mathbf{n}\,dA = 0, \quad A = 1, 2, \ldots, N \tag{1}$$

In problems of this type, it is plausible to neglect diffusion velocities, $\mathbf{u}_A \ll \mathbf{v}_A$, so that Eq. 1 takes the form

$$\int_A \rho_A \mathbf{v} \cdot \mathbf{n}\,dA = 0, \quad A = 1, 2, \ldots, N \tag{2}$$

Under these circumstances, we can combine the equations for nitrogen and oxygen to obtain the following mass balance for *dry air*:

$$\int_{A_1} \rho_{air} \mathbf{v} \cdot \mathbf{n}\,dA + \int_{A_2} \rho_{air} \mathbf{v} \cdot \mathbf{n}\,dA + \int_{A_3} \rho_{air} \mathbf{v} \cdot \mathbf{n}\,dA = 0 \tag{3}$$

Once again, we note that ρ_{air} indicates the density of oxygen and nitrogen as indicated in Eq. 5.36. Since significant amounts of nitrogen and oxygen do not leave the system in Stream #3, the mass balances for dry air and water take the form

Dry Air:
$$-(\dot{m}_{air})_1 + (\dot{m}_{air})_2 = 0 \tag{4}$$

Water:
$$-(\dot{m}_{H_2O})_1 + (\dot{m}_{H_2O})_2 + (\dot{m}_{H_2O})_3 = 0 \tag{5}$$

Here we have used the nomenclature illustrated by

$$\int_{A_1} \rho_{air} |\mathbf{v} \cdot \mathbf{n}| \, dA = (\dot{m}_{air})_1 \tag{6}$$

Eqs. 4 and 5 represent the fundamental balance equations associated with the system illustrated in Figure 5.7a, and we need to use the information that is given to determine $(\dot{m}_{H_2O})_3$.

Since information is given about the *humidity* of the incoming air stream, we need to think about how we can connect the mass flow rates in Eqs. 4 and 5 with the humidity vaguely described by Eq. 5.35 or the point humidity precisely defined by Eq. 5.36. We start by examining the ratio of mass flow rates (water to dry air) given by

$$\frac{(\dot{m}_{H_2O})_1}{(\dot{m}_{air})_1} = \frac{\displaystyle\int_{A_1} \rho_{H_2O} |\mathbf{v} \cdot \mathbf{n}| \, dA}{\displaystyle\int_{A_1} \rho_{air} |\mathbf{v} \cdot \mathbf{n}| \, dA} \tag{7}$$

If the velocity, $\mathbf{v} \cdot \mathbf{n}$, is *uniform* across the entrance, or the species densities, ρ_{H_2O} and ρ_{air}, are *uniform* across the entrance, we can follow the development in Sec. 4.5.1 to conclude that

$$\frac{(\dot{m}_{H_2O})_1}{(\dot{m}_{air})_1} = \frac{\displaystyle\int_{A_1} \rho_{H_2O} |\mathbf{v} \cdot \mathbf{n}| \, dA}{\displaystyle\int_{A_1} \rho_{air} |\mathbf{v} \cdot \mathbf{n}| \, dA} = \frac{\langle\rho_{H_2O}\rangle_1 Q_1}{\langle\rho_{air}\rangle_1 Q_1} = \frac{\langle\rho_{H_2O}\rangle_1}{\langle\rho_{air}\rangle_1} \tag{8}$$

In addition, if we accept the ratio of the *area average species densities* as our measure of the humidity we obtain

$$\frac{\langle\rho_{H_2O}\rangle_1}{\langle\rho_{air}\rangle_1} = \frac{(\dot{m}_{H_2O})}{(\dot{m}_{air})_1} = (\text{humidity})_1 \tag{9}$$

This suggests a particular strategy for solving this mass balance problem. Dividing Eq. 5 by the mass flow rate of dry air in Stream #1 leads to

$$-\frac{(\dot{m}_{H_2O})_1}{(\dot{m}_{air})_1} + \frac{(\dot{m}_{H_2O})_2}{(\dot{m}_{air})_1} + \frac{(\dot{m}_{H_2O})_3}{(\dot{m}_{air})_1} = 0 \qquad (10)$$

and on the basis of the mass balance for "dry air" given by Eq. 4, we can express this result in the form

$$-\frac{(\dot{m}_{H_2O})_1}{(\dot{m}_{air})_1} + \frac{(\dot{m}_{H_2O})_2}{(\dot{m}_{air})_2} + \frac{(\dot{m}_{H_2O})_3}{(\dot{m}_{air})_1} = 0 \qquad (11)$$

Use of equations of the form of Eq. 9 leads to a "humidity balance" equation given by

$$(\text{humidity})_1 = (\text{humidity})_2 + \frac{(\dot{m}_{H_2O})_3}{(\dot{m}_{air})_1} \qquad (12)$$

Our objective in this example is to determine how much liquid water is removed from the air, and Eq. 12 provides this information in the form

$$(\dot{m}_{H_2O})_3 = (\dot{m}_{air})_1 \left[(\text{humidity})_1 - (\text{humidity})_2 \right] \qquad (13)$$

Here, it becomes clear that the solution to this problem requires that we determine the humidity in Streams #1 and #2. This motivates us to make use of Eq. 5.37 that we list here as

$$\text{point humidity} = \frac{MW_{H_2O}\, p_{H_2O}}{MW_{air} \left(p - p_{H_2O} \right)} \qquad (14)$$

In addition, we are given information about the percent relative humidity defined by Eq. 5.38 and listed here as

$$\% \text{ relative humidity} = \frac{p_{H_2O}}{p_{H_2O,\, vap}} \times 100 \qquad (15)$$

In Stream #1, the percent relative humidity is 80% and this provides

$$(\% \text{ relative humidity})_1 = \frac{\left(p_{H_2O} \right)_1}{\left(p_{H_2O,\, vap} \right)_1} \times 100 = 80 \qquad (16)$$

which allows us to express the *partial pressure* of water vapor in Stream #1 as

$$\left(p_{H_2O} \right)_1 = 0.80 \times \left(p_{H_2O,\, vap} \right)_1 \qquad (17)$$

The *vapor pressure* of water, in both Stream #1 and Stream #2, can be determined by Antoine's equation (see Sec A.3 of Appendix A) that provides

$$\left(p_{H_2O, vap} \right)_1 = p_{H_2O, vap} \, (T = 80F) = 3.484 \text{ kPa} \qquad (18a)$$

$$\left(p_{H_2O, vap} \right)_2 = p_{H_2O, vap} \, (T = 15C) = 1.705 \text{ kPa} \qquad (18b)$$

Returning to Eq. 17 and making use of Eq. 18a leads to

$$\left(p_{H_2O} \right)_1 = 0.80 \times \left[p_{H_2O, vap} \, (T = 80F) \right] = 2,787 \text{ Pa} \qquad (19)$$

and we are now ready to determine the humidity in Stream #1.

We are given that the total pressure in the system is equivalent to 755 mmHg and this leads to a pressure given by

$$p = \frac{755 \text{ mmHg}}{760 \text{ mmHg/atm}} \times \frac{101.3 \times 10^3 \text{ Pa}}{\text{atm}} = 100,634 \text{ Pa} \qquad (20)$$

Use of this result in Eq. 14 leads to the humidity in Stream #1 given by

$$(\text{humidity})_1 = \frac{MW_{H_2O} \left(p_{H_2O} \right)_1}{MW_{air} \left[p - \left(p_{H_2O} \right)_1 \right]}$$

$$= \frac{18.015 \,\text{g H}_2\text{O/mol}}{28.84 \,\text{g dry air/mol}} \times \frac{2,795 \text{ Pa}}{100,634 \text{ Pa} - 2,795 \text{ Pa}}$$

$$= 0.01784 \text{ g H}_2\text{O/g dry air} \qquad (21)$$

We are given that Stream #2 is in equilibrium with Stream #3, thus the relative humidity in Stream #2 is 100% and the partial pressure of water vapor can be determined in terms of a *process equilibrium relation* as

$$(p_{H_2O})_2 = (p_{H_2O, vap})_3 = p_{H_2O, vap}\big|_{T=15C} = 1,705 \text{ Pa} \qquad (22)$$

This allows us to repeat the type of calculation illustrated by Eq. 21 to obtain

$$(\text{humidity})_2 = \frac{MW_{H_2O} \left(p_{H_2O} \right)_2}{MW_{air} \left[p - \left(p_{H_2O} \right)_2 \right]}$$

$$= 0.01076 \text{g H}_2\text{O/g dry air} \qquad (23)$$

Finally, we return to Eq. 13 to determine the mass flow rate of water leaving the air conditioning unit according to

$$\left(\dot{m}_{H_2O} \right)_3 = 1,000 \text{ kg dry air/h} \left[0.01784 - 0.01076 \right] \frac{\text{kg H}_2\text{O}}{\text{kg dry air}}$$

$$= 7.08 \text{ kg H}_2\text{O/h} \qquad (24)$$

EXAMPLE 5.8   USE OF AIR TO DRY WET SOLIDS

In Figure 5.8a, we have illustrated a co-current air dryer. The solids entering the dryer contain 20% water on a mass basis and the mass flow rate of
the wet solids entering the dryer is 1,000 lb_m/h. The dried solids contain
5% water on a mass basis, and the temperature of the solid stream leaving
the dryer is 65°C. The complete *design* of the dryer is a complex process that
requires a knowledge of the flow rate of the dry air entering the dryer. This can
be determined by a macroscopic mass balance analysis.

FIGURE 5.8a   Air dryer.

As the air and the wet solids pass through the dryer, an equilibrium condition is approached and mass transfer of water from the wet solids to the air
stream diminishes. As an example, we assume that the air leaving the dryer
is in equilibrium with the solids leaving the dryer, and this allows us to determine the *maximum amount* of water that can be removed from the wet solids.
For this type of approximation, the dryer becomes an *equilibrium stage*.

To construct a control volume for the analysis of this system, we need only
make cuts where information is *given* and *required* and this leads to the control
volume shown in Figure 5.8b.

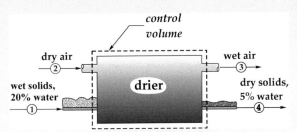

FIGURE 5.8b   Control volume.

We begin this problem with the species macroscopic *mole* balance for a
steady-state process in the absence of chemical reaction. In addition, we note
that the species velocity, $\mathbf{v}_A$, can be replaced with the mass average velocity, $\mathbf{v}$,
at entrances and exits in order to obtain

$$\int_A c_A \mathbf{v} \cdot \mathbf{n}\, dA = 0 \tag{1}$$

It is important to remember that the molar concentration can be expressed as

$$c_A = y_A c, \quad \text{gas streams} \tag{2}$$

where y_A is the mole fraction of species A and c is the total molar concentration. The form given by Eq. 2 is especially useful in the analysis of the air stream; however, for the wet solids stream, it is convenient to work in terms of *mass* rather than *moles* and make use of

$$c_A = \frac{\rho_A}{MW_A}, \quad \text{wet solid streams} \tag{3}$$

If we let species A be water and apply Eq. 1 to the control volume shown in Figure 5.8b, we obtain

$$\underbrace{\int_{A_1} \left(\rho_{H_2O}/MW_{H_2O} \right) \mathbf{v} \cdot \mathbf{n}\, dA}_{\substack{\textit{molar flow rate of water} \\ \textit{entering with the solid}}} + \underbrace{\int_{A_4} \left(\rho_{H_2O}/MW_{H_2O} \right) \mathbf{v} \cdot \mathbf{n}\, dA}_{\substack{\textit{molar flow rate of water} \\ \textit{leaving with the solid}}} +$$

Water:

$$\underbrace{\int_{A_2} y_{H_2O}\, c\, \mathbf{v} \cdot \mathbf{n}\, dA}_{\substack{\textit{molar flow rate of water} \\ \textit{entering with the air}}} + \underbrace{\int_{A_3} y_{H_2O}\, c\, \mathbf{v} \cdot \mathbf{n}\, dA}_{\substack{\textit{molar flow rate of water} \\ \textit{leaving with the air}}} = 0 \tag{4}$$

Here, we have used Eqs. 2 and 3 in order to arrange the fluxes in forms that are convenient, *but not necessary* for this particular problem, and we can express those fluxes in terms of averaged quantities to obtain

Water:

$$-\frac{\langle \rho_{H_2O} \rangle_1 Q_1}{MW_{H_2O}} + \frac{\langle \rho_{H_2O} \rangle_4 Q_4}{MW_{H_2O}}$$

$$-(y_{H_2O})_2\, \dot{M}_2 + (y_{H_2O})_3\, \dot{M}_3 = 0 \tag{5}$$

In this representation of the macroscopic mole balance for water, we have drawn upon the analysis presented in Sec. 4.5. Specifically, we have imposed the following assumptions

Gas streams: $\qquad\qquad\qquad c\,\mathbf{v} \cdot \mathbf{n} = \text{constant} \qquad\qquad (6a)$

Wet solid streams: $\qquad\qquad\quad \mathbf{v} \cdot \mathbf{n} = \text{constant} \qquad\qquad (6b)$

in which "constant" means constant across the area of the entrances and exits. The three phases contained in the wet solid streams are illustrated in Figure 5.8c for stream #1. The *total density* in these streams consists of the density of the solid, the water, and the air, and this density can be written explicitly as

$$\langle \rho \rangle = \langle \rho_{solid} \rangle + \langle \rho_{H_2O} \rangle + \langle \rho_{air} \rangle \tag{7}$$

The mass fraction of water in the wet solids is defined by

$$\langle \omega_{H_2O} \rangle = \frac{\langle \rho_{H_2O} \rangle}{\langle \rho \rangle} \tag{8}$$

and use of this representation in Eq. 5 leads to

Water:
$$-\frac{\langle \omega_{H_2O} \rangle_1 \dot{m}_1}{MW_{H_2O}} + \frac{\langle \omega_{H_2O} \rangle_4 \dot{m}_4}{MW_{H_2O}}$$
$$- (y_{H_2O})_2 \dot{M}_2 + (y_{H_2O})_3 \dot{M}_3 = 0 \tag{9}$$

Here we have identified the mass flow rates of the wet solid streams according to

$$\dot{m}_1 = \langle \rho \rangle_1 Q_1, \quad \dot{m}_4 = \langle \rho \rangle_4 Q_4 \tag{10}$$

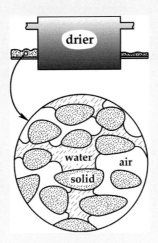

FIGURE 5.8c Wet solids entering the dryer.

A little thought (see Problem 5.33) will indicate that a mass balance for the *solid material* leads to

Solid material:
$$\left[1 - \langle \omega_{H_2O} \rangle_1 \right] \dot{m}_1 = \left[1 - \langle \omega_{H_2O} \rangle_4 \right] \dot{m}_4 \tag{11}$$

and this result can be used in Eq. 9 to obtain

Water:
$$(y_{H_2O})_2 \dot{M}_2 = (y_{H_2O})_3 \dot{M}_3 + \left\{ \frac{\langle \omega_{H_2O} \rangle_4 - \langle \omega_{H_2O} \rangle_1}{\left[1 - \langle \omega_{H_2O} \rangle_4 \right] MW_{H_2O}} \right\} \dot{m}_1 \tag{12}$$

A molar balance for the air will allow us to eliminate $\dot{M}_3$ from this result and the calculation of $\dot{M}_2$ easily follows (see Problem 5.34).

5.6 CONTINUOUS EQUILIBRIUM STAGE PROCESSES

In Sec. 5.5, we considered systems that consisted of a single contacting process in which *equilibrium conditions* were assumed to exist at the exit streams. Knowing when the condition of equilibrium is a reasonable approximation, requires a detailed study of the heat and mass transfer processes that are taking place. These details will be studied in subsequent courses where it will be shown that the condition of equilibrium is a reasonable approximation for many mass transfer processes. Our first example of an equilibrium stage was the batch liquid-liquid extraction process illustrated in Figure 5.4. In that case, one could repeat the extraction process to obtain an arbitrarily small value of the concentration of species A as indicated by Eq. 5.45. In this section, we wish to illustrate how this same type of multi-stage extraction process can be achieved for a steady-state process. In Figure 5.9, we have illustrated an arrangement of mixer-settlers that can be used to reduce the concentration of species A in the organic stream to an arbitrarily small value. Rather than working with the details illustrated in this figure, we will represent the mixer-settler unit as a single box so that our counter-current extraction process takes the form illustrated in Figure 5.10. Here, we note that the nomenclature used to identify the incoming and outgoing streams is different than that used in Figure 5.6 for a single liquid-liquid extraction unit. In this case, the number of the unit is used to identify the *outgoing streams*, and this simplification is necessary for an efficient treatment of a system containing N units.

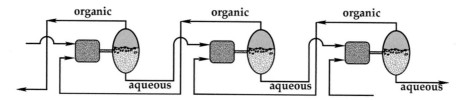

FIGURE 5.9   Multi-stage extraction process.

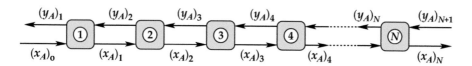

FIGURE 5.10   Schematic representation of a multi-stage extraction process.

5.6.1 SEQUENTIAL ANALYSIS-ALGEBRAIC

There are a variety of *problem statements* associated with a system of the type illustrated in Figure 5.10. For example, one might be given the following: (1) The number of stages, N. (2) The mole fractions for the *inlet conditions*, $(x_A)_o$ and $(y_A)_{N+1}$. (3) The equilibrium relation applicable to each stage. (4) The molar flow rates of the two phases, $\dot{M}_\beta$ and $\dot{M}_\gamma$. Given this information one would then be asked to determine the *outlet conditions*, $(y_A)_1$ and $(x_A)_N$.

Alternatively, one might be given: (1) The mole fractions for the *inlet streams*, $(x_A)_o$ and $(y_A)_{N+1}$. (2) The equilibrium relation. (3) The molar flow rates of the two phases, $\dot{M}_\beta$ and $\dot{M}_\gamma$. In this case, one would be asked to determine the *number of stages, N*, that would be required to achieve a *desired composition* in one of the *outlet streams*.

In the following paragraphs, we develop an algebraic solution for the cascade of N equilibrium stages illustrated in Figure 5.10, and we illustrate how this result can be applied to several different problem statements. We begin our analysis of the cascade of equilibrium stages with the single unit illustrated in Figure 5.11 in which the obvious control volume has been identified. For this case, the mole balance for species A takes the form

$$\underbrace{(x_A)_1\,\dot{M}_\beta + (y_A)_1\,\dot{M}_\gamma}_{\substack{\text{molar flow of species } A \\ \text{out of the control volume}}} = \underbrace{(x_A)_o\,\dot{M}_\beta + (y_A)_2\,\dot{M}_\gamma}_{\substack{\text{molar flow of species } A \\ \text{into the control volume}}} \qquad (5.66)$$

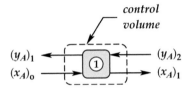

FIGURE 5.11   Single unit extraction process.

Here, we impose the special condition given by

Special condition: $(x_A)_o = 0$ (5.67)

So that Eq. 5.66 takes the form

$$(x_A)_1\,\dot{M}_\beta + (y_A)_1\,\dot{M}_\gamma = (y_A)_2\,\dot{M}_\gamma \qquad (5.68)$$

At this point, our objective is to develop a relation between $(y_A)_1$ and $(y_A)_2$, and we begin by arranging Eq. 5.68 in the form

$$(y_A)_1 + (x_A)_1 \left(\dot{M}_\beta / \dot{M}_\gamma \right) = (y_A)_2 \qquad (5.69)$$

Here, we note that the process equilibrium relation is given by

Process equilibrium relation: $(y_A)_1 = K_{eq,A}\,(x_A)_1$ (5.70)

and use of this result allows us to simplify the mole balance for species A to obtain

$$(y_A)_1 + (y_A)_1 \left(\dot{M}_\beta / \dot{M}_\gamma\, K_{eq,A} \right) = (y_A)_2 \qquad (5.71)$$

We now define the *absorption factor* according to

$$A = \left(\dot{M}_\beta / \dot{M}_\gamma\, K_{eq,A} \right) \qquad (5.72)$$

in order to develop the following relation between $(y_A)_1$ and $(y_A)_2$

One equilibrium stage:
$$(y_A)_1 = \frac{(y_A)_2}{1+A} \qquad (5.73)$$

Here, we have chosen to arrange the macroscopic mole balance for species A in terms of the mole fraction in the γ-phase. However, the choice is arbitrary and we could just as well have set up the analysis in terms of x_A instead of y_A.

Having developed an expression for the exit mole fraction, $(y_A)_1$, in terms of the entrance mole fraction, $(y_A)_2$, for a single equilibrium stage, we now wish to relate $(y_A)_1$ to $(y_A)_3$. To accomplish this, we examine the two equilibrium stages illustrated in Figure 5.12.

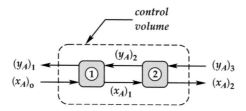

FIGURE 5.12   Two-unit extraction process.

In this case, the molar balance for species A can be expressed as

$$\underbrace{(x_A)_2\,\dot{M}_\beta + (y_A)_1\,\dot{M}_\gamma}_{\substack{\text{molar flow of species } A \\ \text{out of the control volume}}} = \underbrace{(x_A)_o\dot{M}_\beta + (y_A)_3\dot{M}_\gamma}_{\substack{\text{molar flow of species } A \\ \text{into the control volume}}} \qquad (5.74)$$

and we continue to impose the special condition

Special Condition:
$$(x_A)_o = 0 \qquad (5.75)$$

so that Eq. 5.74 takes the form

$$(x_A)_2\,\dot{M}_\beta + (y_A)_1\,\dot{M}_\gamma = (y_A)_3\,\dot{M}_\gamma \qquad (5.76)$$

At this point, our objective is to determine $(y_A)_1$ in terms of $(y_A)_3$, and we begin by arranging this result in the form

$$(y_A)_1 + (x_A)_2\left(\dot{M}_\beta/\dot{M}_\gamma\right) = (y_A)_3 \qquad (5.77)$$

Here we note that the process equilibrium relation is given by

Process equilibrium relation: $(y_A)_2 = K_{eq,A}\,(x_A)_2 \qquad (5.78)$

and use of this result allows us to simplify the mole balance for species A to

$$(y_A)_1 + (y_A)_2 \left(\dot{M}_\beta / \dot{M}_\gamma \, K_{eq,A} \right) = (y_A)_3 \tag{5.79}$$

We now use Eq. 5.73 to determine $(y_A)_2$ so that Eq. 5.79 provides the following result

$$(y_A)_1 + (y_A)_1 \left[A(1+A) \right] = (y_A)_3 \tag{5.80}$$

which can be solved for $(y_A)_1$ to obtain

Two equilibrium stages: $\qquad (y_A)_1 = \dfrac{(y_A)_3}{1 + A + A^2} \tag{5.81}$

We are now ready to determine $(y_A)_1$ when three equilibrium stages are employed as illustrated in Figure 5.13.

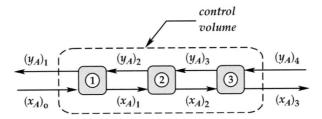

FIGURE 5.13 Three-unit extraction process.

In this case, the molar balance for species A can be expressed as

$$\underbrace{(x_A)_3 \, \dot{M}_\beta + (y_A)_1 \, \dot{M}_\gamma}_{\substack{\text{molar flow of species } A \\ \text{out of the control volume}}} = \underbrace{(x_A)_0 \, \dot{M}_\beta + (y_A)_4 \, \dot{M}_\gamma}_{\substack{\text{molar flow of species } A \\ \text{into the control volume}}} \tag{5.82}$$

and we continue to impose the condition

Special condition: $\qquad (x_A)_0 = 0 \tag{5.83}$

so that Eq. 5.82 takes the form

$$(x_A)_3 \, \dot{M}_\beta + (y_A)_1 \, \dot{M}_\gamma = (y_A)_4 \, \dot{M}_\gamma \tag{5.84}$$

At this point, our objective is to determine $(y_A)_1$ as a function of $(y_A)_4$, and we begin by arranging this result in the form

$$(y_A)_1 + (x_A)_3 \left(\dot{M}_\beta / \dot{M}_\gamma \right) = (y_A)_4 \tag{5.85}$$

Here, we note that the process equilibrium relation is given by

Process equilibrium relation: $\qquad (y_A)_3 = K_{eq,A} \, (x_A)_3 \tag{5.86}$

and use of this result allows us to simplify the mole balance for species A to

$$(y_A)_1 + (y_A)_3 \left(\dot{M}_\beta / \dot{M}_\gamma \, K_{eq,A} \right) = (y_A)_4 \tag{5.87}$$

We now use Eq. 5.80 to eliminate $(y_A)_3$ so that Eq. 5.87 provides the following result

$$(y_A)_1 + (y_A)_1 \, A\left[1 + A(1 + A) \right] = (y_A)_4 \tag{5.88}$$

which can be solved for $(y_A)_1$ as a function of $(y_A)_4$ and the absorption coefficient.

Three equilibrium stages: $(y_A)_1 = \dfrac{(y_A)_4}{1 + A + A^2 + A^3}$ (5.89)

Here, we are certainly in a position to deduce that N equilibrium stages would produce the result given by

N equilibrium stages: $(y_A)_1 = \dfrac{(y_A)_{N+1}}{1 + A + A^2 + A^3 + \cdots + A^N}$ (5.90)

This type of *sequential analysis* can be used to analyze any multi-stage process such as the one illustrated in Figure 5.10; however, one must remember that there are three important simplifications associated with this result. These three simplifications are: (1) The total molar flow rates, $\dot{M}_\beta$ and $\dot{M}_\gamma$, are treated as constants. (2) A linear equilibrium relation is applicable. (3) The mole fraction of species A entering the system in the β-phase is zero, i.e., $(x_A)_o = 0$. This latter constraint can be easily removed and that will be left as an exercise for the student.

If the absorption factor and the number of stages are specified for the system illustrated in Figure 5.10, one can easily calculate the mole fraction of species A at the exit of the system, $(y_A)_1$, in terms of the mole fraction at the entrance, $(y_A)_{N+1}$. If $(y_A)_1$ is specified as some fraction of $(y_A)_{N+1}$ and the number of stages are specified, the required absorption factor can be determine using the implicit expression for A given by (see Appendix B for the solution procedure)

$$1 + A + A^2 + A^3 + \cdots + A^N = \frac{(y_A)_{N+1}}{(y_A)_1} \tag{5.91}$$

In general, it is convenient to use this result along with the macroscopic balance around the *entire cascade* illustrated in Figure 5.10. This is given by

$$\underbrace{(x_A)_N \, \dot{M}_\beta + (y_A)_1 \, \dot{M}_\gamma}_{\substack{\text{molar flow of species } A \\ \text{out of the control volume}}} = \underbrace{(x_A)_o \, \dot{M}_\beta + (y_A)_{N+1} \, \dot{M}_\gamma}_{\substack{\text{molar flow of species } A \\ \text{into the control volume}}} \tag{5.92}$$

and it is often convenient to arrange this result in the form

$$(x_A)_N = \frac{(y_A)_{N+1} - (y_A)_1}{\dot{M}_\beta / \dot{M}_\gamma} \tag{5.93}$$

Here, we have continued to impose the condition that $(x_A)_o = 0$. In order to keep this initial study as simple as possible.

There are several types of problems that can be explored using the analysis leading to Eq. 5.91 and Eq. 5.93, and we list three of the types as follows:

Type I: Given the *inlet* mole fractions, $(x_A)_o$ and $(y_A)_{N+1}$, the system parameters, and the *desired* value of $(y_A)_1$, we would like to determine the number of stages, N.

Type II: Given the *inlet* mole fractions, $(x_A)_o$ and $(y_A)_{N+1}$, the system parameters, and the number of stages, N, we would like to determine the value of $(y_A)_1$.

Type III: Given the *inlet* mole fractions, $(x_A)_o$ and $(y_A)_{N+1}$, the system parameters $K_{eq,A}$, $\dot{M}_\gamma$, the number of stages, N, and a *desired* value of $(y_A)_1$, we would like to determine the molar flow rate of the β-phase, $\dot{M}_\beta$.

EXAMPLE 5.9   ABSORPTION OF ACETONE FROM AIR INTO WATER

In order to extract acetone from air using water, one can apply a contacting device such as the one illustrated in Figure 5.7. The cascade of equilibrium stages can be represented by Figure 5.10; however, the analogous system illustrated in Figure 5.9a is often used to more closely represent the gas-liquid contacting process under consideration.

There are various problems associated with the design and operation of this unit and we begin with the first type of problem identified in the previous paragraph.

Type I: Given the *inlet* mole fractions, $(x_A)_o$ and $(y_A)_{N+1}$, the system parameters, and the *desired* value of $(y_A)_1$, we would like to determine the number of stages, N.

In this case, we are given the inlet mole fractions, $(x_A)_o = 0$ and $(y_A)_{N+1} = 0.010$, and the system parameters, $K_{eq,A} = 2.53$, $\dot{M}_\beta = 90$ kmol/h and $\dot{M}_\gamma = 30$ kmol/h. For these particular values, we want to know how many stages, N, are required to reduce the exit mole fraction of acetone to $(y_A)_1 = 0.001$. In this case, we express Eq. 5.91 as

$$1 + A + A^2 + A^3 + \cdots + A^N = \frac{(y_A)_{N+1}}{(y_A)_1} = \frac{0.010}{0.001} = 10.0 \tag{1}$$

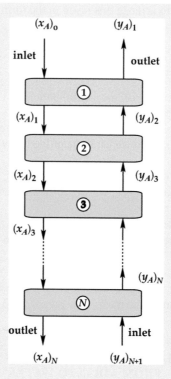

$(x_A)_0$ $(y_A)_1$

inlet   outlet

①

$(x_A)_1$ $(y_A)_2$

②

$(x_A)_2$ $(y_A)_3$

③

$(x_A)_3$ $(y_A)_N$

Ⓝ

outlet   inlet

$(x_A)_N$ $(y_A)_{N+1}$

FIGURE 5.9a Multi-stage gas-liquid contacting device.

in which the value of the absorption coefficient is given by $A = \dot{M}_\beta / \dot{M}_\gamma K_{eq,A} = 1.186$. This leads to the values listed in Table 5.9a where we see that five stages are *insufficient* to achieve the desired result, $(y_A)_1 = 0.001$. In addition, the use of six stages reduces the exit mole fraction of acetone in the air stream to $(y_A)_1 = 0.00081$ which is less than the desired result. For this type of situation, it is the responsibility of the chemical engineer to make a judgment based on safety considerations, environmental constraints, requirements

TABLE 5.9a

Number of Stages, N, Versus $(y_A)_{N+1}/(y_A)_1$

Number of Stages, N	$1 + A + A^2 + A^3 +, ..., + A^N$
1	2.186
2	3.592
3	5.260
4	7.238
5	9.584
6	12.366

for other processing units, and economic optimization. Such matters are covered in a future course on process design, and we have alluded to some of these concerns in Sec. 1.1.

Type II: Given the *inlet* mole fractions, $(x_A)_o$ and $(y_A)_{N+1}$, the system parameters, and the number of stages, N, we would like to determine the value of $(y_A)_1$.

In this case we consider an *existing unit* in which there are 7 stages. The *inlet* mole fractions are given by $(x_A)_o = 0$ and $(y_A)_8 = 0.010$, and the parameters associated with the system are specified as $K_{eq,A} = 2.53$, $\dot{M}_\beta = 90$ kmol/h and $\dot{M}_\gamma = 30$ kmol/h. In order to determine the mole fraction in the γ-phase (air) leaving the cascade, we make use of Eq. 5.91 to express $(y_A)_1$ as

$$(y_A)_1 = \frac{(y_A)_8}{1 + A + A^2 + A^3 + A^4 + A^5 + A^6 + A^7} \tag{2}$$

This leads to the following value of the mole fraction of acetone leaving the top of the cascade illustrated in Figure 5.9a:

$$(y_A)_1 = \frac{0.010}{15.67} = 0.00064 \tag{3}$$

In addition to calculating the exit mole fraction of acetone in the air stream, one may want to determine the mole fraction of acetone in the water stream leaving the cascade. This is obtained from Eq. 5.93 which provides

$$(x_A)_7 = \frac{(y_A)_8 - (y_A)_1}{\dot{M}_\beta / \dot{M}_\gamma} = \frac{0.010 - 0.00064}{(90 \text{ kmol/h})/(30 \text{ kmol/h})} = 0.0031 \tag{4}$$

Sometimes, it is possible to change the operating characteristics of a cascade to achieve a desired result and this situation is considered in the following example.

Type III: Given the *inlet* mole fractions, $(x_A)_o$ and $(y_A)_{N+1}$, the system parameters $K_{eq,A}$, $\dot{M}_\gamma$, the number of stages, N, and a *desired* value of $(y_A)_1$, we would like to determine the molar flow rate of the β-phase, $\dot{M}_\beta$.

In this case, we consider an *existing unit* in which there are 6 stages. The *inlet* mole fractions are given by $(x_A)_o = 0$ and $(y_A)_7 = 0.010$, and the specified parameters associated with the system are $K_{eq,A} = 2.53$ and $\dot{M}_\gamma = 30$ kmol/h. The desired value of the mole fraction of acetone in the exit air stream is $(y_A)_1 = 0.0005$. We begin the analysis with Eq. 5.91 that leads to

$$1 + A + A^2 + A^3 + A^4 + A^5 + A^6 = \frac{(y_A)_7}{(y_A)_1} = \frac{0.010}{0.0005} = 20.0 \tag{5}$$

Since $\dot{M}_\beta$ is an adjustable parameter, we need only solve this *implicit equation* for the absorption coefficient in order to determine the molar flow rate. Use of one of the iterative methods described in Appendix B leads to

$$A = 1.342 \tag{6}$$

and from the definition of the absorption coefficient given by Eq. 5.72 we determine the molar flow rate of the β-phase to be

$$\dot{M}_\beta = A \, \dot{M}_\gamma \, K_{eq,A} = 101.9 \text{ kmol/h} \tag{7}$$

Adjusting a molar flow rate to achieve a specific separation is a convenient operational technique provided that the auxiliary equipment required to change the flow rate is readily available.

5.6.2 SEQUENTIAL ANALYSIS-GRAPHICAL

The algebraic solution to the cascade of equilibrium stages illustrated in Figure 5.10 can also be represented in terms of a graph. In this case, we consider the cascade illustrated in Figure 5.14 in which the index n is bounded according to $1 \le n \le N$. The macroscopic mole balance for species A associated with this control volume can be expressed as

$$\underbrace{(x_A)_n \, \dot{M}_\beta + (y_A)_1 \, \dot{M}_\gamma}_{\substack{\textit{molar flow of species A} \\ \textit{out of the control volume}}} = \underbrace{(x_A)_o \, \dot{M}_\beta + (y_A)_{n+1} \, \dot{M}_\gamma}_{\substack{\textit{molar flow of species A} \\ \textit{into the control volume}}} \tag{5.94}$$

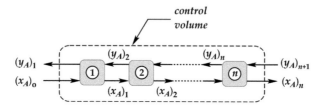

FIGURE 5.14   Cascade of equilibrium stages.

In this case, we want to develop an expression for $(y_A)_{n+1}$ thus we arrange Eq. 5.94 in the form

$$(y_A)_{n+1} = (x_A)_n \left(\dot{M}_\beta / \dot{M}_\gamma \right) + \left[(y_A)_1 - (x_A)_o \left(\dot{M}_\beta / \dot{M}_\gamma \right) \right], \quad n = 1, 2, \dots, N \tag{5.95}$$

This is sometime called the "operating line" and it can be used in conjunction with the "equilibrium line"

$$(y_A)_n = K_{eq,A} (x_A)_n, \quad n = 1, 2, \dots, N \tag{5.96}$$

to provide a graphical representation of the solution developed using the sequential analysis. To be precise, we note that Eqs. 5.95 and 5.96 represents a series of *operating points* and *equilibrium points*; however, the construction of *operating lines* and *equilibrium lines* is a useful graphical tool.

To illustrate the graphical construction associated with Eqs. 5.95 and 5.96, we consider the Type I problem discussed in Sec. 5.6.1 and repeated here as

Type I: Given the *inlet* mole fractions, $(x_A)_o$ and $(y_A)_{N+1}$, the system parameters, and the *desired* value of $(y_A)_1$, we would like to determine the number of stages, N.

In this case, we are given the *inlet* mole fractions, $(x_A)_o = 0$ and $(y_A)_{N+1} = 0.010$, and the system parameters, $K_{eq,A} = 2.25$, $\dot{M}_\beta = 97.5$ kmol/h and $\dot{M}_\gamma = 30$ kmol/h. For these particular values, we want to know how many stages, N, are required to reduce the exit mole fraction of acetone to $(y_A)_1 = 0.001$. The *equilibrium line* associated with Eq. 5.96 is given directly by

Equilibrium line: $$y_A = 2.25 x_A \qquad (5.97)$$

while the construction of the operating line associated with Eq. 5.95 requires that we assume the value of $(y_A)_1$ in order to obtain

Operating line: $$y_A = 3.25\, x_A + 0.001 \qquad (5.98)$$

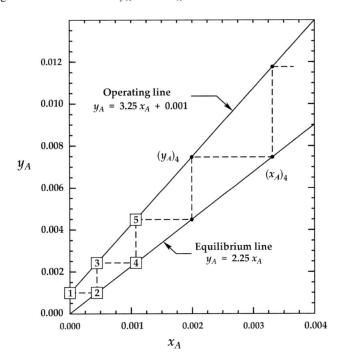

FIGURE 5.15   Graphical analysis of a cascade of equilibrium stages.

This approach is comparable to our use of Eq. 5.91 in Example 5.9 where the analysis of a Type I problem required that we specify $(y_A)_1$ in our search for the number of stages associated with $1 + A + \cdots + A^N$.

Given the operating line and the equilibrium line, we can construct the *operating points* and the *equilibrium points* and this is done in Figure 5.15. Some of these points are identified by numbers such as $\boxed{4}$ while others are identified by solid dots such as •. In order to connect the *graphical analysis* shown in Figure 5.15 with the *sequential analysis* given in Sec. 5.6.1, we recall Figure 5.11 in the form given below by Figure 5.16. There we have clearly identified Points #1, #2, and #3 in the graphical analysis with those same pairs of values in the sequential analysis. In Figure 5.15 and in Figure 5.16, we see that Point #1 represents a point on the operating line, that Point #2 represents a point on the equilibrium line, and that Point #3 represents a second point on the operating line.

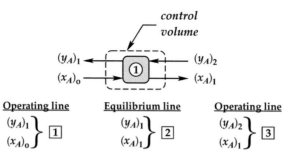

FIGURE 5.16 Mole balance around unit #1.

To continue this type of comparison, we recall Figure 5.12 in the form given below by Figure 5.17. There we have clearly identified Points #3, #4, and #5 in the graphical analysis with those same pairs of values in the sequential analysis. In Figures 5.15 and 5.17, we see that Point #3 represents a point on the operating line, that Point #4 represents a point on the equilibrium line, and that Point #5 represents a second point on the operating line.

The sequential analysis presented in Sec. 5.6.1 has the advantage of providing a clear set of equations that describe the mass transfer process occurring in a cascade of equilibrium stages. The graphical analysis is certainly less accurate,

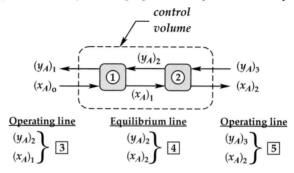

FIGURE 5.17 Mole balance around units #1 and #2.

but it can illustrate quite effectively how the system responds to changes in the operating conditions. As an example, we can *reduce* the molar flow rate of the β-phase (water) and examine the effect that will have on the separation process. For the case in which species A is transferred from the γ-phase to the β-phase, the reduction of $\dot{M}_\beta$ will surely make the separation process less efficient. This change is clearly indicated in Figure 5.18 where the effect of reducing $\dot{M}_\beta$ by 21% diminishes the efficiency of the cascade significantly. The graphical representation of a cascade of equilibrium stages is explored more thoroughly in Example 5.10.

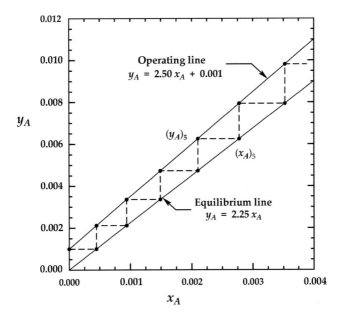

FIGURE 5.18 Reduced molar flow rate for the β-phase.

EXAMPLE 5.10 GRAPHICAL ANALYSIS OF THE ABSORPTION PROCESS

Here, we extend Example 5.9 to explore the use of graphical methods to analyze the absorption of acetone from air into water. The molar flow rate of the gas mixture (air-acetone) entering the cascade (at stage N) is $\dot{M}_\gamma = 30$ kmol/h and the mole fraction of acetone is given by $(y_A)_{N+1} = 0.010$. The molar flow rate of the liquid (water) entering the cascade (at stage #1) is $\dot{M}_\beta = 90$ kmol/h and the mole fraction of acetone in this entering liquid stream is zero, i.e., $(x_A)_o = 0$. In this case, the slope of the operating line and the equilibrium coefficient are given by

$$\dot{M}_\beta/\dot{M}_\gamma = 3.0, \quad K_{eq,A} = 2.53 \tag{1}$$

and we can use these values to determine an operating line and an equilibrium line. Following the development given in Sec. 5.6.2, we construct the graph illustrated in Figure 5.10a.

There we find that five stages are *not sufficient* to produce the desired result, i.e., $(y_A)_{N+1} = 0.010$, and one is forced to make a choice of adding a sixth stage or accepting a somewhat reduced separation.

If we reduce the water flow rate from 90 kmol/h to 70 kmol/h, the slope of the operating line is reduced from 3.00 to 2.33, and the situation illustrated in Figure 5.10a changes to that illustrated in Figure 5.10b. An overall mole balance indicates that $(x_A)_N = 0.00386$ and the graphical construction indicates that slightly more than 18 equilibrium stages are required.

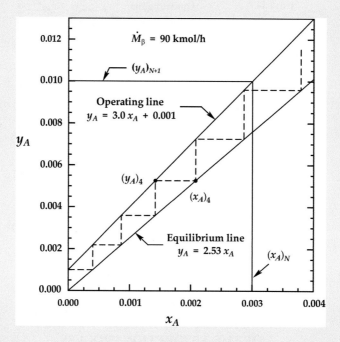

FIGURE 5.10a   Graphical analysis.

This means that we need 19 stages to accomplish the objective of reducing the mole fraction in the γ-phase to $(y_A)_1 = 0.001$. If the water flow rate is further reduced to 67 kmol/h, we see in Figure 5.10c that the equilibrium line and the operating line intersect at $x_A = 0.00337$. This is referred to as a *pinch point* within the cascade of equilibrium stages, and an infinite number of equilibrium stages are required to reach this point. Consequently, it is impossible to reach the desired design specifications for the cascade, i.e., $(y_A)_{N+1} = 0.010$ and $(x_A)_N = 0.0403$.

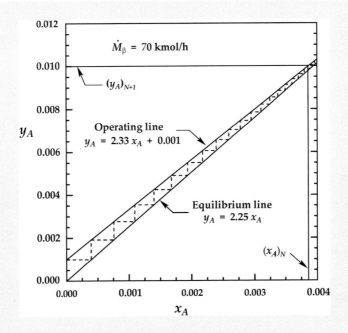

FIGURE 5.10b Influence of a reduction in the molar flow rate of water.

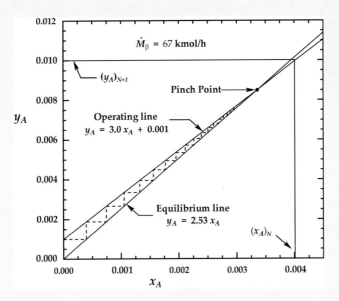

FIGURE 5.10c Pinch point for a cascade of equilibrium stages.

The sequential analysis and the graphical analysis of continuous equilibrium stage processes provide a relatively clear illustration of the physical processes involved; however, more complex systems are routinely encountered in industrial practice. In those cases, powerful numerical methods are extremely useful.

5.7 PROBLEMS

Note: Problems marked with the symbol ⌨ will be difficult to solve without the use of computer software. Physical property data are available in Appendix A.

SECTION 5.1

5.1. Show that the mole fraction in an ideal gas mixture can be expressed as $y_A = p_A/p$.

5.2. Demonstrate that the volume percent of a mixture is the same as the mole percent for an ideal gas mixture.

5.3. Assuming ideal gas behavior, determine the average molecular mass of a mixture made of equal amounts of mass of chlorine, argon, and ammonia.

SECTION 5.2

5.4. A liquid mixture of hydrocarbons has 40% by weight of cyclohexane, 40% of benzene, and 20% toluene. Assuming that volumes are additive compute the following:

    a. species densities of the components in the mixture.
    b. overall density of the mixture
    c. concentrations of the components in mol/m^3
    d. mole fractions of the components in the mixture.

SECTION 5.3

5.5. Determine the vapor pressure, in Pascal, of ethyl ether at 25°C and at 30°C. Estimate the heat of vaporization of ethyl ether using these two vapor pressures and the Clausius-Clapeyron equation.

5.6. Determine the vapor pressure of methanol at 25°C and compare it to that of ethanol at the same temperature. Consider the ethanol-methanol system to be an ideal solution in the liquid phase and an ideal gas mixture in the vapor phase. Determine the mole fraction of methanol in the vapor phase when the liquid phase mole fraction is 0.50. If the liquid phase is allowed to slowly evaporate, will it become richer in methanol or ethanol? Here you are asked to provide an intuitive answer concerning the composition of the liquid phase during the process of distillation. In Chapter 8, a precise analysis of the process will be presented.

5.7. Determine the vapor-liquid equilibrium curve of a binary mixture of cyclo-hexane and acetone. Plot the mole fraction of acetone in the vapor phase versus the mole fraction of acetone in the liquid phase at one atmosphere (760 mmHg).

5.8. Use Eqs. 5.26 and 5.27 order to derive Eq. 5.28.

5.9. Demonstrate that Eq. 5.31 is valid for an ideal system containing three compo-nents when an appropriate constraint is imposed.

5.10. Given the general representations for the chemical potentials

$$(\mu_A)_{gas} = G(T, p, y_A, y_B, \text{etc.}), \quad (\mu_A)_{liquid} = F(T, p, x_A, x_B, \text{etc.}) \tag{1}$$

make use of Eq. 5.24 to develop Henry's law given by Eqs. 5.31 and 5.32. Use a Taylor series expansion for $\mu_{A, gas}$ (see Problem 5.31) to obtain

$$(\mu_A)_{gas} = G\Big|_{y_A=0} + \frac{\partial G}{\partial y_A}\Big|_{y_A=0} (y_A) + \frac{\partial^2 G}{\partial y_A^2}\Big|_{y_A=0} (y_A)^2 + \cdots \tag{2}$$

Since the chemical potential of species A is zero when there is no species A present, this simplifies to

$$(\mu_A)_{gas} = \frac{\partial G}{\partial y_A}\Big|_{y_A=0} (y_A) + \frac{\partial^2 G}{\partial y_A^2}\Big|_{y_A=0} (y_A)^2 + \cdots \tag{3}$$

Restricting this development to dilute solutions of species A, we can impose $y_A \ll 1$ in order to express the chemical potential in the gas phase as

$$(\mu_A)_{gas} = \frac{\partial G}{\partial y_A}\Big|_{y_A=0} (y_A) \tag{4}$$

Develop a similar representation for the liquid phase and show how these special representations for the chemical potential lead to Henry's law given by Eq. 5.31.

SECTION 5.4

5.11. An equimolar mixture of ethanol and ethyl ether is kept in a closed container at 103 KPa and 95°C. The temperature of the container is slowly reduced to the dew point of the mixture. Determine:

a. What is the dew point temperature of the mixture?
b. What is the pressure of the container at the dew point temperature of the mixture?
c. What is the composition of the first drop of liquid at the dew point?

5.12. A liquid mixture of n-hexane (mole fraction equal to 0.32) and n-heptane is heated until it begins boiling. Find the bubble point at $p = 760$ mmHg. What are the mole fractions in the vapor when the mixture begins to boil?

5.13. A vapor mixture of benzene and toluene is slowly cooled inside a constant volume vessel. Initially, the pressure inside the vessel is 300 mmHg and the temperature is 70°C. As the vessel is cooled, the pressure inside the vessel decreases. Assume the vapor behaves like an ideal gas and take the dew point of the mixture to be 60°C. What is the mole fraction of benzene in the initial vapor mixture?

5.14. Given Eq. 5.36 as the definition of the point humidity, explore the possible definitions for the area averaged humidity and the volume averaged humidity. Refer to Example 5.6 for guidance.

5.15. Derive Eq. 5.37 from Eq. 5.36.

5.16. Consider a day when the percent relative humidity is 70%, the temperature is 80°F and the barometric pressure is 1 atm. What is the humidity, the mole fraction of water in the air, and the dew point of the air?

5.17. A mole of air is sampled from the atmosphere when the atmospheric pressure is 765 mmHg, the temperature is 25°C, and relative humidity is 75%. The sample of air is placed inside a closed container and heated to 135°C and then compressed to 2 atm. What are the relative humidity, the humidity, and the mole fraction of water in the compressed air?

5.18. A humidifier is used to introduce moisture into air supplied to an office building during winter days. Outside air at atmospheric pressure and 5°C is introduced into the heating system at a rate of 100 m³/min, on a dry air basis. The relative humidity of the outside air is 95%, and the heating system delivers warm air into the building at 20°C. How much water must be introduced into the warm air, in kg/min, the keep the relative humidity inside the building at 75%?

5.19. The modified mass fraction, Ω_A, is defined by

$$\Omega_A = \rho_A \bigg/ \sum_{B=1}^{B=N-1} \rho_B$$

Use this definition to develop relations analogous to Eqs. 5.40 and 5.42.

SECTION 5.5

5.20. Demonstrate that if the batch process illustrated in Figure 5.4 is repeated N times, the concentration of species A in the organic phase is given by

$$(c_{A\gamma})_N = (c_{A\gamma})_o / (1+A)^N \tag{1}$$

in which A is the absorption factor defined by

$$A = V_\beta / \kappa_{eq,A} \, V_\gamma \tag{2}$$

Here V_γ represents the original volume of the organic phase, and V_β represents the volume of the aqueous phase used in each of the N steps in the process.

5.21. The result indicated by Eq. 1 in Problem 5.20 is based on the condition that the concentration of species A is zero in the aqueous phase used in the extraction process. Repeat the analysis of the batch extraction process assuming that the concentration of species A in the original aqueous phase is $(c_{A\beta})_o$.

5.22. In the analysis of the batch liquid-liquid extraction process illustrated in Figure 5.4, the equilibrium relation was given as

Equilibrium relation: $c_{A\gamma} = \kappa_{eq,A} \, (c_{A\beta})$ (1)

Illustrate how $\kappa_{eq,A}$ is related to the Henry's law equilibrium coefficient given by $K_{eq,A}$ in Eq. 5.31. Use $y_{A\gamma}$ to represent the mole fraction of species A in the organic phase and $x_{A\beta}$ to represent the mole fraction of species A in the aqueous phase.

5.23. In order to justify the simplification indicated by Eq. 5.54, we need estimates of $\dot{M}_1 - \dot{M}_4$ and $\dot{M}_2 - \dot{M}_3$. If we let species B represent the organic phase (the γ-phase), and we assume that none of this species is transferred to the aqueous phase (the β-phase), a mole balance for species B takes the form

Species B: $-(y_B)_2 \, \dot{M}_2 + (y_B)_3 \, \dot{M}_3 = 0$ (1)

Use this species mole balance along with the definition

$$(y_B)_3 = (y_B)_2 + \Delta(y_B)$$ (2)

to obtain an estimate of $\dot{M}_2 - \dot{M}_3$. Use this estimate to identify the conditions that are required in order that the molar flow rates of the γ-phase are constrained by $\dot{M}_2 - \dot{M}_3 \ll \dot{M}_3$

5.24. Small amounts of an inorganic salt contained in an organic fluid stream can be removed by contacting the stream with pure water as illustrated in Figure 5-24.

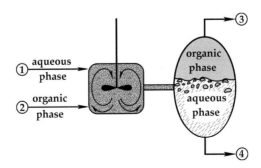

FIGURE 5-24   Liquid-liquid extraction.

The process requires that the organic and aqueous streams be contacted in a mixer that provides a large surface area for mass transfer, and then separated in a settler. If the mixer is efficient, the two phases will be in equilibrium as they leave the settler and you are to *assume* that this is the case for this problem. You are given the following information:

a. Organic stream flow rate: 1,000 lb_m/min
b. Specific gravity of the organic fluid: $\rho_{org}/\rho_{H_2O} = 0.87$
c. Salt concentration in the organic stream entering the mixer:
$(c_A)_{org} = 0.0005$ mol/L
d. Equilibrium relation for the inorganic salt: $(c_A)_{org} = \kappa_{eq,A} (c_A)_{aq}$
where $\kappa_{eq,A} = 1/60$

Here $(c_A)_{aq}$ represents the *salt concentration* in the aqueous phase that is in equilibrium with the salt concentration in the organic phase, $(c_A)_{org}$. In this problem, you are asked to determine the mass flow rates of the water stream that will reduce the salt concentration in the organic stream to 0.1, 0.01, and 0.001 times the original salt concentration. The aqueous and organic phases are to be considered completely immiscible, i.e., only salt is transferred between the two phases. In addition, the amount of material transferred is so small that the volumetric flow rates of the two streams can be considered constant.

5.25. In this problem, we examine the process of recovering fission materials from spent nuclear fuel rods. This is usually referred to as *reprocessing* of the fuel to recover plutonium (Pu) and the active isotope of uranium (U$_{235}$). Reprocessing can be done by separation of the soluble isotope nitrates from a solution in nitric acid by a solvent such as a 30% solution of tributyl phosphate (TBP) in dodecane ($C_{12}H_{26}$) in which the nitrates are preferentially soluble. Industrial reprocessing of nuclear fuels is done by countercurrent operation of many liquid-liquid separation stages. These separation stages consist of well-mixed contacting tanks, where $UO_2(NO_3)_2$ is exchanged between two immiscible liquid phases, and separation tanks, where the organic and aqueous phases are separated. A schematic of a separation stage is shown in Figure 5-25a.

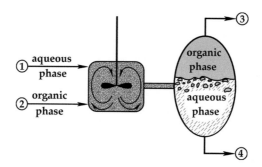

FIGURE 5-25a   Liquid-liquid separation stage for reprocessing.

In this process, an aqueous solution of uranyl nitrate $[UO_2(NO_3)_2]$ is one of the feed streams to the separation stage, and the mass flow rate of the aqueous feed phase is, $\dot{m}_1 = 400 \, kg/h$. The second feed stream is an organic solution of TBP in dodecane $(C_{12}H_{26})$ which we assume to be a single component. The organic and inorganic phases are assumed to be immiscible, thus only the uranil nitrate is transferred from one stream to the other. The process specifications are indicated in Figure 5-25b, and for this problem it is the *mass flow rates* that you are asked to determine.

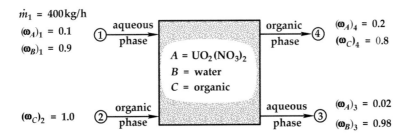

$\dot{m}_1 = 400 \, kg/h$
$(\omega_A)_1 = 0.1$
$(\omega_B)_1 = 0.9$

① aqueous phase

$A = UO_2(NO_3)_2$
$B = $ water
$C = $ organic

organic phase ④
$(\omega_A)_4 = 0.2$
$(\omega_C)_4 = 0.8$

$(\omega_C)_2 = 1.0$ ② organic phase

aqueous phase ③
$(\omega_A)_3 = 0.02$
$(\omega_B)_3 = 0.98$

FIGURE 5-25b Specified stream variables.

5.26. A gas stream consisting of air with a small amount of acetone is purified by contacting with a water stream in the contacting device illustrated in Figure 5.7. The inlet gas stream (Stream #2) contains one percent acetone and has a molar flow rate of 30 kmol/h. The molar flow rate of the pure water stream (Stream #1) is 90 kmol/h. The process operates at constant temperature and pressure, and the process equilibrium relation is given by

Process equilibrium relation: $\qquad (y_A)_3 = K_{eq,A}(x_A)_4 \qquad\qquad$ (1)

where $K_{eq,A} = 2.53$. Assume that water and air are immiscible and that the molar flow rates entering and leaving the equilibrium stage are constant. In this problem, you are asked to

    I. Determine the absorption factor for this equilibrium stage.
    II. Determine the mole fractions of acetone in the two exit streams.

5.27. The concept of an equilibrium stage is a very useful tool for the design of multi-component separations, and a typical equilibrium stage for a distillation column is shown in Figure 5-27.

A liquid stream, S1, flowing downward encounters a vapor stream, S2, flowing upward. We assume that the vapor and liquid streams exchange mass inside the equilibrium stage until they are in equilibrium with each other. Equilibrium is determined by a ratio of the molar fractions of each component in the liquid and vapor streams according to

Equilibrium relation: $\qquad K_A = \dfrac{(y_A)_4}{(x_A)_3} \qquad A = 1, 2, ..., N$

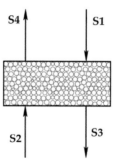

FIGURE 5-27   Sketch of an equilibrium stage process.

The streams leaving the stage, S3 (liquid), and S4 (vapor) are in equilibrium with each other and therefore satisfy the above relation. The ratio of the molar flow rates of the output streams is a function of the energy balance within the stage. In this problem, we assume that the ratio of the liquid output molar flow rate to the vapor output molar flow rate, $\dot{M}_3/\dot{M}_4$, is given. Assuming that the compositions, i.e. the mole fractions of the components and the molar flow rates of the input streams, S1, and S2, are known, and the equilibrium constant for one of the components is given, develop the mass balances for a two component vapor-liquid equilibrium stage.

5.28. A single stage, binary distillation process is illustrated in Figure 5-28. The total molar flow rate entering the unit is $\dot{M}_1$ and the mole fraction of species A in this liquid stream is $(x_A)_1$. Heat is supplied in order to generate a vapor stream, and the ratio, $\dot{M}_2/\dot{M}_3 = \beta$, depends on the rate at which heat is supplied. At the vapor-liquid interface, we can assume local thermodynamic equilibrium (see Eq. 5.24) in order to express the vapor-phase mole fraction in terms of the liquid-phase mole fraction according to

Equilibrium relation: $y_A = \dfrac{\alpha_{AB}\, x_A}{1 + x_A(\alpha_{AB} - 1)}$,   at the vapor-liquid interface

Here, α_{AB} represents the relative volatility. If the distillation process is *slow enough*, one can assume that the vapor and the liquid leaving the distillation unit are in equilibrium; however, at this point in our studies we do not know what is meant by *slow enough*. In order to proceed with an *approximate solution* to this problem, we replace the equilibrium relation with a *process* equilibrium relation given by

$$(y_A)_2 = \frac{\alpha_{AB}\, (x_A)_3}{1 + (x_A)_3(\alpha_{AB} - 1)},\quad \text{between the exit streams}$$

Use of this relation means that we are treating the system shown in Figure 5-28 as an *equilibrium stage*. Given a detailed study of mass transfer in a subsequent course, one can make a judgment concerning the conditions that are required in order that this *process equilibrium relation* be satisfactory. For the present, you are asked to use the above relation to derive an *implicit expression* for $(y_A)_2$ in terms of $(x_A)_1$ and examine three special cases: $\alpha_{AB} \to 0$, $\beta \to 0$, $\alpha_{AB} = 1$.

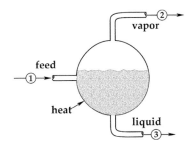

FIGURE 5-28   Single stage binary distillation.

5.29. A saturated solution of calcium hydroxide enters a boiler as shown in Figure 5-29 and a fraction, φ, of the water entering the boiler is vaporized. Under these circumstances, a portion of the calcium hydroxide precipitates and you would like to determine the mass fraction of this suspended solid calcium hydroxide in the liquid stream leaving the boiler. Assume that no calcium hydroxide leaves in the vapor stream, that none accumulates in the boiler, and that the temperature of the liquid entering and leaving the boiler is a constant. Assume that the solid calcium hydroxide leaving the boiler is in *equilibrium* with the dissolved carbon dioxide, i.e., the boiler is an *equilibrium stage*. The solubility is often expressed as;

Equilibrium relation: solubility $= S = g$ of $Ca(OH)_2$ /g of H_2O

however, a more precise description can be constructed.

In this problem, you are asked to develop a general solution for the mass fraction of the suspended solid in the liquid stream leaving the boiler in terms of φ and S. For $\varphi = 0.50, 0.21$, and 0.075, determine the mass fraction of suspended solid when $S = 2.5 \times 10^{-3}$.

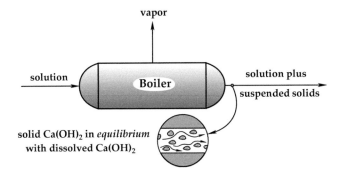

FIGURE 5-29   Precipitation of calcium hydroxide in a boiler.

5.30. In Problem 5.29, an equilibrium relation between solid calcium hydroxide and dissolved calcium hydroxide was given by

Equilibrium relation: $\text{solubility} = S = \text{g of } Ca(OH)_2/\text{g of } H_2O$ (1)

In Sec. 5.3.1, we expressed the general gas/liquid equilibrium relation for species A in terms of the *chemical potential* and for a solid/liquid system we would express Eq. 5.24 as

Equilibrium relation: $(\mu_A)_{solid} = (\mu_A)_{liquid}$ (2)

For the process considered in Problem 5.29, we assume that the solid phase is pure calcium hydroxide so that Eq. 2 takes the form

$$(\mu_A^\circ)_{solid} = (\mu_A)_{liquid}$$ (3)

The description of phase equilibrium phenomena in terms of the chemical potential will be the subject of a subsequent course in thermodynamics; however, at this point one can illustrate how Eqs. 1 and 3 are related.

The development begins with a general representation for the chemical potential at some fixed temperature and pressure. This is given by

$$(\mu_A)_{liquid} = F(T, p, x_A, x_B, \text{ etc.})$$

where x_A is the mole fraction of species A (calcium hydroxide) in the liquid phase. A Taylor series expansion about $x_A = 0$ leads to (see Problem 5.31)

$$(\mu_A)_{liquid} = F\Big|_{x_A=0} + \frac{\partial F}{\partial x_A}\Big|_{x_A=0}(x_A) + \frac{\partial^2 F}{\partial x_A^2}\Big|_{x_A=0}(x_A)^2 + \cdots$$ (4)

The first term in this expansion is zero and when the mole fraction of species A is small compared to one, $x_A \ll 1$, we can make use of a linear form of Eq. 4 given by

$$(\mu_A)_{liquid} = \frac{\partial F}{\partial x_A}\Big|_{x_A=0}(x_A)$$ (5)

In this problem, you are asked to use Eq. 3 along with Eq. 5 and the approximation

$$x_A = \frac{c_A}{c_B}, \quad \text{when } c_A \ll c_B$$ (6)

to derive Eq. 1. Here c_A represents the molar concentration of calcium hydroxide and c_B represents the molar concentration of water. In terms of species A and species B, it will be convenient to express the solubility in the form

Equilibrium relation: $\text{solubility} = S = \rho_A/\rho_B$ (7)

and note that this can be related to the molar form by use of $c_A MW_A = \rho_A$ and $c_B MW_B = \rho_B$.

5.31. In the previous problem, we made use of a Taylor series expansion to obtain a simplified expression for the chemical potential as a function of the mole fraction. A Taylor series expansion is a powerful tool for predicting the value of a function at some position $x = b$ when information about the function is *only* available at $x = a$. If we think about the definite integral given by

$$f(b) = f(a) + \int_{x=a}^{x=b} (df/dx)\, dx \tag{1}$$

we see that we can determine $f(b)$, given $f(a)$, only if we know the derivative, df/dx *everywhere* between a and b. To see how we can use Eq. 1 to determine $f(b)$ using *only* information at $x = a$, we first make use of the change of variable, or *transformation*, defined by

$$\eta = x - b \tag{2}$$

This leads to the relations

$$dx = d\eta, \quad df/dx = df/d\eta \tag{3}$$

which can be used to express Eq. 1 in the form

$$f(b) = f(a) + \int_{\eta=a-b}^{\eta=0} (df/d\eta)\, d\eta \tag{4}$$

Remember that the rule for differentiating a product

$$\frac{d}{d\eta}(UV) = U\frac{dV}{d\eta} + V\frac{dU}{d\eta} \tag{5}$$

is the basis for the technique known as *integration by parts* where it is employed in the form

$$U\frac{dV}{d\eta} = \frac{d}{d\eta}(UV) - V\frac{dU}{d\eta} \tag{6}$$

If we make use of the two representations

$$U = \frac{df}{d\eta}, \quad V = \eta \tag{7}$$

Eq. 6 provides the following identity for the derivative of f with respect to η:

$$\frac{df}{d\eta} = \frac{d}{d\eta}\left[\frac{df}{d\eta}\eta\right] - \eta\frac{d}{d\eta}\left[\frac{df}{d\eta}\right] \qquad (8)$$

Use this result in Eq. 4 to produce the second term in the Taylor series expansion for $f(b)$ about $f(a)$. Repeat this entire procedure to extend the representation for $f(b)$ to obtain the third term in the Taylor series expansion and thus illustrate how one obtains a representation for $f(b)$ that involves only the function and its derivatives evaluated at $x = a$. Keep in mind that the only mathematical tools used in this derivation are the definition of the definite integral and the rule for differentiating a product[††].

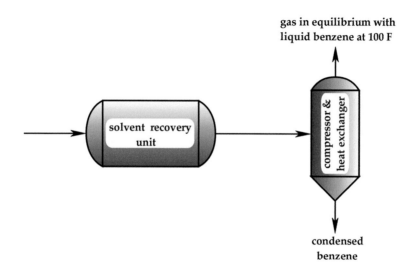

gas in equilibrium with
liquid benzene at 100 F

solvent recovery
unit

compressor &
heat exchanger

condensed
benzene

FIGURE 5-32 Recovery-condenser system.

5.32. A gas mixture leaves a solvent recovery unit as illustrated in Figure 5-32.

The partial pressure of benzene in this stream is 80 mmHg and the total pressure is 750 mmHg. The volumetric analysis of the gas, on a benzene-free basis, is 15% CO_2, 4% O_2, and the remainder is nitrogen. This gas is compressed to 5 atm and cooled to 100°F. Calculate the percentage of benzene condensed in the process. Assume that CO_2, O_2, and N_2 are insoluble in benzene, thus the liquid phase is pure benzene.

[††] Stein, S.K. and Barcellos, A. 1992, *Calculus and Analytic Geometry*, McGraw-Hill, Inc., New York, pages 134 and 226.

5.33. Derive the form of the solid phase mass balance given by Eq. 11 in Example 5.8.

5.34. In this problem, we wish to complete the analysis given in Example 5.8 in order to determine the total molar flow rate of fresh air entering the dryer. To accomplish this, we must first derive the macroscopic mole balance for air which can then be applied to the dryer illustrated in Example 5.8.

Part I. Consider air to consist of nitrogen and oxygen and determine under what conditions the mole balances for these two components can be added to obtain the special form

$$\frac{d}{dt}\int_V c_{air}\,dV + \int_A c_{air}\,\mathbf{v}\cdot\mathbf{n}\,dA = 0 \tag{1}$$

in which $c_{air} = c_{N_2} + c_{O_2}$. Begin the analysis with the macroscopic mole balances given by

$$\frac{d}{dt}\int_V c_A\,dV + \int_A c_A\mathbf{v}_A\cdot\mathbf{n}\,dA = \int_V R_A\,dV \tag{2}$$

$$\frac{d}{dt}\int_V c_B\,dV + \int_A c_B\mathbf{v}_B\cdot\mathbf{n}\,dA = \int_V R_B\,dV \tag{3}$$

and identify the simplifications that are necessary in order to obtain Eq. 1.

Part II. Use Eq. 1 to determine the *molar flow* rate of the incoming air indicated by $\dot{M}_2$ in Eq. 5 of Example 5.8. Clearly identify the process equilibrium relation that must be imposed in order to solve this problem.

SECTION 5.6

5.35. A sequential analysis of the multi-stage extraction process illustrated in Figure 5.10 led to the relation between $(y_A)_1$ and $(y_A)_{N+1}$ given by Eq. 5.90. That result was based on the special condition given by $(x_A)_o = 0$. In this problem, you are asked to develop a general form of Eq. 5.90 that is applicable for any value of $(x_A)_o$.

5.36. In Example 5.9, what value of $\dot{M}_\gamma$ can be used instead of $\dot{M}_\gamma = 30$ kmol/h so that the exit condition of $(y_A)_1 = 0.001$ is satisfied exactly?

5.37. In Example 5.9, an implicit equation for the absorption factor was given by

$$1 + A + A^2 + A^3 + A^4 + A^5 + A^6 = 20.0$$

The solution for A can be obtained by the methods described in Appendix B. Use at least one of the following methods to determine the value of the absorption coefficient:

a. The bisection method
b. The false position method

    c. Newton's method

    d. Picard's method

    e. Wegstein's method

5.38. Point #2 in Figure 5.15 is represented by an equation given in Sec. 5.6.1. Identify the equation.

5.39. Point #3 in Figure 5.15 is represented by an equation given in Sec. 5.6.1. Identify the equation.

5.40. Point #4 in Figure 5.15 is represented by an equation given in Sec. 5.6.1. Identify the equation.

5.41. Point #5 in Figure 5.15 is represented by an equation given in Sec. 5.6.1. Identify the equation.

5.42. In Example 5.10, the β-phase mole fraction entering the Nth stage was listed as $(x_A)_N = 0.00386$. Indicate how this mole fraction was determined and verify that the molar flow rate of acetone leaving in the liquid stream remains unchanged because of the change in $\dot{M}_\beta$.

5.43. Verify that $(x_A)_N = 0.00386$ for the conditions associated with Figure 5.10b in Example 5.10.

5.44. Verify that $(x_A)_N = 0.0403$ for the conditions associated with Figure 5.10c in Example 5.10.

5.45. Consider the process described in Example 5.10 for $\dot{M}_\beta = 90$ kmol/h and $\dot{M}_\gamma = 30$ kmol/h. Assume that the mole fraction of acetone in the air (γ-phase) entering the system is specified as $(y_A)_6 = 0.010$ and assume that the mole fraction of acetone in the water (β-phase) entering the system is specified as $(x_A)_o = 0$. Given that there are five equilibrium stages, reconstruct Figure 5.10a, and use the reconstructed figure to determine $(y_A)_1$.

5.46. Consider the process described in Example 5.10 for $\dot{M}_\beta = 70$ kmol/h and $\dot{M}_\gamma = 30$ kmol/h. Assume that the mole fraction of acetone in the air (γ-phase) entering the system is specified as $(y_A)_6 = 0.010$ and assume that the mole fraction of acetone in the water (β-phase) entering the system is specified as $(x_A)_o = 0$. Given that there are five equilibrium stages, reconstruct Figure 5.10b and use the reconstructed figure to determine $(y_A)_1$.

5.47. Consider the process described in Example 5.10 for $\dot{M}_\beta = 67$ kmol/h and $\dot{M}_\gamma = 30$ kmol/h. Assume that the mole fraction of acetone in the air (γ-phase) entering the system is specified as $(y_A)_6 = 0.010$ and assume that the mole fraction of acetone in the water (β-phase) entering the system is specified as $(x_A)_o = 0$. Given that there are five equilibrium stages, reconstruct Figure 5.10c and use the reconstructed figure to determine $(y_A)_1$.

6 Stoichiometry

Up to this point, we have used various forms of the two axioms associated with the principle of conservation of mass. The species mass balance for a fixed control volume

Axiom I: $$\frac{d}{dt}\int_V \rho_A dV + \int_A \rho_A \mathbf{v}_A \cdot \mathbf{n}dA = \int_V r_A dV, \quad A = 1, 2, \dots, N \qquad (6.1)$$

has been used to solve problems dealing with liquid and solid systems when it was convenient to work with the species mass density ρ_A or the mass fraction ω_A. Those problems did not involve chemical reactions, thus the net mass rate of production of species A owing to chemical reactions was zero, i.e., $r_A = 0$. When it was convenient to work with the species molar concentration c_A or the mole fraction x_A, we used the molar form of Eq. 6.1 with $R_A = 0$.

In this chapter, we begin our study of material balances *with chemical reactions*, and we continue this study throughout the remainder of the text. Aris[*] pointed out that stoichiometry is essentially the "bookkeeping" of atomic species, and we use this bookkeeping to develop constraints on the net molar rates of production, R_A, R_B, $R_C, \dots, R_N$. These stoichiometric constraints reduce the degrees of freedom and they represent a key aspect of macroscopic balance analysis for systems with chemical reaction. The net global rates of production can be determined experimentally without any detailed knowledge of the chemical kinetics, and we show how this is done in Example 6.2. Knowledge of the net global rates of production allows us to specify the flow sheets referred to in Sec. 1.2; however, they do not allow us to specify the size of the units illustrated in Figure 1.5. To determine the *size of a chemical reactor*, we must know how the net rates of production depend on the concentration of the participating chemical species and this matter is explored in Chapter 9.

6.1 CHEMICAL REACTIONS

In the presence of chemical reactions, the *total* mass balance is obtained directly from Eq. 6.1 by summing that result over all N species and imposing the second axiom

Axiom II: $$\sum_{A=1}^{A=N} r_A = 0 \qquad (6.2)$$

---

[*] Aris, R. 1965, page 7, *Introduction to the Analysis of Chemical Reactors*, Prentice-Hall, Inc., Englewood Cliffs, New Jersey.

DOI: 10.1201/9781003283751-6   **179**

This leads to the total mass balance given by

$$\frac{d}{dt} \int_V \rho dV + \int_A \rho \mathbf{v} \cdot \mathbf{n} \, dA = 0 \tag{6.3}$$

For problems involving a gas phase and the use of an equation of state (like the ideal gas law), the molar form of Eq. 6.1 is more convenient and can be written as

Axiom I: $$\frac{d}{dt} \int_V c_A dV + \int_A c_A \mathbf{v}_A \cdot \mathbf{n} \, dA = \int_V R_A \, dV, \quad A = 1, 2, ..., N \tag{6.4}$$

Here, R_A is the net molar rate of production of species A per unit volume owing to chemical reactions. This is related to r_A by

$$R_A = r_A / MW_A \tag{6.5}$$

and Eq. 6.2 provides a constraint on the net molar rates of production given by

Axiom II: $$\sum_{A=1}^{A=N} MW_A R_A = 0 \tag{6.6}$$

Here, we note that MW_A represents the *molecular mass* of species A and that we have chosen a nomenclature based on the traditional phrase, *molecular weight*. It is important to remember that r_A and R_A represent both the *creation* of species A (when r_A and R_A are *positive*) and the *consumption* of species A (when r_A and R_A are *negative*). For systems involving chemical reactions, Eq. 6.4 is preferred over Eq. 6.1 for two reasons. To begin with, chemical reaction rates are traditionally expressed in terms of molar concentrations, c_A, c_B, etc., and one needs to determine how R_A is related to these molar concentrations. For example, if species A is undergoing an irreversible decomposition, the net molar rate of production might be expressed as

$$R_A = -\frac{k c_A^2}{1 + k' c_A}, \quad \textit{irreversible decomposition} \tag{6.7}$$

where k and k' are coefficients to be determined by experiment. In this case, the negative sign indicates that species A is being *consumed* by the chemical reaction. One can use Eq. 6.7 along with Eq. 6.4 to predict the behavior of a system, i.e., *to design a system*. Chemical reaction rate equations such as Eq. 6.7 are considered in Chapter 9.

In the absence of specific chemical kinetic information about R_A, one can only use Eq. 6.4 *to analyze a system* in terms of the flow rates, global rates of production, and composition of the streams entering and leaving the system. The second reason that Eq. 6.4 is preferred over Eq. 6.1 is that the net molar rates of production of the various species are easily related to the *atomic structure* of the molecules involved in the chemical reactions. These relations can be constructed in terms of *stoichiometric coefficients* and as an example we consider the special case illustrated in Figure 6.1. Here we have suggested that ethane reacts with oxygen to form carbon dioxide and water, a process that is often referred to as *complete combustion*. The stoichiometry of this process can be *visualized* as

$$\frac{1}{2}C_2H_6 + \frac{7}{4}O_2 \rightarrow \frac{3}{2}H_2O + CO_2 \qquad (6.8)$$

and we call this a *stoichiometric schema*. In general, this stoichiometric schema has no connection with the actual kinetics of the reaction, thus Eq. 6.8 *does not mean* that 1/2 a molecule of C_2H_6 collides with 7/4 of a molecule of O_2 to create 3/2 of a molecule of H_2O and one molecule of CO_2. The actual molecular processes involved in the oxidation of ethane are far more complicated than suggested by Eq. 6.8, and an introduction to these processes is given in Chapter 9. The coefficients in Eq. 6.8 are often deduced by *counting atoms*, and this process is based on the idea that

Axiom II: $\left\{\begin{array}{c} \textit{atomic species are} \\ \textit{neither created nor} \\ \textit{destroyed by} \\ \textit{chemical reactions} \end{array}\right\}$ $\qquad (6.9)$

If the process illustrated in Figure 6.1 is carried out with a stoichiometric mixture of ethane and oxygen, one *might* find that the product stream contains *mostly* CO_2 and H_2O, but one might also find small amounts of CO, CH_3OH, C_2H_4, etc. and it is not always obvious that these *small amounts* can be ignored. In fact, we believe that these small amounts should be a matter of *constant concern*.

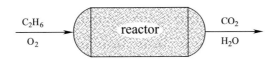

FIGURE 6.1   Combustion reaction.

6.1.1   PRINCIPLE OF STOICHIOMETRIC SKEPTICISM

When a system of the type represented by Figure 6.1 is encountered, it is appropriate to immediately consider the alternative illustrated in Figure 6.2. It is always possible that the "other molecular species" suggested in Figure 6.2 may be present in amounts small enough so that the schema represented by Eq. 6.8 is a satisfactory *approximation*. However, what is meant by *small enough* may be difficult to determine since the "other molecular species" may consist of biocides or carcinogens or other species that could be damaging to the environment even in small amounts. Under certain circumstances, small amounts may produce major consequences, and we want students to react to any proposed *stoichiometric schema* in the manner indicated by

Figure 6.2. This idea will be especially important in terms of our studies of reaction kinetics in Chapter 9 where small amounts of *reactive intermediates* or *Bodenstein products* actually control the macroscopic process suggested in Figure 6.2.

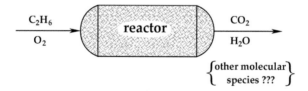

FIGURE 6.2   Incomplete combustion reaction.

One can indeed *postulate* that ethane and oxygen will react to produce carbon dioxide and water, but the postulate needs to be verified by experiment. As an example only, we imagine that the process illustrated in Figure 6.2 is carried out so that ethane is *partially oxidized* to produce ethylene oxide, carbon dioxide, carbon monoxide, and water. Under these circumstances, the stoichiometry of the reaction might be represented by the following undetermined schema:

$$? \, C_2H_6 + ? \, O_2 \rightarrow ? \, CO + ? \, C_2H_4O + ? \, H_2O + CO_2 \tag{6.10}$$

In this case, the stoichiometric coefficients could be found by *counting atoms* to obtain

$$2C_2H_6 + 4O_2 \rightarrow CO + C_2H_4O + 4H_2O + CO_2 \tag{6.11}$$

and one could also *count atoms* to develop a different schema given by

$$2C_2H_6 + \frac{19}{4}O_2 \rightarrow 2CO + \frac{1}{2}C_2H_4O + 5H_2O + CO_2 \tag{6.12}$$

Here it should be clear that we need *more information* to treat the case of *partial oxidation* of ethane, and to organize this additional information efficiently, we need a precise *mathematical representation* of the concept that atomic species are neither created nor destroyed by chemical reactions. It is important to understand that Eqs. 6.11 and 6.12 are *pictures* of the concept that atoms are conserved, and what we need are *equations* describing the concept that atoms are conserved. In this text, we use *arrows* to represent pictures and *equal signs* to represent equations.

6.2   CONSERVATION OF ATOMIC SPECIES

To be precise about the role of atomic species in chemical reactions, we first need to replace the word statement given by Eq. 6.9 with a *word equation* that we write as

$$\text{Axiom II:} \quad \left\{ \begin{array}{l} \text{the molar rate of production} \\ \text{per unit volume of } J\text{-type atoms} \\ \text{owing to chemical reactions} \end{array} \right\} = 0, \quad J = 1, 2, \ldots, T \tag{6.13}$$

Here the nomenclature is intended to suggest *aTomic* species. From this form of Axiom II we need to extract a *mathematical equation*, and in order to do this we *define* the number N_{JA} as

$$N_{JA} = \left\{ \begin{array}{c} \text{number of moles of} \\ \text{J-type atoms per mole} \\ \text{of molecular species } A \end{array} \right\}, \quad J = 1, 2, \ldots, T, \text{ and } A = 1, 2, \ldots, N \quad (6.14)$$

We refer to N_{JA} as the *atomic species indicator* and we identify the array of coefficients associated with N_{JA} as the *atomic matrix*[†]. To illustrate the structure of the atomic matrix, we consider the complete oxidation of ethane illustrated in Figure 6.1. That process provides the basis for the following *visual representation* of the atomic matrix:

$$
\begin{array}{c}
\text{Molecular Species} \quad \rightarrow \quad C_2H_6 \quad O_2 \quad H_2O \quad CO_2 \\
\begin{array}{c} \textit{carbon} \\ \textit{hydrogen} \\ \textit{oxygen} \end{array}
\left[\begin{array}{cccc} 2 & 0 & 0 & 1 \\ 6 & 0 & 2 & 0 \\ 0 & 2 & 1 & 2 \end{array} \right]
\end{array}
\quad (6.15)
$$

This representation *connects* atoms with molecules in a convenient manner, and it is exactly what one uses to count atoms and balance chemical equations. There are two symbols that are useful for representing the *atomic matrix*. The first of these is given by $[N_{JA}]$ which has the obvious connection to Eq. 6.14, while the second is given by A which has the obvious connection to the name of this matrix. In this text, we will encounter both representations for the *atomic matrix* as indicated by

$$[N_{JA}] = \left[\begin{array}{cccc} 2 & 0 & 0 & 1 \\ 6 & 0 & 2 & 0 \\ 0 & 2 & 1 & 2 \end{array} \right], \quad \text{or} \quad A = \left[\begin{array}{cccc} 2 & 0 & 0 & 1 \\ 6 & 0 & 2 & 0 \\ 0 & 2 & 1 & 2 \end{array} \right] \quad (6.16)$$

In order to use the atomic species indicator, N_{JA}, to construct an equation representing the concept that atoms are neither created nor destroyed by chemical reaction, we first recall the definition of R_A

$$R_A = \left\{ \begin{array}{c} \textit{net} \text{ molar rate of production} \\ \text{per unit volume of species } A \\ \text{owing to chemical reactions} \end{array} \right\} \quad (6.17)$$

[†] Amundson, N.R. 1966, page 54, *Mathematical Methods in Chemical Engineering: Matrices and Their Application*, Prentice-Hall, Inc., Englewood Cliffs, New Jersey.

which is consistent with the pictorial representation of R_{CO_2} given earlier in Figure 4.1. Next, we form the product of the atomic species indicator with R_A to obtain

$$N_{JA}R_A = \left\{ \begin{array}{l} \text{number of moles of} \\ J\text{-type atoms per mole} \\ \text{of molecular species } A \end{array} \right\} \left\{ \begin{array}{l} net \text{ molar rate of production} \\ \text{per unit volume of species } A \\ \text{owing to chemical reactions} \end{array} \right\} \quad (6.18)$$

A little thought will indicate that the product of N_{JA} and R_A can be described as,

$$N_{JA}R_A = \left\{ \begin{array}{l} \text{net molar rate of production per unit} \\ \text{volume of } J\text{-type atoms owing to the} \\ \text{molar rate of production of species } A \end{array} \right\} \quad (6.19)$$

and the axiomatic statement given by Eq. 6.13 takes the form[‡]

Axiom II: $$\sum_{A=1}^{A=N} N_{JA}\, R_A = 0, \quad J = 1, 2, \ldots, T \quad (6.20)$$

This equation represents a precise mathematical statement that atomic species are neither created nor destroyed by chemical reactions, and it provides a set of T equations that *constrain* the N net rates of production, R_A, $A = 1, 2, \ldots, N$. While Axiom II provides T equations, the equations are not necessarily independent. The number of independent equations is given by the *rank of the atomic matrix* and we will be careful to indicate that rank when specific processes are examined. If ions are involved in the reactions, one must impose the condition of *conservation of charge* as described in Appendix E. Some comments concerning *heterogeneous reactions* are given in Appendix F.

The net rate of production of species A indicated by R_A can also be expressed in terms of the *creation* and *consumption* of species A according to

$$R_A = \left\{ \begin{array}{l} \text{molar rate of } creation \text{ of} \\ \text{species } A \text{ per unit volume} \\ \text{owing to chemical reactions} \end{array} \right\} - \left\{ \begin{array}{l} \text{molar rate of } consumption \text{ of} \\ \text{species } A \text{ per unit volume} \\ \text{owing to chemical reactions} \end{array} \right\} \quad (6.21)$$

Here we need to think carefully about the description of R_A given by Eq. 6.17 where we have used the word *net* to represent the sum of the *creation* of species A and the *consumption* of species A. This means that Eqs. 6.17 and 6.21 are equivalent descriptions of R_A and the reader is free to choose which ever set of words is most appealing.

[‡] Truesdell, C. and Toupin, R. 1960, The Classical Field Theories, in *Handbuch der Physik*, Vol. III, Part 1, edited by S. Flugge, Springer-Verlag, New York, page 473.

If we make use of the *atomic matrix* and the *column matrix of the net rates of production*, we can express Axiom II as

$$
\text{Axiom II:}
\begin{bmatrix}
N_{11} & N_{12} & N_{13} & \cdots & N_{1,N-1,} & N_{1N} \\
N_{21} & N_{22} & . & \cdots & N_{2,N-1} & N_{2N} \\
N_{31} & N_{32} & . & \cdots & . & . \\
. & . & . & \cdots & . & . \\
. & . & . & \cdots & . & . \\
N_{T1} & N_{T2} & . & \cdots & N_{T,N-1} & N_{TN}
\end{bmatrix}
\begin{bmatrix}
R_1 \\ R_2 \\ R_3 \\ . \\ R_{N-1} \\ R_N
\end{bmatrix}
=
\begin{bmatrix}
0 \\ 0 \\ 0 \\ . \\ 0
\end{bmatrix}
\qquad (6.22)
$$

Everything we need to know about the conservation of atomic species is contained in this linear matrix equation; however, we need this information in *different forms* that will be developed in this chapter. In our development, we will find *patterns* associated with the atomic matrix and these patterns will be connected to the physical problems under consideration.

6.2.1 AXIOMS AND THEOREMS

In dealing with axioms and proved theorems, it is important to accept the idea that the choice is not necessarily unique. From the authors' perspective, Eq. 6.20 and Eq. 6.22 represent the preferred form of the axiom indicating that atoms are neither created nor destroyed by chemical reactions. We have identified both as Axiom II; however, we have also identified Eq. 6.2 as Axiom II. Equation 6.2 indicates that *mass is conserved* during chemical reactions while Eq. 6.20 indicates that *atoms are conserved* during chemical reactions. Surely, both of these are not independent axioms, thus we should be able to derive one from the other. In the following paragraphs, we show how the *axiom* given by Eqs. 6.20 can be used to prove Eq. 6.2 as a *theorem*.

To carry out this proof, we first multiply Eqs. 6.20 by the atomic mass of the Jth atomic species leading to

$$
AW_J \sum_{A=1}^{A=N} N_{JA} R_A = 0, \quad J = 1, 2, \ldots, T \qquad (6.23)
$$

Here, we have used AW_J to represent the *atomic mass* of species J in the same manner that we have used MW_A to represent the *molecular mass* of species A. We now sum Eq. 6.23 over all *atomic species* to obtain

$$
\sum_{J=1}^{J=T} AW_J \sum_{A=1}^{A=N} N_{JA} R_A = 0 \qquad (6.24)
$$

Since the sum over J is independent of the sum over A, we can place the sum over J inside the sum over A leading to the form

$$\sum_{A=1}^{A=N} \sum_{J=1}^{J=T} A W_J N_{JA} R_A = 0 \qquad (6.25)$$

We now note that R_A is independent of the process of summing over all J, thus we can take R_A outside of the first sum and express Eq. 6.25 as

$$\sum_{A=1}^{A=N} R_A \sum_{J=1}^{J=T} A W_J N_{JA} = 0 \qquad (6.26)$$

At this point, we need only recognize that the molecular mass of species A is defined by

$$MW_A = \sum_{J=1}^{J=T} A W_J N_{JA} \qquad (6.27)$$

in order to express Eq. 6.26 as

$$\sum_{A=1}^{A=N} R_A MW_A = 0 \qquad (6.28)$$

Use of the definition given by Eq. 6.5 leads to the following proved theorem:

Proved theorem: $\qquad \sum_{A=1}^{A=N} r_A = 0 \qquad (6.29)$

This result was identified as Axiom II by Eq. 6.2 and earlier by Eq. 4.11 in our initial exploration of the axioms for the mass of multicomponent systems. It should be clear from this development that one person's axiom might be another person's proved theorem. For example, if Eq. 6.29 is taken as Axiom II, one can prove Eq. 6.20 as a theorem. The proof is the object of Problem 6.4, and it requires the constraint that the net rates of production are independent of the atomic masses, AW_J. We prefer Eq. 6.20 as Axiom II since it can be used to prove Eq. 6.29 without imposing any constraints; however, one must accept the idea that different people state the laws of physics in different ways.

6.2.2   LOCAL AND GLOBAL FORMS OF AXIOM II

Up to this point, we have discussed the *local form* of Axiom II, i.e., the form that applies at a point in space. However, when Axiom II is used to analyze the reactors shown in Figures 6.1 and 6.2, we will make use of an integrated form of Eq. 6.20 that applies to the control volume illustrated in Figure 6.3.

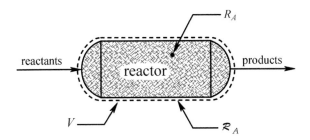

FIGURE 6.3 Local and global rates of production.

Here, we have illustrated the *local* rate of production for species A, designated by R_A, and the *global* rate of production for species A, designated by $\mathcal{R}_A$. The latter is defined by

$$\mathcal{R}_A = \int_V R_A \, dV = \left\{ \begin{array}{l} net \text{ macroscopic molar rate} \\ \text{of production of species } A \\ \text{owing to chemical reactions} \end{array} \right\} \tag{6.30}$$

and we often use an abbreviated description given by

$$\mathcal{R}_A = \left\{ \begin{array}{l} \text{global rate of} \\ \text{production of} \\ \text{species } A \end{array} \right\} \tag{6.31}$$

When dealing with a problem that involves the global rate of production, we need to form the volume integral of Eq. 6.20 to obtain

$$\int_V \sum_{A=1}^{A=N} N_{JA} \, R_A \, dV = 0, \quad J = 1, 2, \ldots, T \tag{6.32}$$

The integral can be taken inside the summation operation, and we can make use of the fact that the elements of N_{JA} are independent of space to obtain

$$\sum_{A=1}^{A=N} N_{JA} \int_V R_A \, dV = 0, \quad J = 1, 2, \ldots, T \tag{6.33}$$

Use of the definition of the global rate of production for species A given by Eq. 6.30 leads to the following global form of Axiom II:

Axiom II (global form): $\displaystyle \sum_{A=1}^{A=N} N_{JA} \, \mathcal{R}_A = 0, \quad J = 1, 2, \ldots, T \tag{6.34}$

Here, one must remember that $\mathcal{R}_A$ has units of *moles per unit time* while R_A has units of *moles per unit time per unit volume*, thus the physical interpretation of these two quantities is different as illustrated in Figure 6.3. In our study of complex systems described in Chapter 7, we will routinely encounter global rates of production and Axiom II (global form) will play a key role in the analysis of those systems.

6.2.3 Solutions of Axiom II

In the previous paragraphs, we have shown that Eq. 6.20 and Eq. 6.22 represent the fundamental concept that atomic species are conserved during chemical reactions. In addition, we made use of the concept that atomic species are conserved by *counting atoms* or *balancing chemical equations* (see Eqs. 6.8, 6.11, and 6.12). The fact that the process of counting atoms is *not unique* for the partial oxidation of ethane is a matter of considerable interest that will be explored carefully in this chapter.

In order to develop a better understanding of Axiom II, we carry out the matrix multiplication indicated by Eq. 6.22 for a system containing three (3) atomic species and six (6) molecular species. This leads to the following set of three (3) equations containing six (6) net rates of production:

Atomic Species 1:

$$N_{11}R_1 + N_{12}R_2 + N_{13}R_3 + N_{14}R_4 + N_{15}R_5 + N_{16}R_6 = 0 \qquad (6.35a)$$

Atomic Species 2:

$$N_{21}R_1 + N_{22}R_2 + N_{23}R_3 + N_{24}R_4 + N_{25}R_5 + N_{26}R_6 = 0 \qquad (6.35b)$$

Atomic Species 3:

$$N_{31}R_1 + N_{32}R_2 + N_{33}R_3 + N_{34}R_4 + N_{35}R_5 + N_{36}R_6 = 0 \qquad (6.35c)$$

This 3×6 system of equations always has the *trivial solution* $R_A = 0$, for $A = 1, 2, ..., 6$, and the necessary and sufficient condition for a *non-trivial solution* to exist is that the *rank* of the atomic matrix be less than the number of molecular species. For this special case of three atomic species and six molecular species, we express this condition as[§]

Non-trivial solution (Special): $r = rank\,[N_{JA}] < 6$ \qquad (6.36)

By *rank* we mean explicitly the *row rank* which represents the *number of linearly independent equations* contained in Eqs. 6.35. It is possible that all three of Eqs. 6.35 are independent and the rank associated with the atomic matrix is three, i.e., $r = rank = 3$. On the other hand, it is possible that one of the three equations is a *linear combination* of the other two equations and the rank is two, i.e., $r = rank = 2$. The general condition concerning the rank of the atomic matrix in Eq. 6.22 is given by

Non-trivial solution (General): $r = rank\,[N_{JA}] < N$ \qquad (6.37)

[§] Kolman, B. 1997, *Introductory Linear Algebra*, 6th Edition, Prentice-Hall, Upper Saddle River, New Jersey.

When the rank is equal to $N-1$, we have a special case of Eq. 6.22 that leads to a *single independent stoichiometric reaction*. In that special case, the $N-1$ net rates of production can be specified in terms of R_N and Eq. 6.22 can be expressed as

$$R_A = v_{AN}\, R_N \qquad\qquad\qquad (6.38a)$$

$$R_B = v_{BN}\, R_N \qquad\qquad\qquad (6.38b)$$

$$R_C = v_{CN}\, R_N \qquad\qquad\qquad (6.38c)$$

$$\cdot$$
$$\cdot$$

$$R_{N-1} = v_{N-1N}\, R_N \qquad\qquad\qquad (6.38n{-}1)$$

Here v_{AN}, v_{BN}, etc., are often referred to as *stoichiometric coefficients*; however, the authors prefer to identify these quantities as *elements of the pivot matrix* as indicated in Example 6.1.

6.2.4 STOICHIOMETRIC EQUATIONS

It is crucial to understand that Eqs. 6.38 are based on the concept that atoms are neither created nor destroyed by chemical reactions. The bookkeeping associated with the conservation of atoms is known as stoichiometry, thus it is appropriate to refer to the equations given by Eqs. 6.38 as stoichiometric equations. In addition, it is appropriate to identify Eqs. 6.38 as a case in which there is a *single* independent net rate of production, and that this *single* independent net rate of production is identified as R_N. In order to be clear about stoichiometry and chemical kinetics, we place Eq. 6.7 side-by-side with Eq. 6.38a to obtain

$$R_A = -\frac{k\,c_A^2}{1+k'c_A}, \text{chemical kinetics} \qquad\qquad (6.39a)$$

$$R_A = v_{AN}\, R_N, \text{stoichiometry} \qquad\qquad (6.39b)$$

Here we note that the symbol R_A in Eq. 6.39a has *exactly the same physical significance* as R_A in Eq. 6.39b. In both cases R_A is defined by Eq. 6.17. However, the description of the right-hand side of these two representations of R_A is quite different. The right-hand side of Eq. 6.39a is a *chemical kinetic relation* while the right-hand side of Eq. 6.39b is a *stoichiometric relation*. The *chemical kinetic representation* depends on the complex processes that occur when molecules dissociate, active complexes are formed, and various molecular fragments coalesce to form products. The *stoichiometric representation* is based solely on the concept that atoms are conserved. The chemical kinetic representation may depend on temperature, pressure, and the presence of catalysts, while the stoichiometric representation remains *invariant* depending only on the conservation of atoms. In this chapter, and throughout most of the text, we will deal with stoichiometric relations based on Axiom II. In Chapter 9, we will examine chemical reaction rate equations, and in that treatment, we will be very careful to identify *stoichiometric constraints* that are associated with *elementary stoichiometry*.

EXAMPLE 6.1   COMPLETE COMBUSTION OF ETHANE

In this example, we consider the complete combustion of ethane, thus the molecular species under consideration are identified as (see Figure 6.1)

$$C_2H_6, \quad O_2, \quad H_2O, \quad CO_2$$

One form of the atomic matrix for this group of molecular species can be visualized as

$$
\begin{array}{cccc}
\text{Molecular Species} \rightarrow & C_2H_6 & O_2 & H_2O & CO_2
\end{array}
$$

$$
\begin{array}{c}
carbon \\
hydrogen \\
oxygen
\end{array}
\left[
\begin{array}{cccc}
2 & 0 & 0 & 1 \\
6 & 0 & 2 & 0 \\
0 & 2 & 1 & 2
\end{array}
\right]
\tag{1}
$$

and for this particular arrangement the atomic matrix is given by

$$
[N_{JA}] =
\begin{bmatrix}
2 & 0 & 0 & 1 \\
6 & 0 & 2 & 0 \\
0 & 2 & 1 & 2
\end{bmatrix}
\tag{2}
$$

A simple calculation (see Problem 6.5) shows that the rank of the matrix is three

$$r = rank\ [N_{JA}] = 3 \tag{3}$$

Thus, we have *three equations* and *four unknowns*. The three homogeneous equations that are analogous to Eqs. 6.35 are given by

$$2R_{C_2H_6} + 0 + 0 + R_{CO_2} = 0 \tag{4a}$$

$$6R_{C_2H_6} + 0 + 2R_{H_2O} + 0 = 0 \tag{4b}$$

$$0 + 2R_{O_2} + R_{H_2O} + 2R_{CO_2} = 0 \tag{4c}$$

while the analogous matrix equation corresponding to Eq. 6.22 takes the form

$$
\begin{bmatrix}
2 & 0 & 0 & 1 \\
6 & 0 & 2 & 0 \\
0 & 2 & 1 & 2
\end{bmatrix}
\begin{bmatrix}
R_{C_2H_6} \\
R_{O_2} \\
R_{H_2O} \\
R_{CO_2}
\end{bmatrix}
=
\begin{bmatrix}
0 \\
0 \\
0
\end{bmatrix}
\tag{5}
$$

It is possible to use intuition and the *picture* given by Eq. 6.8 to express the net rates of production in the form

$$R_{C_2H_6} = -\frac{1}{2} R_{CO_2} \qquad (6a)$$

$$R_{O_2} = -\frac{7}{4} R_{CO_2} \qquad (6b)$$

$$R_{H_2O} = +\frac{3}{2} R_{CO_2} \qquad (6c)$$

however, the use of Eqs. 4 to produce this result is more reliable. Finally, we note that these results for the net rates of production can be expressed in the form of the *pivot theorem* that is described in Sec. 6.4. In terms of the pivot theorem that can be extracted from Eq. 5, we have

$$\underbrace{\begin{bmatrix} R_{C_2H_6} \\ R_{O_2} \\ R_{H_2O} \end{bmatrix}}_{\substack{\text{column matrix} \\ \text{of non-pivot species}}} = \underbrace{\begin{bmatrix} -1/2 \\ -7/4 \\ +3/2 \end{bmatrix}}_{\text{pivot matrix}} \underbrace{\begin{bmatrix} R_{CO_2} \end{bmatrix}}_{\substack{\text{column matrix} \\ \text{of pivot species}}} \qquad (7)$$

This indicates that all the net rates of production are specified if we can determine the net rate of production for carbon dioxide, R_{CO_2}. Indeed, all the rates of production can be determined if we know *any one of the four rates*; however, we have chosen carbon dioxide as the *pivot species* in order to arrange Eqs. 4 and 5 in the forms given by Eqs. 6 and 7.

In the previous example, we illustrated how Axiom II can be used to analyze the stoichiometry for the complete combustion of ethane. The process of complete combustion was described earlier by the single *stoichiometric schema* given by Eq. 6.8 and the coefficients that appeared in that schema are evident in Eqs. 6 and 7 of Example 6.1.

6.2.5   ELEMENTARY ROW OPERATIONS AND
COLUMN/ROW INTERCHANGE OPERATIONS

In working with sets of equations such as those represented by Eq. 6.22, we will employ elementary row operations and column/row operations in order to arrange the equations in a convenient form. Elementary row operations were described earlier in Sec. 4.9.1 and we list them here as they apply to the atomic matrix:

I. Any row in the atomic matrix can be modified by multiplying or dividing by a non-zero scalar without affecting the system of equations.

II. Any row in the atomic matrix can be added or subtracted from another row without affecting the system of equations.

III. Any two rows in the atomic matrix can be interchanged without affecting the system of equations.

The column / row interchange operation that we will use in the treatment of Eq. 6.22 is described as follows:

IV. Any two columns in the atomic matrix can be interchanged without affecting the system of equations *provided that* the corresponding rows of the column matrix of net rates of production are also interchanged.

In terms of Eq. 6.20, this latter operation can be described mathematically as

$$N_{JB}R_B \rightleftarrows N_{JD}R_D, \quad B,D=1,2,...,N, \quad J=1,2,...,T \tag{6.40}$$

We can use these operations to develop *row equivalent* matrices, *row reduced* matrices, *row echelon* matrices, and *row reduced echelon* matrices. In order to illustrate these concepts, we consider the following example of Axiom II:

Axiom II:
$$\begin{bmatrix} 2 & 2 & 0 & 2 & 0 & 4 \\ 6 & 4 & 2 & 4 & 2 & 6 \\ 2 & 2 & 0 & 1 & 1 & 3 \\ 1 & 0 & 1 & 1 & 0 & 0 \end{bmatrix} \begin{bmatrix} R_1 \\ R_2 \\ R_3 \\ R_4 \\ R_5 \\ R_6 \end{bmatrix} = 0 \tag{6.41}$$

Directing our attention to the atomic matrix, we subtract three times the first row from the second row to obtain a *row equivalent* matrix given by

$R2 - 3R1:$
$$\begin{bmatrix} 2 & 2 & 0 & 2 & 0 & 4 \\ 0 & -2 & 2 & -2 & 2 & -6 \\ 2 & 2 & 0 & 1 & 1 & 3 \\ 1 & 0 & 1 & 1 & 0 & 0 \end{bmatrix} \begin{bmatrix} R_1 \\ R_2 \\ R_3 \\ R_4 \\ R_5 \\ R_6 \end{bmatrix} = 0 \tag{6.42}$$

Dividing the first row by two will create a coefficient of one in the first row of the first column. This operation leads to

$$
R1/2: \quad
\begin{bmatrix}
1 & 1 & 0 & 1 & 0 & 2 \\
0 & -2 & 2 & -2 & 2 & -6 \\
2 & 2 & 0 & 1 & 1 & 3 \\
1 & 0 & 1 & 1 & 0 & 0
\end{bmatrix}
\begin{bmatrix}
R_1 \\ R_2 \\ R_3 \\ R_4 \\ R_5 \\ R_6
\end{bmatrix}
= 0
\qquad (6.43)
$$

Multiplication of the second row by $-1/2$ provides

$$
R2(-1/2): \quad
\begin{bmatrix}
1 & 1 & 0 & 1 & 0 & 2 \\
0 & 1 & -1 & 1 & -1 & 3 \\
2 & 2 & 0 & 1 & 1 & 3 \\
1 & 0 & 1 & 1 & 0 & 0
\end{bmatrix}
\begin{bmatrix}
R_1 \\ R_2 \\ R_3 \\ R_4 \\ R_5 \\ R_6
\end{bmatrix}
= 0
\qquad (6.44)
$$

Using several elementary row operations, we construct a *row echelon form* of the atomic matrix that is given by

$$
\begin{bmatrix}
1 & 1 & 0 & 1 & 0 & 2 \\
0 & 1 & -1 & 1 & -1 & 3 \\
0 & 0 & 0 & 1 & -1 & 1 \\
0 & 0 & 0 & 0 & 0 & 0
\end{bmatrix}
\begin{bmatrix}
R_1 \\ R_2 \\ R_3 \\ R_4 \\ R_5 \\ R_6
\end{bmatrix}
= 0
\qquad (6.45)
$$

The row of zeros indicates that one of the four equations represented by Eq. 6.41 is not independent, i.e., it is a linear combination of two or more of the other equations. This means that the rank of the atomic matrix represented in Eq. 6.41 is three, $r = rank\,[N_{JA}] = 3$.

We can make further progress toward the *row reduced echelon form* by subtracting row #2 from row #1 to obtain

$$
R1 - R2: \quad
\begin{bmatrix}
1 & 0 & 1 & 0 & 1 & -1 \\
0 & 1 & -1 & 1 & -1 & 3 \\
0 & 0 & 0 & 1 & -1 & 1 \\
0 & 0 & 0 & 0 & 0 & 0
\end{bmatrix}
\begin{bmatrix}
R_1 \\ R_2 \\ R_3 \\ R_4 \\ R_5 \\ R_6
\end{bmatrix}
= 0
\qquad (6.46)
$$

In this form, the first two columns contain only a single entry along the diagonal, and we would like the third column to have this characteristic. Use of the following column/row interchange

$$N_{J3}\,R_3 \rightleftarrows N_{J4}\,R_4, \quad J=1,2,3 \tag{6.47}$$

provides a step in that direction given by

$$\begin{bmatrix} 1 & 0 & 0 & 1 & 1 & -1 \\ 0 & 1 & 1 & -1 & -1 & 3 \\ 0 & 0 & 1 & 0 & -1 & 1 \\ 0 & 0 & 0 & 0 & 0 & 0 \end{bmatrix} \begin{bmatrix} R_1 \\ R_2 \\ R_3 \\ R_4 \\ R_5 \\ R_6 \end{bmatrix} = 0 \tag{6.48}$$

We now subtract row three from row two in order to obtain the following *row reduced echelon form*:

$$R2 - R3: \qquad \begin{bmatrix} 1 & 0 & 0 & 1 & 1 & -1 \\ 0 & 1 & 0 & -1 & 0 & 2 \\ 0 & 0 & 1 & 0 & -1 & 1 \\ 0 & 0 & 0 & 0 & 0 & 0 \end{bmatrix} \begin{bmatrix} R_1 \\ R_2 \\ R_3 \\ R_4 \\ R_5 \\ R_6 \end{bmatrix} = 0 \tag{6.49}$$

The last row of zeros produces the null equation that we express as

$$0\times R_1 + 0\times R_2 + 0\times R_4 + 0\times R_3 + 0\times R_5 + 0\times R_6 = 0 \tag{6.50}$$

Thus, we can discard that row to obtain

$$\text{Axiom II:} \qquad \begin{bmatrix} 1 & 0 & 0 & | & 1 & 1 & -1 \\ 0 & 1 & 0 & | & -1 & 0 & 2 \\ 0 & 0 & 1 & | & 0 & -1 & 1 \end{bmatrix} \begin{bmatrix} R_1 \\ R_2 \\ R_4 \\ R_3 \\ R_5 \\ R_6 \end{bmatrix} = 0 \tag{6.51}$$

This form has the attractive feature that the *submatrix* located to the left of the dashed line is a *unit matrix*, and this is a useful result for solving sets of equations. Finally, it is crucial to understand that any atomic matrix can always be expressed in *row reduced echelon form*, and uniqueness proofs are available[**][††].

[**] Noble, B. 1969, *Applied Linear Algebra*, Prentice-Hall, Inc., Englewood Cliffs, New Jersey.
[††] Kolman, B. 1997, *Introductory Linear Algebra*, Sixth Edition, Prentice-Hall, Upper Saddle River, New Jersey.

EXAMPLE 6.2   EXPERIMENTAL DETERMINATION
OF THE RATE OF PRODUCTION OF ETHYLENE

Here, we consider the experimental determination of a *global* rate of production for the steady-state, catalytic dehydrogenation of ethane as illustrated in Figure 6.2a. We *assume* (see Sec. 6.1.1) that the reaction produces ethylene and hydrogen, thus only C_2H_6, C_2H_4, and H_2 are present in the system. We are given that the feed is pure ethane and the feed flow rate is 100 kmol/min. The product Stream #2 is subject to a measurement indicating that the molar flow rate of hydrogen in that stream is 30 kmol/min, and we wish to use this information to determine the *global rate of production* for ethylene. For steady-state conditions, the axiom given by Eq. 6.4 takes the form

$$\text{Axiom I:} \qquad \int_A c_A \mathbf{v}_A \cdot \mathbf{n}\, dA = \int_V R_A dV, \quad A \Rightarrow C_2H_6, C_2H_4, H_2 \qquad (1)$$

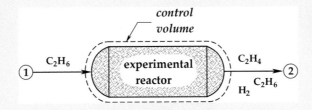

FIGURE 6.2a   Experimental reactor.

Application of this result to the control volume illustrated in Figure 6.2a provides the following three equations:

$$\text{Ethane:} \qquad -(y_{C_2H_6})_1 \dot{M}_1 + (y_{C_2H_6})_2 \dot{M}_2 = \mathcal{R}_{C_2H_6} \qquad (2)$$

$$\text{Ethylene:} \qquad -(y_{C_2H_4})_1 \dot{M}_1 + (y_{C_2H_4})_2 \dot{M}_2 = \mathcal{R}_{C_2H_4} \qquad (3)$$

$$\text{Hydrogen:} \qquad -(y_{H_2})_1 \dot{M}_1 + (y_{H_2})_2 \dot{M}_2 = \mathcal{R}_{H_2} \qquad (4)$$

Here we have used $\mathcal{R}_A$ to represent the global net rate of production for species A that is defined by (see Eq. 6.30)

$$\mathcal{R}_A = \int_V R_A dV, \quad A \Rightarrow C_2H_6, C_2H_4, H_2 \qquad (5)$$

The units of the global rate of production, $\mathcal{R}_A$, are *moles/time* while the units of the rate of production, R_A, are *moles/(time $\times$ volume)*, and one must be careful to note this difference.

At the entrance and exit of the control volume, we have two constraints on the mole fractions given by

Stream #1: $\qquad\qquad\qquad (y_{C_2H_6})_1 + (y_{C_2H_4})_1 + (y_{H_2})_1 = 1$ $\qquad$ (6)

Stream #2: $\qquad\qquad\qquad (y_{C_2H_6})_2 + (y_{C_2H_4})_2 + (y_{H_2})_2 = 1$ $\qquad$ (7)

For this particular process, the global form of Axiom II can be expressed as

Axiom II: $\qquad\qquad\qquad \sum_{A=1}^{A=N} N_{JA}\, \mathcal{R}_A = 0, \quad J \Rightarrow C, H$ $\qquad$ (8)

The visual representation of the *atomic matrix* is given by

$$
\begin{array}{r}
\text{Molecular Species} \rightarrow C_2H_6 \quad C_2H_4 \quad H_2 \\[4pt]
\begin{array}{r}
carbon \\ hydrogen
\end{array}
\begin{bmatrix} 2 & 2 & 0 \\ 6 & 4 & 2 \end{bmatrix}
\end{array}
\qquad (9)
$$

and we express the explicit form of this matrix as

$$
A = \begin{bmatrix} 2 & 2 & 0 \\ 6 & 4 & 2 \end{bmatrix}, \quad \text{or} \quad [N_{JA}] = \begin{bmatrix} 2 & 2 & 0 \\ 6 & 4 & 2 \end{bmatrix}
\qquad (10)
$$

Use of this result for the atomic matrix with Eq. 9 leads to

$$
\begin{bmatrix} 2 & 2 & 0 \\ 6 & 4 & 2 \end{bmatrix}
\begin{bmatrix} \mathcal{R}_{C_2H_6} \\ \mathcal{R}_{H_2} \\ \mathcal{R}_{C_2H_4} \end{bmatrix}
= \begin{bmatrix} 0 \\ 0 \end{bmatrix}
\qquad (11)
$$

At this point, we can follow the development in Sec. 6.2.5 to obtain

$$
\begin{bmatrix} 1 & 0 & 1 \\ 0 & 1 & -1 \end{bmatrix}
\begin{bmatrix} \mathcal{R}_{C_2H_6} \\ \mathcal{R}_{H_2} \\ \mathcal{R}_{C_2H_4} \end{bmatrix}
= \begin{bmatrix} 0 \\ 0 \end{bmatrix}
\qquad (12)
$$

in which C_2H_4 has been chosen to be the *pivot species* (see Sec. 6.4). Carrying out the matrix multiplication leads to

$$
\mathcal{R}_{C_2H_6} = -\mathcal{R}_{C_2H_4}
\qquad (13a)
$$

$$
\mathcal{R}_{H_2} = \mathcal{R}_{C_2H_4}
\qquad (13b)
$$

in which $\mathcal{R}_{C_2H_4}$ is to be determined experimentally. A degree of freedom analysis will show that a unique solution is available and we can summarize the various equations as

Ethane mole balance: $-100\ \text{kmol/min} + (y_{C_2H_6})_2\ \dot{M}_2 = \mathcal{R}_{C_2H_6}$ (14)

Ethylene mole balance: $(y_{C_2H_4})_2\ \dot{M}_2 = \mathcal{R}_{C_2H_4}$ (15)

Hydrogen mole balance: $30\ \text{kmol/min} = \mathcal{R}_{H_2}$ (16)

Stream #1: $(y_{C_2H_6})_1 = 1,\quad (y_{C_2H_4})_1 = 0,\quad (y_{H_2})_1 = 0$ (17)

Stream #2: $(y_{C_2H_6})_2 + (y_{C_2H_4})_2 + (y_{H_2})_2 = 1$ (18)

Axiom II constraint: $\mathcal{R}_{C_2H_6} = -\mathcal{R}_{C_2H_4}$ (19)

Axiom II constraint: $\mathcal{R}_{H_2} = \mathcal{R}_{C_2H_4}$ (20)

The solution to Eqs. 14–20 is given by

$$\dot{M}_2 = 130\ \text{kmol/min} \tag{21}$$

$$(y_{C_2H_6})_2 = \frac{7}{13},\quad (y_{C_2H_4})_2 = \frac{3}{13},\quad (y_{H_2})_2 = \frac{3}{13} \tag{22}$$

$$\mathcal{R}_{C_2H_4} = 30\ \text{kmol/min} \tag{23}$$

Here, we see how the experimental system illustrated in Figure 6.2a can be used to determine the global rate of production for ethylene, $\mathcal{R}_{C_2H_4}$.

In this example, we have made use of the global form of Axiom II given by Eq. 6.34 as opposed to the local form given by Eq. 6.20. In addition, we can integrate the local form given by Eq. 6.6 to obtain

$$\sum_{A=1}^{A=N} MW_A \mathcal{R}_A = 0 \tag{24}$$

This form of Axiom II reminds us that, in general, moles *are not conserved* and they are certainly not conserved in this specific example.

In the previous example, we illustrated how a net rate of production could be determined experimentally for the case of a *single independent stoichiometric reaction*. When this condition exists for an N-component system, we can express $N-1$ rates of production in terms of a single rate of production, R_N. For the complete combustion of ethane described in Example 6.1, there are four molecular species, and the rates of production for C_2H_6, O_2, H_2O can be related to the rate of production for CO_2. For the rate of production of ethylene described in Example 6.2, we have another example of a single independent reaction. In more complex systems, the stoichiometry is represented by multiple independent stoichiometric reactions, and we consider such a case in the following example.

EXAMPLE 6.3   PARTIAL OXIDATION OF CARBON

Carbon and oxygen can react to form carbon monoxide and carbon dioxide, thus, the reaction involves *four molecular species* and *two atomic species* as indicated by

Molecular Species: $C, \; O_2, \; CO, \; CO_2$ (1)

Atomic Species: C *and* O   (2)

A visual representation of the atomic matrix for this system is given by

$$
\text{Molecular Species} \;\rightarrow\;
\begin{array}{cccc}
C & O_2 & CO & CO_2
\end{array}
$$

$$
\begin{array}{c}
carbon \\
oxygen
\end{array}
\begin{bmatrix}
1 & 0 & 1 & 1 \\
0 & 2 & 1 & 2
\end{bmatrix}
\qquad (3)
$$

and this can be used with Eq. 6.22 to obtain

Axiom II:
$$
\begin{bmatrix}
1 & 0 & 1 & 1 \\
0 & 2 & 1 & 2
\end{bmatrix}
\begin{bmatrix}
R_C \\
R_{O_2} \\
R_{CO} \\
R_{CO_2}
\end{bmatrix}
=
\begin{bmatrix}
0 \\
0
\end{bmatrix}
\qquad (4)
$$

A simple calculation shows that the rank of the atomic matrix is two

$$
r = rank \, [N_{JA}] = 2 \qquad (5)
$$

Thus, we have *two equations* and *four unknowns*. Here we note that the atomic matrix can be expressed in row reduced echelon form (see Eq. 6.51) leading to

Axiom II:
$$
\begin{bmatrix}
1 & 0 & 1 & 1 \\
0 & 1 & 1/2 & 1
\end{bmatrix}
\begin{bmatrix}
R_C \\
R_{O_2} \\
R_{CO} \\
R_{CO_2}
\end{bmatrix}
=
\begin{bmatrix}
0 \\
0
\end{bmatrix}
\qquad (6)
$$

and the homogeneous system of equations corresponding to this form is given by

$$
R_C + 0 + R_{CO} + R_{CO_2} = 0 \qquad (7a)
$$

$$
0 + R_{O_2} + \frac{1}{2} R_{CO} + R_{CO_2} = 0 \qquad (7b)
$$

Given *two* equations and *four* rates of production, it is clear that we must *determine* two rates of production in order to determine all the rates of production. We will associate these two rates with two *pivot species*, and if we choose the pivot species to be carbon monoxide and carbon dioxide the rates of production for carbon and oxygen are given by

$$R_C = -R_{CO} - R_{CO_2} \tag{8a}$$

$$R_{O_2} = -\frac{1}{2} R_{CO} - R_{CO_2} \tag{8b}$$

Here, we have followed the same style used in Example 6.2 and placed the pivot species on the right-hand side of Eqs. 8. In matrix notation, this result can be expressed as (see Sec. 6.4)

$$\begin{bmatrix} R_C \\ R_{O_2} \end{bmatrix} = \begin{bmatrix} -1 & -1 \\ -1/2 & -1 \end{bmatrix} \begin{bmatrix} R_{CO} \\ R_{CO_2} \end{bmatrix} \tag{9}$$

$$\underbrace{\phantom{\begin{bmatrix} R_C \\ R_{O_2} \end{bmatrix}}}_{\substack{\text{column matrix} \\ \text{of non-pivot species}}} \qquad \underbrace{\phantom{\begin{bmatrix} -1 & -1 \\ -1/2 & -1 \end{bmatrix}}}_{\text{pivot matrix}} \underbrace{\phantom{\begin{bmatrix} R_{CO} \\ R_{CO_2} \end{bmatrix}}}_{\substack{\text{column matrix} \\ \text{of pivot species}}}$$

in which the 2×2 matrix is referred to as the *pivot matrix* since it is the matrix that maps the net rates of production of the *pivot species* onto the net rates of production of the *non-pivot species*. Other possibilities can be constructed by using different pivot species and the development of these has been left as an exercise for the student.

The partial oxidation of carbon is an especially simple example of multiple independent stoichiometric equations, i.e., *rank* $[N_{JA}] < N - 1$. The partial oxidation of ethane, illustrated in Eqs. 6.10–6.12, provides a more challenging problem.

EXAMPLE 6.4   PARTIAL OXIDATION OF ETHANE

As an example only, we imagine that the process illustrated in Figure 6.2 is carried out so that ethane is *partially oxidized* to produce ethylene oxide, carbon dioxide, carbon monoxide and water. Thus, the molecular species involved in the process are assumed to be

Molecular species: C_2H_6, O_2, H_2O,   CO, CO_2, C_2H_4O \hfill (1)

and the rates of production for these species are constrained by

Axiom II: \qquad\qquad $\sum\limits_{A=1}^{A=6} N_{JA} R_A = 0$, $J \Rightarrow C, H, O$ \hfill (2)

A visual representation of the *atomic matrix* is given by

Molecular Species $\rightarrow$ C_2H_6 O_2 H_2O CO CO_2 C_2H_4O

$$
\begin{array}{l}
\textit{carbon} \\
\textit{hydrogen} \\
\textit{oxygen}
\end{array}
\begin{bmatrix}
2 & 0 & 0 & 1 & 1 & 2 \\
6 & 0 & 2 & 0 & 0 & 4 \\
0 & 2 & 1 & 1 & 2 & 1
\end{bmatrix}
\tag{3}
$$

and the matrix representation of Eq. 2 takes the form

Axiom II:
$$
\begin{bmatrix}
2 & 0 & 0 & 1 & 1 & 2 \\
6 & 0 & 2 & 0 & 0 & 4 \\
0 & 2 & 1 & 1 & 2 & 1
\end{bmatrix}
\begin{bmatrix}
R_{C_2H_6} \\
R_{O_2} \\
R_{H_2O} \\
R_{CO} \\
R_{CO_2} \\
R_{C_2H_4O}
\end{bmatrix}
=
\begin{bmatrix}
0 \\
0 \\
0
\end{bmatrix}
\tag{4}
$$

By a series of elementary row operations, we can transform the atomic matrix to the row reduced echelon form so that Eq. 4 can be expressed as (see Sec. 6.5.2)

Axiom II:
$$
\begin{bmatrix}
1 & 0 & 0 & 1/2 & 1/2 & 1 \\
0 & 1 & 0 & 5/4 & 7/4 & 1 \\
0 & 0 & 1 & -3/2 & -3/2 & -1
\end{bmatrix}
\begin{bmatrix}
R_{C_2H_6} \\
R_{O_2} \\
R_{H_2O} \\
R_{CO} \\
R_{CO_2} \\
R_{C_2H_4O}
\end{bmatrix}
=
\begin{bmatrix}
0 \\
0 \\
0
\end{bmatrix}
\tag{5}
$$

Here, we see that the rank of the atomic matrix is three, $r = 3$, thus the rank is less than N which is equal to six. Since the rank $< N$ a non-trivial solution exists consisting of three independent equations given by

$$
R_{C_2H_6} = -\frac{1}{2} R_{CO} - \frac{1}{2} R_{CO_2} - R_{C_2H_4O}
\tag{6a}
$$

$$
R_{O_2} = -\frac{5}{4} R_{CO} - \frac{7}{4} R_{CO_2} - R_{C_2H_4O}
\tag{6b}
$$

$$
R_{H_2O} = \frac{3}{2} R_{CO} + \frac{3}{2} R_{CO_2} + R_{C_2H_4O}
\tag{6c}
$$

Here, we have chosen CO, CO_2, and C_2H_4O as the *pivot species* with the idea that the rates of production for these species will be determined experimentally. Given the rates of production for the *pivot species*, Eqs. 6 can be used to determine the rates of production for the *non-pivot species*, C_2H_6, O_2, and H_2O.

In this section, we have illustrated how Axiom II, given by Eqs. 6.20 or by Eq. 6.22, is used to *constrain* the net rates of production. When we have a *single independent stoichiometric reaction*, such as the complete combustion of ethane, one need only measure a single net rate of production in order to determine all the net rates of production. This case is illustrated in Eq. 6.8 and discussed in detail in Example 6.2. When we have *multiple independent stoichiometric reactions*, such as the partial oxidation of carbon (Example 6.3) or the partial oxidation of ethane (Example 6.4), we need to measure more than one net rate of production in order to determine all the net rates of production.

6.2.6 MATRIX PARTITIONING

Axiom II provides an example of the multiplication of a $T \times N$ matrix with a $1 \times N$ column matrix. Multiplication of matrices can also be represented in terms of *submatrices*, provided that one is careful to follow the rules of matrix multiplication. As an example, we consider the following matrix equation

$$\begin{bmatrix} a_{11} & a_{12} & a_{13} & a_{14} & a_{15} \\ a_{21} & a_{22} & a_{23} & a_{24} & a_{25} \\ a_{31} & a_{32} & a_{33} & a_{34} & a_{35} \end{bmatrix} \begin{bmatrix} b_1 \\ b_2 \\ b_3 \\ b_4 \\ b_5 \end{bmatrix} = \begin{bmatrix} c_1 \\ c_2 \\ c_3 \end{bmatrix} \tag{6.52}$$

which conforms to the rule that the number of columns in the first matrix is equal to the number of rows in the second matrix. Equation 6.52 represents the three individual equations given by

$$a_{11} b_1 + a_{12} b_2 + a_{13} b_3 + a_{14} b_4 + a_{15} b_5 = c_1 \tag{6.53a}$$

$$a_{21} b_1 + a_{22} b_2 + a_{23} b_3 + a_{24} b_4 + a_{25} b_5 = c_2 \tag{6.53b}$$

$$a_{31} b_1 + a_{32} b_2 + a_{33} b_3 + a_{34} b_4 + a_{35} b_5 = c_3 \tag{6.53c}$$

which can also be expressed in compact form according to

$$A B = C \tag{6.54}$$

Here, the matrices A, B, and C are defined explicitly by

$$A = \begin{bmatrix} a_{11} & a_{12} & a_{13} & a_{14} & a_{15} \\ a_{21} & a_{22} & a_{23} & a_{24} & a_{25} \\ a_{31} & a_{32} & a_{33} & a_{34} & a_{35} \end{bmatrix} \quad B = \begin{bmatrix} b_1 \\ b_2 \\ b_3 \\ b_4 \\ b_5 \end{bmatrix} \quad C = \begin{bmatrix} c_1 \\ c_2 \\ c_3 \end{bmatrix} \tag{6.55}$$

In addition to the matrix multiplication that we have used up to this point, matrix multiplication can also be carried out in terms of *partitioned matrices*.

If we wish to obtain a *column partition* of the matrix A in Eq. 6.52, we must also create a row partition of matrix B in order to conform to the rules of matrix multiplication that are discussed in Appendix C. This column/row partition takes the form

$$
\begin{bmatrix}
a_{11} & a_{12} & a_{13} & a_{14} & a_{15} \\
a_{21} & a_{22} & a_{23} & a_{24} & a_{25} \\
a_{31} & a_{32} & a_{33} & a_{34} & a_{35}
\end{bmatrix}
\begin{bmatrix}
b_1 \\ b_2 \\ b_3 \\ b_4 \\ b_5
\end{bmatrix}
=
\begin{bmatrix}
c_1 \\ c_2 \\ c_3
\end{bmatrix}
\tag{6.56}
$$

and the *submatrices* are identified explicitly according to

$$
A_{11} =
\begin{bmatrix}
a_{11} & a_{12} & a_{13} \\
a_{21} & a_{22} & a_{23} \\
a_{31} & a_{32} & a_{33}
\end{bmatrix}
\quad
B_1 =
\begin{bmatrix}
b_1 \\ b_2 \\ b_3
\end{bmatrix}
\quad
A_{12} =
\begin{bmatrix}
a_{14} & a_{15} \\
a_{24} & a_{25} \\
a_{34} & a_{35}
\end{bmatrix}
\quad
B_2 =
\begin{bmatrix}
b_4 \\ b_5
\end{bmatrix}
\tag{6.57}
$$

Use of these representations in Eq. 6.56 leads to

$$
\begin{bmatrix} A_{11} & A_{12} \end{bmatrix}
\begin{bmatrix}
B_1 \\ B_2
\end{bmatrix}
= C
\tag{6.58}
$$

and matrix multiplication in terms of the submatrices provides

$$
A_{11}B_1 + A_{12}B_2 = C
\tag{6.59}
$$

To verify that this result is identical to Eq. 6.53, we use Eqs. 6.57 and the third of Eqs. 6.55 to obtain

$$
\begin{bmatrix}
a_{11} & a_{12} & a_{13} \\
a_{21} & a_{22} & a_{23} \\
a_{31} & a_{32} & a_{33}
\end{bmatrix}
\begin{bmatrix}
b_1 \\ b_2 \\ b_3
\end{bmatrix}
+
\begin{bmatrix}
a_{14} & a_{15} \\
a_{24} & a_{25} \\
a_{34} & a_{35}
\end{bmatrix}
\begin{bmatrix}
b_4 \\ b_5
\end{bmatrix}
=
\begin{bmatrix}
c_1 \\ c_2 \\ c_3
\end{bmatrix}
\tag{6.60}
$$

Carrying out the matrix multiplication on the left-hand side of this result leads to

$$
\begin{bmatrix}
a_{11}b_1 & a_{12}b_2 & a_{13}b_3 \\
a_{21}b_1 & a_{22}b_2 & a_{23}b_3 \\
a_{31}b_1 & a_{32}b_2 & a_{33}b_3
\end{bmatrix}
+
\begin{bmatrix}
a_{14}b_4 & a_{15}b_5 \\
a_{24}b_4 & a_{25}b_5 \\
a_{34}b_4 & a_{35}b_5
\end{bmatrix}
=
\begin{bmatrix}
c_1 \\ c_2 \\ c_3
\end{bmatrix}
\tag{6.61}
$$

At this point, we add the two matrices on the left-hand side of this result following the rules for matrix addition given in Sec. 2.6 in order to obtain

$$\begin{bmatrix} a_{11}b_1 & a_{12}b_2 & a_{13}b_3 & a_{14}b_4 & a_{15}b_5 \\ a_{21}b_1 & a_{22}b_2 & a_{23}b_3 & a_{24}b_4 & a_{25}b_5 \\ a_{31}b_1 & a_{32}b_2 & a_{33}b_3 & a_{34}b_4 & a_{35}b_5 \end{bmatrix} = \begin{bmatrix} c_1 \\ c_2 \\ c_3 \end{bmatrix} \tag{6.62}$$

This is an alternate representation of Eq. 6.52 that immediately leads to Eqs. 6.53. A more detailed discussion of matrix multiplication and partitioning is given in Appendix C; however, the results in this section are sufficient for our treatment of Axiom II. In the following paragraphs we learn that partitioning of the atomic matrix leads to especially useful forms of Axiom II.

6.3 PIVOTS AND NON-PIVOTS

In the previous section, we illustrated that the number of constraining equations associated with Axiom II is equal to the *rank* = r of the atomic matrix which is less than or equal to the number of atomic species, T. Because of this, the number of pivot species must be equal to $N - r$ and the number of the non-pivot species must be equal to the rank of the atomic matrix, r. The choice of pivot species and non-pivot species is not completely arbitrary, since it is a *necessary condition* that all the atomic species be present in at least one non-pivot species. In this section, we consider the issue of pivots and non-pivots in terms of an example and some analysis using matrix partitioning that was discussed in Sec. 6.2.6.

**EXAMPLE 6.5   PRODUCTION OF
BUTADIENE FROM ETHANOL[††].**

Ethanol produced by fermentation of natural sugars from grain can be used in the production of butadiene which is a basic feedstock for the production of synthetic rubber. The following molecular species are involved in the production of butadiene C_4H_6 from ethanol C_2H_5OH:

$$C_2H_5OH, \quad C_2H_4, \quad H_2O, \quad CH_3CHO, \quad H_2, \quad C_4H_6$$

Since ethanol is the reactant, it is reasonable to arrange the atomic matrix in the form

Molecular Species $\rightarrow$	C_2H_5OH	C_2H_4	H_2O	CH_3CHO	H_2	C_4H_6	
carbon	2	2	0	2	0	4	
hydrogen	6	4	2	4	2	6	(1)
oxygen	1	0	1	1	0	0	

[††] Kvisle, S., Aguero, A. and Sneeded, R.P.A. 1988, Transformation of ethanol into 1,3-butadiene over magnesium oxide/silica catalysts, Appl. Catal. **43**, 117–121.

If we assume that the rank of the matrix is three ($r = rank = 3$), we are confronted with six unknowns and three equations, thus the non-pivot species are represented by C_2H_5OH, C_2H_4, and H_2O. The atomic matrix can be expressed explicitly by

$$A = \begin{bmatrix} 2 & 2 & 0 & 2 & 0 & 4 \\ 6 & 4 & 2 & 4 & 2 & 6 \\ 1 & 0 & 1 & 1 & 0 & 0 \end{bmatrix} \tag{2}$$

and a series of elementary row operations leads to the *row echelon form* given by

$$A^{\cdot} = \begin{bmatrix} 1 & 1 & 0 & 1 & 0 & 2 \\ 0 & 1 & -1 & 1 & -1 & 3 \\ 0 & 0 & 0 & 1 & -1 & 1 \end{bmatrix} \tag{3}$$

This indicates that the rank of the atomic matrix is three ($rank = 3$) and we have three independent equations to determine the six rates of production. The calculation indicated in Eq. 6.22 is given by

$$\begin{bmatrix} 1 & 1 & 0 & 1 & 0 & 2 \\ 0 & 1 & -1 & 1 & -1 & 3 \\ 0 & 0 & 0 & 1 & -1 & 1 \end{bmatrix} \begin{bmatrix} R_{C_2H_5OH} \\ R_{C_2H_4} \\ R_{H_2O} \\ R_{CH_3CHO} \\ R_{H_2} \\ R_{C_4H_6} \end{bmatrix} = \begin{bmatrix} 0 \\ 0 \\ 0 \end{bmatrix} \tag{4}$$

At this point, we see that the atomic matrix is not in the row reduced echelon form; however, we can obtain this form by means of a *column/row* interchange. In terms of Eq. 6.20 we express a judicious choice of a column/row interchange as

$$N_{JC}\, R_C \rightleftarrows N_{JD}\, R_D, \quad B, D \Rightarrow H_2O, CH_3CHO, \quad J \Rightarrow C, H, O \tag{5}$$

in order to express Eq. 4 in the form

$$\begin{bmatrix} 1 & 1 & 1 & 0 & 0 & 2 \\ 0 & 1 & 1 & -1 & -1 & 3 \\ 0 & 0 & 1 & 0 & -1 & 1 \end{bmatrix} \begin{bmatrix} R_{C_2H_5OH} \\ R_{C_2H_4} \\ R_{CH_3CHO} \\ R_{H_2O} \\ R_{H_2} \\ R_{C_4H_6} \end{bmatrix} = \begin{bmatrix} 0 \\ 0 \\ 0 \end{bmatrix} \tag{6}$$

Here, we note that our original choice of non-pivot species, C_2H_5OH, C_2H_4, and H_2O, has been changed by the application of Eq. 5 that leads to the non-pivot species represented by C_2H_5OH, C_2H_4, and CH_3CHO.

At this point, we make use of some routine elementary row operations to obtain the desired *row reduced echelon form*

$$
\begin{bmatrix}
1 & 0 & 0 & 1 & 1 & -1 \\
0 & 1 & 0 & -1 & 0 & 2 \\
0 & 0 & 1 & 0 & -1 & 1
\end{bmatrix}
\begin{bmatrix}
R_{C_2H_5OH} \\
R_{C_2H_4} \\
R_{CH_3CHO} \\
R_{H_2O} \\
R_{H_2} \\
R_{C_4H_6}
\end{bmatrix}
=
\begin{bmatrix}
0 \\
0 \\
0
\end{bmatrix}
\tag{7}
$$

Given this representation of Axiom II, we can apply a *column/row partition* illustrated by

$$
\left[\begin{array}{ccc:ccc}
1 & 0 & 0 & 1 & 1 & -1 \\
0 & 1 & 0 & -1 & 0 & 2 \\
0 & 0 & 1 & 0 & -1 & 1
\end{array}\right]
\begin{bmatrix}
R_{C_2H_5OH} \\
R_{C_2H_4} \\
R_{CH_3CHO} \\
\hdashline
R_{H_2O} \\
R_{H_2} \\
R_{C_4H_6}
\end{bmatrix}
=
\begin{bmatrix}
0 \\
0 \\
0
\end{bmatrix}
\tag{8}
$$

which immediately leads to

$$
\underbrace{\begin{bmatrix}
1 & 0 & 0 \\
0 & 1 & 0 \\
0 & 0 & 1
\end{bmatrix}}_{\substack{\text{non-pivot} \\ \text{submatrix}}}
\begin{bmatrix}
R_{C_2H_5OH} \\
R_{C_2H_4} \\
R_{CH_3CHO}
\end{bmatrix}
+
\underbrace{\begin{bmatrix}
1 & 1 & -1 \\
-1 & 0 & 2 \\
0 & -1 & 1
\end{bmatrix}}_{\substack{\text{pivot} \\ \text{submatrix}}}
\begin{bmatrix}
R_{H_2O} \\
R_{H_2} \\
R_{C_4H_6}
\end{bmatrix}
=
\begin{bmatrix}
0 \\
0 \\
0
\end{bmatrix}
\tag{9}
$$

Here the non-pivot submatrix is the *unit matrix* that maps a column matrix onto itself as indicated by

$$
\underbrace{\begin{bmatrix}
1 & 0 & 0 \\
0 & 1 & 0 \\
0 & 0 & 1
\end{bmatrix}}_{\substack{\text{non-pivot} \\ \text{submatrix}}}
\begin{bmatrix}
R_{C_2H_5OH} \\
R_{C_2H_4} \\
R_{CH_3CHO}
\end{bmatrix}
=
\begin{bmatrix}
R_{C_2H_5OH} \\
R_{C_2H_4} \\
R_{CH_3CHO}
\end{bmatrix}
\tag{10}
$$

Substitution of this result into Eq. 9 provides the following simple form

$$
\begin{bmatrix} R_{C_2H_5OH} \\ R_{C_2H_4} \\ R_{CH_3CHO} \end{bmatrix} = - \underbrace{\begin{bmatrix} 1 & 1 & -1 \\ -1 & 0 & 2 \\ 0 & -1 & 1 \end{bmatrix}}_{\substack{pivot \\ submatrix}} \begin{bmatrix} R_{H_2O} \\ R_{H_2} \\ R_{C_4H_6} \end{bmatrix} \tag{11}
$$

From this we extract a representation for the column matrix of *non-pivot species* in terms of the *pivot matrix* of stoichiometric coefficients and the column matrix of *pivot species*. This representation is given by

$$
\underbrace{\begin{bmatrix} R_{C_2H_5OH} \\ R_{C_2H_4} \\ R_{CH_3CHO} \end{bmatrix}}_{\substack{column\ matrix \\ of\ non\text{-}pivot\ species}} = \underbrace{\begin{bmatrix} -1 & -1 & 1 \\ 1 & 0 & -2 \\ 0 & 1 & -1 \end{bmatrix}}_{pivot\ matrix} \underbrace{\begin{bmatrix} R_{H_2O} \\ R_{H_2} \\ R_{C_4H_6} \end{bmatrix}}_{\substack{column\ matrix \\ of\ pivot\ species}} \tag{12}
$$

This is a special case of the *pivot theorem* in which we see that the net rates of production of the pivot species are mapped onto the net rates of production of the non-pivot species by the pivot matrix. The matrix multiplication indicated in Eq. 12 can be carried out to obtain

$$
R_{C_2H_5OH} = -R_{H_2O} - R_{H_2} + R_{C_4H_6} \tag{13a}
$$

$$
R_{C_2H_4} = R_{H_2O} + 0 - 2R_{C_4H_6} \tag{13b}
$$

$$
R_{CH_3CHO} = 0 + R_{H_2} - R_{C_4H_6} \tag{13c}
$$

In this example, we have provided a template for the solution of Eq. 6.22 in which the mathematical steps are always the same, i.e., (I) Begin with the atomic matrix given by Eq. 2 and develop the row reduced echelon form given by Eq. 3. (II) Partition Axiom II as indicated by Eqs. 8 and 9. (III) Carry out the matrix multiplication to obtain the net rates of production for the non-pivot species indicated by Eqs. 13.

In the design of a butadiene production unit, the representation for the net rate of production of the non-pivot species is a crucial part of the analysis. In Chapter 7, we will apply this type of analysis to several processes.

6.3.1 RANK OF THE ATOMIC MATRIX

In the previous paragraphs, we have seen several examples of the atomic matrix when the rank of that matrix was less than the number of molecular species, i.e., $r < N$. Here we wish to illustrate two special cases in which $r = N$ and $r < N$.

First, we consider a reactor containing only methyl chloride $CH_3 Cl$, ethyl chloride $C_2H_5 Cl$, and chlorine Cl_2. We illustrate the atomic matrix by

$$
\begin{array}{c}
\text{Molecular Species} \;\to\; \begin{array}{ccc} CH_3Cl & C_2H_5Cl & Cl_2 \end{array} \\
\begin{array}{l}
carbon \\
hydrogen \\
chlorine
\end{array}
\begin{bmatrix}
1 & 2 & 0 \\
3 & 5 & 0 \\
1 & 1 & 2
\end{bmatrix}
\end{array}
\tag{6.63}
$$

and use elementary row operations to obtain the *row reduced echelon form* given by

$$
A^{*} = \begin{bmatrix}
1 & 0 & 0 \\
0 & 1 & 0 \\
0 & 0 & 1
\end{bmatrix}
\tag{6.64}
$$

Use of this result in Axiom II provides

$$
\begin{bmatrix}
1 & 0 & 0 \\
0 & 1 & 0 \\
0 & 0 & 1
\end{bmatrix}
\begin{bmatrix}
R_{CH_3Cl} \\
R_{C_2H_5Cl} \\
R_{Cl_2}
\end{bmatrix}
=
\begin{bmatrix}
0 \\
0 \\
0
\end{bmatrix}
\tag{6.65}
$$

and we immediately obtain the *trivial solution* given by[§§].

$$
R_{CH_3Cl} = 0
\tag{6.66a}
$$

$$
R_{C_2H_5Cl} = 0
\tag{6.66b}
$$

$$
R_{Cl_2} = 0
\tag{6.66c}
$$

This indicates that *no net rate of production can occur* in a reactor containing only methyl chloride, ethyl chloride, and chlorine. If we allow molecular hydrogen to be present in the reactor, our system is described by

$$
\begin{array}{c}
\text{Molecular Species} \to \begin{array}{cccc} CH_3Cl & C_2H_5Cl & Cl_2 & H_2 \end{array} \\
\begin{array}{l}
carbon \\
hydrogen \\
chlorine
\end{array}
\begin{bmatrix}
1 & 2 & 0 & 0 \\
3 & 5 & 0 & 2 \\
1 & 1 & 2 & 0
\end{bmatrix}
\end{array}
\tag{6.67}
$$

and the atomic matrix takes the form

$$
A = \begin{bmatrix}
1 & 2 & 0 & 0 \\
3 & 5 & 0 & 2 \\
1 & 1 & 2 & 0
\end{bmatrix}
\tag{6.68}
$$

[§§] Kolman, B. 1997, page 59, *Introductory Linear Algebra*, 6th Edition, Prentice-Hall, Upper Saddle River, New Jersey.

The *row reduced echelon form* of this atomic matrix can be expressed as

$$A^* = \begin{bmatrix} 1 & 0 & 0 & 4 \\ 0 & 1 & 0 & -2 \\ 0 & 0 & 1 & -1 \end{bmatrix} \tag{6.69}$$

and from Axiom II we obtain

$$\begin{bmatrix} 1 & 0 & 0 & 4 \\ 0 & 1 & 0 & -2 \\ 0 & 0 & 1 & -1 \end{bmatrix} \begin{bmatrix} R_{CH_3Cl} \\ R_{C_2H_5Cl} \\ R_{Cl_2} \\ R_{H_2} \end{bmatrix} = \begin{bmatrix} 0 \\ 0 \\ 0 \end{bmatrix} \tag{6.70}$$

In this case, we have $r = rank = 3$ and $N = 4$, and the net rates of production for methyl chloride, ethyl chloride, and chlorine can be represented in terms of the net rate of production of hydrogen. The *column/row partition* of Eq. 6.70 is illustrated by

$$\begin{bmatrix} 1 & 0 & 0 & | & 4 \\ 0 & 1 & 0 & | & -2 \\ 0 & 0 & 1 & | & -1 \end{bmatrix} \begin{bmatrix} R_{CH_3Cl} \\ R_{C_2H_5Cl} \\ R_{Cl_2} \\ \hline R_{H_2} \end{bmatrix} = \begin{bmatrix} 0 \\ 0 \\ 0 \end{bmatrix} \tag{6.71}$$

and this immediately leads to

$$\begin{bmatrix} 1 & 0 & 0 \\ 0 & 1 & 0 \\ 0 & 0 & 1 \end{bmatrix} \begin{bmatrix} R_{CH_3Cl} \\ R_{C_2H_5Cl} \\ R_{Cl_2} \end{bmatrix} + \begin{bmatrix} 4 \\ -2 \\ -1 \end{bmatrix} \begin{bmatrix} R_{H_2} \end{bmatrix} = \begin{bmatrix} 0 \\ 0 \\ 0 \end{bmatrix} \tag{6.72}$$

This can be solved for the column matrix of non-pivot species according to

$$\underbrace{\begin{bmatrix} R_{CH_3Cl} \\ R_{C_2H_5Cl} \\ R_{Cl_2} \end{bmatrix}}_{\substack{\text{column matrix} \\ \text{of non-pivot species}}} = \underbrace{\begin{bmatrix} -4 \\ 2 \\ 1 \end{bmatrix}}_{\text{pivot matrix}} \underbrace{\begin{bmatrix} R_{H_2} \end{bmatrix}}_{\substack{\text{column matrix} \\ \text{of pivot species}}} \tag{6.73}$$

Here, we have a *non-trivial solution* in which the three rates of production of the *non-pivot species* are specified in terms of the rate of production of the *pivot species*, R_{H_2}. One should always keep in mind that the *null solution* is still possible for this case, i.e., all four net rates of production may be zero depending on the conditions in our hypothetical reactor.

6.4   AXIOMS AND THEOREMS

To summarize our studies of stoichiometry, we note that atomic species are neither created nor destroyed by chemical reactions. In terms of the *atomic matrix* and the *column matrix of net rates of production*, this concept can be expressed as

Axiom II: $$A\,R = 0 \qquad (6.74)$$

As indicated in the previous section, the *row reduced echelon form* of the atomic matrix can always be developed, thus we can express Eq. 6.74 as

Row reduced echelon form: $$A^*\,R = 0 \qquad (6.75)$$

The product of the atomic matrix times the column matrix of net rates of production can be partitioned according to (see Sec. 6.2.6)

Column/Row partition: $$\begin{bmatrix} I & W \end{bmatrix} \begin{bmatrix} R_{NP} \\ R_P \end{bmatrix} = 0 \qquad (6.76)$$

Here the *column partition* of A^* provides the *non-pivot submatrix I* and the *pivot submatrix W*, while the *row partition* of R provides the *non-pivot column submatrix R_{NP}* and the *pivot column submatrix R_P*. Carrying out the matrix multiplication indicated by Eq. 6.76 leads to

$$I\,R_{NP} + W\,R_P = 0 \qquad (6.77)$$

and operation of the unit matrix on R_{NP} provides the obvious result given by

$$I\,R_{NP} = R_{NP} \qquad (6.78)$$

At this point, we define the *pivot matrix P* according to

Pivot matrix: $$P = -W \qquad (6.79)$$

and we use this result, along with Eq. 6.78, in Eq. 6.77 to obtain the *pivot theorem*

Pivot theorem: $$R_{NP} = P\,R_P \qquad (6.80)$$

These five concepts represent the foundations of stoichiometry, and they appear in various special forms throughout this chapter and in subsequent chapters. When ionic species are involved, *conservation of charge* must be taken into account as indicated in Appendix E. Heterogeneous reactions can be analyzed using the framework presented in this chapter and the details are discussed in Appendix F. Reactions involving *optical isomers* require some care that is illustrated in Problems 6.32 and 6.33.

In Chapter 7, we will make repeated use of the *global form* of the pivot theorem. To develop the global form, we integrate Eq. 6.80 over the volume V to obtain

Global pivot theorem: $$\mathcal{R}_{NP} = P\,\mathcal{R}_P \qquad (6.81)$$

A summary of the matrices presented in this chapter is given in Sec. 9.4 along with a discussion of the several matrices used in the study of chemical reaction kinetics.

6.5 PROBLEMS

Note that problems marked with the symbol ⌨ will be difficult to solve without the use of computer software.

SECTION 6.1

6.1. By "counting atoms" provide at least one version of a balanced chemical equation based on

$$? C_2H_6 + ? O_2 \rightarrow ? CO + ? C_2H_4O + ? H_2O + CO_2$$

that is different from the two examples given in the text.

SECTION 6.2

6.2. Construct an atomic matrix for the following set of components: Sodium hydroxide $NaOH$, methyl bromide CH_3Br, methanol CH_3OH, and sodium bromide $NaBr$.

6.3. Construct an atomic matrix for a system containing the following molecular species: NH_3, O_2, NO, N_2, H_2O, and NO_2.

6.4. Begin with the statement that mass is neither created nor destroyed by chemical reaction

$$\sum_{A=1}^{A=N} r_A = 0$$

and use it to derive Eq. 6.20. Be careful to state any restrictions that might be necessary in order to complete the derivation.

6.5. The rank of a matrix is conveniently determined using the row reduced form of the matrix. Consider the atomic matrix given by Eq. 2 of Example 6.1 and use elementary row operations to develop the row reduced echelon form of that matrix.

6.6. Using Eq. 6.20, show how to obtain Eqs. 4 in Example 6.1.

6.7. Use elementary row operations to express Eq. 5 of Example 6.1 in terms of the row reduced echelon form of the atomic matrix. Indicate how Eqs. 6 are obtained using the row reduced echelon form.

6.8. First, find the rank of the atomic matrix developed in Problem 6.3. Next, choose N_2, H_2O, and NO_2 as the pivot species and develop a solution for R_{NH_3}, R_{O_2}, and R_{NO}.

6.9. Express the atomic matrix in Eq. 12 of Example 6.2 in row reduced echelon form. Use that form to express $\mathcal{R}_{C_2H_6}$ and $\mathcal{R}_{C_2H_4}$ in terms of $\mathcal{R}_{H_2}$.

6.10. Represent Eqs. 13 of Example 6.2 in the form of the pivot theorem illustrated by Eq. 7 of Example 6.1.

6.11. In this problem, you are asked to consider the complete combustion of methanol, thus the molecular species under consideration are

$$CH_3OH, \quad O_2, \quad H_2O, \quad CO_2$$

Develop the atomic matrix in row reduced echelon form, and use Axiom II with CO_2 as the pivot species in order to determine the rates of production, $R_{CH_3OH}, R_{O_2}, R_{H_2O}$ in terms of R_{CO_2}. Express your results in the form analogous to Eq. 7 of Example 6.1.

6.12. Consider the complete combustion of methane to produce water and carbon dioxide. Construct the atomic matrix in row reduced echelon form, and show that Axiom II can be used to express the rates of production of methane, oxygen, and water in terms of the single pivot species, carbon dioxide. Express your results in the form of the pivot theorem illustrated by Eq. 7 of Example 6.1.

6.13. For the molecular species listed in Problem 6.2, determine the ratio of rates of production given by

$$\frac{R_{NaOH}}{R_{NaBr}}, \quad \frac{R_{CH_3Br}}{R_{NaBr}}, \quad \frac{R_{CH_3OH}}{R_{NaBr}}$$

6.14. The production of alumina $NaAlO_2$ from bauxite $Al(OH)_3$ requires sodium hydroxide $NaOH$ as a reactant and yields water H_2O as product. For this system, determine the ratio of the rates of production given by

$$\frac{R_{Al(OH)_3}}{R_{NaAlO_2}}, \quad \frac{R_{NaOH}}{R_{NaAlO_2}}, \quad \frac{R_{H_2O}}{R_{NaAlO_2}}$$

6.15. At $T = 20°C$, the rate of disappearance (i.e., the net rate of production) of methyl bromide in the reaction between methyl bromide and sodium hydroxide (see Problems 6.2 and 6.12) is

$$-R_{CH_3Br} = 0.2 \, \text{mol/m}^3\text{s}$$

The molecular species involved in this reaction are sodium hydroxide ($NaOH$), methyl bromide CH_3Br, methanol CH_3OH, and sodium bromide $NaBr$. In this problem, you are asked to

    a. Determine the rate of production of sodium hydroxide, methanol, and sodium bromide in kmol/m³s.

b. Determine the rates of production for all components in kg/m³s.

c. Show that mass is neither created nor destroyed by this chemical reaction.

d. Verify that atomic species are neither created nor destroyed by this chemical reaction.

6.16. Methanol can be synthesized by reacting carbon monoxide and hydrogen over a catalyst. The mass rate of production of methanol at 400 K is $r_{CH_3OH} = 0.035\,\text{kg/m}^3\text{s}$. Determine the rate of production in kmol/m³s for all components of the synthesis reaction.

6.17. Given an atomic matrix of the form

$$A = \begin{bmatrix} 1 & 2 & 1 & 4 \\ 0 & 2 & 4 & 2 \\ 0 & 1 & 5 & 4 \\ 0 & 1 & 3 & 2 \end{bmatrix}$$

indicate how one can obtain a row reduced matrix of the form

$$A'' = \begin{bmatrix} 1 & 2 & 1 & 4 \\ 0 & 1 & 2 & 1 \\ 0 & 0 & 1 & 1 \\ 0 & 0 & 0 & 0 \end{bmatrix}$$

6.18. For the partial oxidation of carbon described in Example 6.3, explore the use of pivot species other than CO and CO_2. This will lead to five more possibilities in addition to those given by Eqs. 7 in Example 6.3.

6.19. Use elementary row operations to verify that Eq. 4 leads to Eq. 5 in Example 6.4.

6.20. Express Eqs. 6 of Example 6.4 in the form of the pivot theorem as illustrated by Eq. 9 in Example 6.3.

6.21. Meta-cresol sulfonic acid is a reddish-brown liquid that is used as a pharmaceutical intermediate in the manufacture of *disinfectants* and in the manufacture of *friction dust* for disk brakes. This chemical is produced along with water by the sulfonation of meta-cresol with sulfuric acid. In this problem, you are first asked to prove that the rank of the atomic matrix is three. Then show that the net rates of production of meta-cresol C_7H_8O, sulfuric acid H_2SO_4, and meta-cresol sulfonic acid $C_7H_8O_4S$ can be expressed in terms of the net rate of production of water H_2O.

6.22. In Sec. 6.2.6, we illustrated how to construct a *column/row partition* of a matrix equation, and here you are asked to repeat that type of analysis for a *row/row partition* of the following matrix equation:

$$
\begin{bmatrix}
a_{11} & a_{12} & a_{13} \\
a_{21} & a_{22} & a_{23} \\
a_{31} & a_{32} & a_{33} \\
a_{41} & a_{42} & a_{43} \\
a_{51} & a_{52} & a_{53}
\end{bmatrix}
\begin{bmatrix}
b_1 \\
b_2 \\
b_3
\end{bmatrix}
=
\begin{bmatrix}
c_1 \\
c_2 \\
c_3 \\
c_4 \\
c_5
\end{bmatrix}
\tag{1}
$$

If the 5×3 matrix is partitioned according to

$$
\begin{bmatrix}
a_{11} & a_{12} & a_{13} \\
a_{21} & a_{22} & a_{23} \\
a_{31} & a_{32} & a_{33} \\
a_{41} & a_{42} & a_{43} \\
\hline
a_{51} & a_{52} & a_{53}
\end{bmatrix}
\tag{2}
$$

indicate how Eq. 1 must be partitioned. Clearly indicate how the partitioned version of Eq. 1 leads to the detailed results associated with Eq. 1 that are given by

$$
a_{11} b_1 + a_{12} b_2 + a_{13} b_3 = c_1
\tag{3a}
$$

$$
a_{21} b_1 + a_{22} b_2 + a_{23} b_3 = c_2
\tag{3b}
$$

$$
a_{31} b_1 + a_{32} b_2 + a_{33} b_3 = c_3
\tag{3c}
$$

$$
a_{41} b_1 + a_{42} b_2 + a_{43} b_3 = c_4
\tag{3d}
$$

$$
a_{51} b_1 + a_{52} b_2 + a_{53} b_3 = c_5
\tag{3e}
$$

SECTION 6.3

6.23. In Example 6.5, show how to obtain Eq. 3 beginning with the atomic matrix identified in Eq. 2.

6.24. Show how to develop Eq. 6.69 in terms of Eq. 6.68 using the elementary row operations described in Sec. 6.2.5.

6.25. Show how to obtain Eq. 6.73 from Eq. 6.72.

6.26. When methane is partially combusted with oxygen, one finds the following molecular species: CH_4, O_2, CO, CO_2, H_2O, and H_2. Determine the number of independent stoichiometric reactions and comment on the restrictions concerning the choice of pivot and non-pivot species.

SECTION 6.4

6.27. Rucker *et al.*[***] have studied the catalytic conversion of acetylene (C_2H_2) to form benzene (C_6H_6) along with hydrogen (H_2) and ethylene (C_2H_4). For this system, chose benzene and ethylene as the pivot species, determine the rank of the atomic matrix, and apply the pivot theorem to determine the net rates of production of the non-pivot species, $R_{C_2H_2}$ and R_{H_2}.

6.28. The preparation of styrene (C_8H_8) and benzene (C_6H_6) from acetylene (C_2H_2) has been considered by Tanaka *et al.*[†††] and for this system a visual representation of the atomic matrix is given by

$$\text{Molecular Species} \rightarrow \quad \begin{matrix} C_2H_2 & C_6H_6 & C_8H_8 \end{matrix}$$

$$\begin{matrix} carbon \\ oxygen \end{matrix} \begin{bmatrix} 2 & 6 & 8 \\ 2 & 6 & 8 \end{bmatrix}$$

Determine the rank of the atomic matrix and use the pivot theorem to represent the rate of production of the non-pivot species in terms of the rate of production of the pivot species.

6.29. The reaction of acetylene (C_2H_2) with methanol (CH_3OH) to produce methyl ether ($CH_3OC_2H_3$) is sometimes known as Reppe chemistry[‡‡‡]. Determine the rank of the atomic matrix for this system. Choose methyl ether as the pivot species and use the pivot theorem to represent the rates of production of the non-pivot species in terms of the rate of production of the pivot species.

6.30. Given a system containing the molecular species: CH_4, O_2, Cl_2, CH_3Cl, HCl, H_2O, and CO_2, determine the rank of the atomic matrix. Use the pivot theorem to express the net rates of production of the non-pivot species, CH_4, O_2, Cl_2, CH_3Cl in terms of the net rates of production of the pivot species, HCl, H_2O, and CO_2.

6.31. In this problem, we consider the catalytic oxidation (O_2) of ethane (C_2H_6) to produce ethylene (C_2H_4) and acetic acid (CH_3COOH) along with carbon dioxide (CO_2), carbon monoxide (CO) and water (H_2O). This process has been studied experimentally[§§§] in order to determine the factors affecting the *selectivity* (see

---

[***] Rucker, T.G., Logan, M.A., Gentle, T.M., Muetterties, E.L. and Somorjai, G.A. 1986, Conversion of acetylene to Benzene over palladium single-crystal surfaces. 1. The low-pressure stoichiometric and high-pressure catalytic reactions, J. Phys. Chem. **90**, 2703–2708.

[†††] Tanaka, M., Yamamote, M. and Oku, M. 1955, Preparation of styrene and benzene from acetylene and vinyl acetylene, USPO, 272–299.

[‡‡‡] Reppe, W.J. 1892–1969, A German scientist notable for his contributions to the chemistry of acetylene.

[§§§] Sankaranarayanan, T.M., Ingle, R.H., Gaikwad, T.B., Lokhande, S.K., Raja. T., Devi, R.N., Ramaswany, V., and Manikandan, P. 2008, Selective oxidation of ethane over Mo-V-Al-O oxide catalysts: Insight to the factors affecting the selectivity of ethylene and acetic acid and structure-activity correlation studies, Catal. Lett. **121**: 39–51.

Chapter 7) of ethylene and acetic acid. Make use of the pivot theorem to demonstrate that one must measure (as one possibility) the net rates of production for CH_3COOH, C_2H_4, H_2O, and CO in order to predict the net rates of production for C_2H_6, CO_2, and O_2.

6.32. From Figure 6-32, we see that α-butylene has a very different structure than isobutylene. Thus, we expect reactions involving these two molecules might be quite different. However, in terms of stoichiometry, these two molecules are indistinguishable, and this means that we need to be careful in constructing the atomic matrix. For the combustion of a mixture of α-butylene and isobutylene, we can express the atomic matrix as

Molecular Species $\rightarrow$	$C_4H_8^{\alpha}$	$C_4H_8^{iso}$	H_2O	CO_2	CO
carbon	4	4	0	1	1
hydrogen	8	8	2	0	0
oxygen	0	0	1	2	0

Take CO_2 and $C_4H_8^{iso}$ as the pivot species in order to obtain expressions for $R_{C_4H_8}^{\alpha}$, R_{CO} and R_{H_2O} in terms of R_{CO_2} and $R_{C_4H_8}^{iso}$.

α-butylene   isobutylene

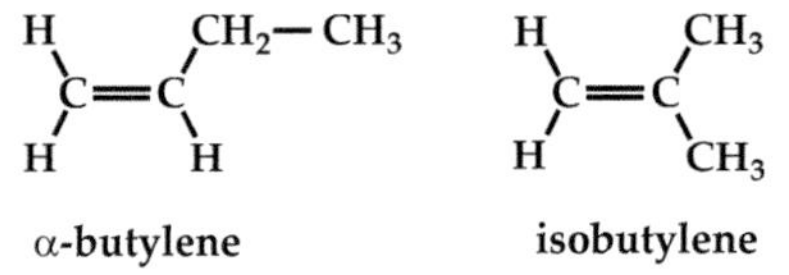

FIGURE 6-32   Isomers of butylene.

6.33. A complex system of optical isomers involves ortho-cresol $C_7H_8O^{OC}$, meta-cresol $C_7H_8O^{MC}$, water H_2O, sulfuric acid H_2SO_4, ortho-cresol-sulfonic acid $C_7H_8O_4S^{OCS}$, and meta-cresol sulfonic acid $C_7H_8O_4S^{MCS}$. Use Axiom II to develop three constraints for the net rates of production associated with these six molecular species.

6.34. Identify $\mathcal{R}_{NP}$, P, and $\mathcal{R}_P$ in Example 6.5.

7 Material Balances for Complex Systems

Most recent paradigm shifts in the mathematical analysis of physical systems are due to the use of computers. In Chapter 4, we encountered the application of matrices in the formulation of material balance problems, and for small matrices, those problems could be solved easily. For large matrices, solutions are difficult to obtain (see Sec. 4.8) and computer-aided calculations are necessary. In this chapter, we consider the *transition* from simple and small systems to complex and large systems. We begin with some moderately complex processes involving reactors, separators, and recycle streams. These systems can be analyzed without the use of computers. In Sec. 7.4, we introduce sequential analysis using iterative methods that require some programming. This sequential analysis forms the basis for process simulators that will be studied in a senior-level design course; however, it is absolutely essential to understand the details presented in this chapter prior to using process simulators for computer-aided design.

In Chapters 4 and 5, we studied multicomponent, multiphase systems *without* chemical reactions, and in Chapter 6, we learned how to analyze multiple, independent stoichiometric reactions in a general manner. We are now ready to study more complex systems with chemical reactors such as the one shown in Figure 7.1. Here we have identified several distinct control volumes, and the choice of the control volumes that provides the most convenient analysis will be examined in this chapter.

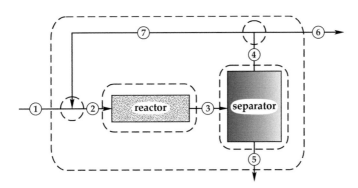

FIGURE 7.1 Reactor and separator with recycle.

In Sec. 4.7.1, we developed a degree-of-freedom analysis for systems with N components, M streams, and no chemical reactions. Here, we extend that analysis to

DOI: 10.1201/9781003283751-7

include chemical reactions in systems for which the governing equations are given by

Axiom I:
$$\frac{d}{dt}\int_V c_A\, dV + \int_A c_A \mathbf{v}_A \cdot \mathbf{n}\, dA = \int_V R_A\, dV, \quad A = 1, 2, \ldots, N \tag{7.1}$$

Axiom II:
$$\sum_{A=1}^{A=N} N_{JA}\, R_A = 0, \quad J = 1, 2, \ldots, T \tag{7.2}$$

When Axiom II is applied to control volumes, we will make use of the species global net rates of production defined by

$$\mathcal{R}_A = \int_V R_A\, dV, \quad A = 1, 2, \ldots, N \tag{7.3}$$

and we will follow the development in Sec. 6.2.2 so that Eq. 7.2 takes the form

Axiom II:
$$\sum_{A=1}^{A=N} N_{JA}\, \mathcal{R}_A = 0, \quad J = 1, 2, \ldots, T \tag{7.4}$$

In terms of the global net rate of production, Axiom I takes the form

Axiom I:
$$\frac{d}{dt}\int_V c_A\, dV + \int_A c_A\, v_A \cdot \mathbf{n}\, dA = \mathcal{R}_A, \quad A = 1, 2, \ldots, N \tag{7.5}$$

These two results are applicable to any fixed control volume and we will use them throughout this chapter to determine molar flow rates, mass flow rates, mole fractions, etc. In addition to solving problems in terms of Eqs. 7.4 and 7.5, one can use those equations to derive *atomic species balances*. This is done in Appendix D where we illustrate how to solve problems in terms of the T atomic species rather than in terms of the N molecular species.

7.1　MULTIPLE REACTIONS: CONVERSION, SELECTIVITY AND YIELD

Most chemical reaction systems of industrial interest produce one or more primary or desirable products and one or more secondary or undesirable products. For example, benzene (C_6H_6) and propylene (C_3H_6) undergo reaction in the presence of a catalyst to form both the *desired product*, isopropyl benzene or cumene (C_9H_{12}) and the *undesired product*, p-diisopropyl benzene ($C_{12}H_{18}$). This situation is illustrated in Figure 7.2 in which the reactants, benzene and propylene, also appear in the exit stream because the reaction does not go to completion. In addition, the undesirable product, p-diisopropyl benzene appears in the exit stream. It is important to remember that this process should be considered in terms of the principle of *stoichiometric skepticism* described in Sec. 6.1.1.

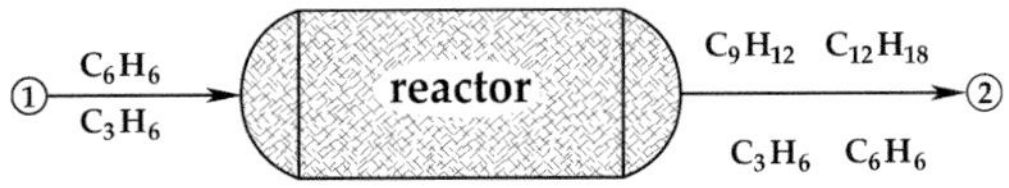

FIGURE 7.2   Production of cumene.

The analysis of this reactor is based on Axioms I and II. For a single entrance (Stream #1) and a single exit (Stream #2), Axiom I takes the form

Axiom I: $-(\dot{M}_A)_1 + (\dot{M}_A)_2 = \mathcal{R}_A, \quad A = 1, 2, 3, 4$ (7.6)

For a system containing two atomic species and four molecular species, the global form of Axiom II is given by

Axiom I: $$\sum_{A=1}^{A=4} N_{JA} \mathcal{R}_A = 0, \quad J = 1, 2$$ (7.7)

In both the experimental study of the reactor shown in Figure 7.2 and in the operation of that reactor, it is useful to have a number of defined quantities that characterize the performance. The first of these defined quantities is called the *conversion* which is given by

Conversion of reactant $A = \dfrac{\left(\begin{array}{c}\text{Total molar rate of}\\ \text{consumption of species } A\end{array}\right)}{\left(\begin{array}{c}\text{Molar flow rate of}\\ \text{species } A \text{ in the feed}\end{array}\right)}$ (7.8)

Since $\mathcal{R}_A$ represents the net molar *rate of production* of species A, we can express the conversion as

Conversion of reactant $A = \dfrac{-\mathcal{R}_A}{(\dot{M}_A)_1}$ (7.9)

An experimental determination of the conversion of reactant A requires the measurement of the molar flow rate of species A into and out of the reactor illustrated in Figure 7.2. If the reaction of benzene and propylene does not go to completion, one might obtain a result given by

Conversion of $C_6H_6 = \dfrac{-\mathcal{R}_{C_6H_6}}{(\dot{M}_{C_6H_6})_1} = \dfrac{(\dot{M}_{C_6H_6})_1 - (\dot{M}_{C_6H_6})_2}{(\dot{M}_{C_6H_6})_1} = 0.68$ (7.10)

This indicates that 68% of the incoming benzene is consumed in the reaction, but it does not indicate how much of this benzene reacts to form the *desired product*, cumene, and how much reacts to form the *undesired product*, p-diisopropyl benzene. In an efficient reactor, the conversion would be close to one and the amount of undesired product would be small.

A second defined quantity, the *selectivity*, indicates how one product (cumene for example) is favored over another product (p-diisopropyl benzene for example), and this quantity is defined by

$$\text{Selectivity of } \mathbf{D/U} = \frac{\text{Total molar rate of production of } \mathbf{Desired} \text{ product}}{\text{Total molar rate of production of } \mathbf{Undesired} \text{ product}} = \frac{\mathcal{R}_D}{\mathcal{R}_U}$$

$$(7.11)$$

For the system illustrated in Figure 7.2, the desired product is cumene C_9H_{12}, and the undesired product is p-diisopropyl benzene $C_{12}H_{18}$. If the selectivity for this pair of molecules is 0.85, we have

$$\text{Selectivity of } (C_9H_{12}/C_{12}H_{18}) = \frac{\mathcal{R}_{C_9H_{12}}}{\mathcal{R}_{C_{12}H_{18}}} = 0.85 \qquad (7.12)$$

and this suggests rather *poor performance* of the reactor since the rate of production of the undesirable product is greater than the rate of production of the desired product. In an efficient reactor, the selectivity would be *large compared to one*.

A third defined quantity is the *yield* of a reactor which is the ratio of the *rate of production* of a product to the *rate of consumption* of a reactant. We express the yield for a general case as

$$\text{Yield of } (A/B) = \frac{\text{Total rate of production of species } A}{\text{Total rate of consumption of species } B} = \frac{\mathcal{R}_A}{-\mathcal{R}_B} \qquad (7.13)$$

If we choose the product to be p-diisopropyl benzene and the reactant to be propylene, the yield takes the form

$$\text{Yield of } (C_{12}H_{18}/C_3H_6) = \frac{\text{Total rate of production of p-diisopropyl benzene}}{\text{Total rate of consumption of propylene}}$$

$$= \frac{\mathcal{R}_{C_{12}H_{18}}}{-\mathcal{R}_{C_3H_6}} \qquad (7.14)$$

If the yield of p-diisopropyl benzene is 0.15, we have the result given by

$$\text{Yield of } (C_{12}H_{18}/C_3H_6) = \frac{\mathcal{R}_{C_{12}H_{18}}}{-\mathcal{R}_{C_3H_6}}$$

$$= \frac{\mathcal{R}_{C_{12}H_{18}}}{(\dot{M}_{C_3H_6})_1 - (\dot{M}_{C_3H_6})_2} = 0.15 \qquad (7.15)$$

In an efficient reactor, the yield of an *undesirable* product would be small compared to one, while the yield of a *desirable product* would be close to one.

The conversion, selectivity, and yield of a reactor must be determined experimentally and these quantities will depend on the operating conditions of the reactor (i.e., temperature, pressure, type of catalyst, etc.). In addition, the value of these quantities will depend on their definitions, and within the chemical engineering literature one encounters a variety of definitions. To avoid errors, the definitions of *conversion*, *selectivity* and *yield* must be given in precise form and we have done this in the preceding paragraphs.

EXAMPLE 7.1   PRODUCTION OF ETHYLENE

Ethylene, C_2H_4, is one of the most useful molecules in the petrochemical industry (see Figure 1.7) since it is the building block for poly-ethylene, ethylene glycol, and many other chemical compounds used in the production of polymers. Ethylene can be produced by catalytic dehydrogenation of ethane, C_2H_6, as shown in Figure 7.1a. There we have indicated that the stream leaving the reactor contains the *desired product*, ethylene, C_2H_4, in addition to hydrogen, H_2, methane, CH_4, propylene, C_3H_6, and some un-reacted ethane, C_2H_6. As in every example of this type, the reader needs to consider the principle of *stoichiometric skepticism* discussed in Sec. 6.1.1 since small amounts of unidentified molecular species are always present in the output of a reactor.

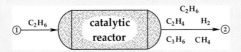

FIGURE 7.1a   Catalytic production of ethylene.

The experiment illustrated in Figure 7.1a has been performed in which the molar flow rate entering the reactor (Stream #1) is 100 mol/s of ethane. From measurements of the effluent stream (Stream #2), the following information regarding the *conversion* of ethane, the *selectivity* of ethylene relative to propylene, and the *yield* of ethylene is available:

$$C = \text{Conversion of } C_2H_6 = \frac{-\mathcal{R}_{C_2H_6}}{(\dot{M}_{C_2H_6})_1} = 0.2 \tag{1}$$

$$S = \text{Selectivity of } C_2H_4/C_3H_6 = \frac{\mathcal{R}_{C_2H_4}}{\mathcal{R}_{C_3H_6}} = 5 \tag{2}$$

$$Y = \text{Yield of } C_2H_4/C_2H_6 = \frac{\mathcal{R}_{C_2H_4}}{-\mathcal{R}_{C_2H_6}} = 0.75 \tag{3}$$

The global net rates of production of individual species are constrained by the stoichiometry of the system that can be expressed in terms of Axiom II. The atomic matrix for this system is given by

Molecular Species $\rightarrow H_2 \quad CH_4 \quad C_2H_6 \quad C_2H_4 \quad C_3H_6$

$$\begin{array}{c} carbon \\ hydrogen \end{array} \begin{bmatrix} 0 & 1 & 2 & 2 & 3 \\ 2 & 4 & 6 & 4 & 6 \end{bmatrix} \tag{4}$$

and the elements of this matrix are the entrees in $[N_{JA}]$ that can be expressed as

$$[N_{JA}] = \begin{bmatrix} 0 & 1 & 2 & 2 & 3 \\ 2 & 4 & 6 & 4 & 6 \end{bmatrix} \tag{5}$$

Use of this atomic matrix with Axiom II as given by Eq. 6.22 leads to

Axiom II: $\qquad \begin{bmatrix} 0 & 1 & 2 & 2 & 3 \\ 2 & 4 & 6 & 4 & 6 \end{bmatrix} \begin{bmatrix} \mathcal{R}_{H_2} \\ \mathcal{R}_{CH_4} \\ \mathcal{R}_{C_2H_6} \\ \mathcal{R}_{C_2H_4} \\ \mathcal{R}_{C_3H_6} \end{bmatrix} = \begin{bmatrix} 0 \\ 0 \end{bmatrix} \tag{6}$

At this point, the atomic matrix can be expressed in *row reduced echelon form* leading to the global pivot theorem given by Eq. 6.81. We express this result in the form

Global pivot theorem: $\qquad \begin{bmatrix} \mathcal{R}_{H_2} \\ \mathcal{R}_{CH_4} \end{bmatrix} = \begin{bmatrix} 1 & 2 & 3 \\ -2 & -2 & -3 \end{bmatrix} \begin{bmatrix} \mathcal{R}_{C_2H_6} \\ \mathcal{R}_{C_2H_4} \\ \mathcal{R}_{C_3H_6} \end{bmatrix} \tag{7}$

and we carry out the matrix multiplication to obtain

$$\mathcal{R}_{H_2} = \mathcal{R}_{C_2H_6} + 2\,\mathcal{R}_{C_2H_4} + 3\,\mathcal{R}_{C_3H_6} \tag{8}$$

$$\mathcal{R}_{CH_4} = -2\,\mathcal{R}_{C_2H_6} - 2\,\mathcal{R}_{C_2H_4} - 3\,\mathcal{R}_{C_3H_6} \tag{9}$$

The five net rates of production that appear in Eqs. 6–9 are represented in the degree of freedom analysis given in Table 7.1a. There we see that there are zero degrees of freedom and we have a solvable problem.

At this point, we direct our attention to Eq. 8 and make use of the information provided in Eqs. 1, 2, and 3 in order to express the net molar rate of production of hydrogen as

$$\mathcal{R}_{H_2} = \left[-1 + 2Y + 3\frac{Y}{S} \right] C \, (\dot{M}_{C_2H_6})_1 \tag{10}$$

Moving on to Eq. 9, we again utilize Eqs. 1, 2, and 3 in order to obtain

$$\mathcal{R}_{CH_4} = \left[2 - 2Y - \frac{3Y}{S} \right] C \, (\dot{M}_{C_2H_6})_1 \tag{11}$$

TABLE 7.1a

Degrees-of-freedom for Production of Ethylene from Ethane

Stream & Net Rate of Production Variables

compositions (5 species & 2 streams)	$N \times M = 5 \times 2 = 10$
flow rates (2 streams)	$M = 2$
net rates of production (5 species)	$N = 5$
Generic Degrees of Freedom (**A**)	17
Number of Independent Balance Equations	
mass/mole balance equations (5 species)	$N = 5$
Number of Constraints for Compositions (2 streams)	$M = 2$
Number of Constraints for Reactions (2 atomic species)	$T = 2$
Generic Specifications and Constraints (**B**)	9

Specified Stream Variables		
compositions	4	Stream #1
flow rates	1	Stream #1
Constraints for Compositions		0
Auxiliary Constraints	3	Eqs. 1, 2, and 3
Particular Specifications and Constraints (**C**)		8
Degrees of Freedom (**A** − **B** − **C**)		0

Given the experimental values of the conversion, yield and selectivity indicated by Eqs. 1–3, along with the molar flow rate of Stream #1,

$$(\dot{M}_{C_2H_6})_1 = 100 \text{ mol/s} \tag{12}$$

we can solve Eqs. 10 and 11 to determine the global rates of production for methane and hydrogen.

$$\mathcal{R}_{H_2} = 19.0 \text{ mol/s}, \quad \mathcal{R}_{CH_4} = 1.0 \text{ mol/s} \tag{13}$$

Given these global rates of production for CH_4 and H_2, the values for the global rates of production of the other species can be easily calculated and this is left as an exercise for the student.

7.2 COMBUSTION REACTIONS

Computation of the rate of production or consumption of chemical species during combustion is an important part of chemical engineering practice. Efficient use of irreplaceable fossil energy resources is ecologically responsible and economically sound. Fuels are burned in combustion chambers using air as the source of oxygen, O_2, as illustrated in Figure 7.3. The fuel enters the combustion chamber at Stream #1 and air is supplied via Stream #2. Since air is 79% nitrogen, N_2, and nitrogen can form NOX as part of the combustion reaction, it is good practice to use only the amount of air that is needed for the reaction. On the other hand, the need to burn the fuels completely requires excess air to displace the reaction equilibrium. Matters are complicated further by the presence of water in the air stream. In this simplified treatment of combustion, we will ignore the complexities associated with NOX in the exit stream and water vapor in the entrance stream.

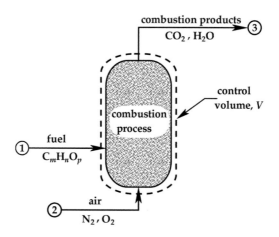

FIGURE 7.3   Combustion process.

7.2.1 Theoretical Air

The first step in analyzing a combustion process is to determine the rate at which air must be supplied in order to achieve complete combustion. This is called the *theoretical air* requirement for the combustion process and it is based on the fuel composition. Most fuels consist of many different hydrocarbons; however, in this simple example we consider the fuel to consist of a single molecule represented by $C_mH_nO_p$. Examples are propane ($(m=3, n=8, p=0)$), carbon monoxide ($(m=1, n=0, p=1)$), and methanol ($(m=1, n=4, p=1)$). Total combustion is the conversion of all the fuel to CO_2 and H_2O. Our tools for this analysis are Axioms I and II as represented by Eqs. 7.4 and 7.5 with the global rate of production defined by Eq. 7.3. We first express the atomic matrix as

$$\text{Molecular species} \rightarrow C_mH_nO_p \quad O_2 \quad CO_2 \quad H_2O$$

$$\begin{matrix} carbon \\ hydrogen \\ oxygen \end{matrix} \begin{bmatrix} m & 0 & 1 & 0 \\ n & 0 & 0 & 2 \\ p & 2 & 2 & 1 \end{bmatrix} \qquad (7.16)$$

and make use of this representation to express Axiom II in the form

Axiom II:
$$\sum_{A=1}^{A=N} N_{JA}\, \mathcal{R}_A = \begin{bmatrix} m & 0 & 1 & 0 \\ n & 0 & 0 & 2 \\ p & 2 & 2 & 1 \end{bmatrix} \begin{bmatrix} \mathcal{R}_{C_m H_n O_p} \\ \mathcal{R}_{O_2} \\ \mathcal{R}_{CO_2} \\ \mathcal{R}_{H_2O} \end{bmatrix} = \begin{bmatrix} 0 \\ 0 \\ 0 \end{bmatrix} \tag{7.17}$$

The atomic matrix can be expressed in *row reduced echelon form* according to

$$[N_{JA}]^* = \begin{bmatrix} 1 & 0 & 0 & \dfrac{2}{n} \\ 0 & 1 & 0 & \dfrac{4m+n-2p}{2n} \\ 0 & 0 & 1 & -\dfrac{2m}{n} \end{bmatrix} \tag{7.18}$$

This indicates that the rank is three, and all the species production rates can be expressed in terms of a single *pivot species* that we choose to be water. Use of Eq. 7.18 in Eq. 7.17 allows us to express Axiom II in the form

$$\begin{bmatrix} 1 & 0 & 0 & \dfrac{2}{n} \\ 0 & 1 & 0 & \dfrac{4m+n-2p}{2n} \\ 0 & 0 & 1 & -\dfrac{2m}{n} \end{bmatrix} \begin{bmatrix} \mathcal{R}_{C_m H_n O_p} \\ \mathcal{R}_{O_2} \\ \mathcal{R}_{CO_2} \\ \mathcal{R}_{H_2O} \end{bmatrix} = \begin{bmatrix} 0 \\ 0 \\ 0 \end{bmatrix} \tag{7.19}$$

This equation can be partitioned according to the development in Example 6.5 or the development in Sec. 6.4 in order to obtain

$$\begin{bmatrix} 1 & 0 & 0 \\ 0 & 1 & 0 \\ 0 & 0 & 1 \end{bmatrix} \begin{bmatrix} \mathcal{R}_{C_m H_n O_p} \\ \mathcal{R}_{O_2} \\ \mathcal{R}_{CO_2} \end{bmatrix} + \begin{bmatrix} \dfrac{2}{n} \\ \dfrac{4m+n-2p}{2n} \\ -\dfrac{2m}{n} \end{bmatrix} [\mathcal{R}_{H_2O}] = \begin{bmatrix} 0 \\ 0 \\ 0 \end{bmatrix} \tag{7.20}$$

This immediately leads to a special case of the global pivot theorem (see Eq. 6.81)

Global pivot theorem:
$$\begin{bmatrix} \mathcal{R}_{C_m H_n O_p} \\ \mathcal{R}_{O_2} \\ \mathcal{R}_{CO_2} \end{bmatrix} = -\begin{bmatrix} -\dfrac{2}{n} \\ \dfrac{4m+n-2p}{2n} \\ \dfrac{2m}{n} \end{bmatrix} [\mathcal{R}_{H_2O}] \tag{7.21}$$

which provides the species global net rates of production in terms of the global net rate of production of water. These are given by

$$\mathcal{R}_{C_mH_nO_p} = -\frac{2}{n} R_{H_2O}, \quad \mathcal{R}_{O_2} = -\frac{4m+n-2p}{2n} \mathcal{R}_{H_2O}, \quad \mathcal{R}_{CO_2} = \frac{2m}{n} \mathcal{R}_{H_2O} \qquad (7.22)$$

In this particular problem, it is convenient to represent the global net rate of production of fuel (which is negative) in terms of the global net rate of production of oxygen (which is negative), thus we use the first two of Eq. 7.22 to obtain

$$\mathcal{R}_{O_2} = \frac{4m+n-2p}{4} \mathcal{R}_{C_mH_nO_p} \qquad (7.23)$$

At this point, we have completed our analysis of Axiom II and we are ready to apply Axiom I using the control volume illustrated in Figure 7.3. Application of the steady-state form of Eq. 7.5 for both the fuel, $C_mH_nO_p$, and the oxygen, O_2, leads to

Fuel: $\qquad\qquad -(\dot{M}_{C_mH_nO_p})_1 + (\dot{M}_{C_mH_nO_p})_3 = \mathcal{R}_{C_mH_nO_p} \qquad (7.24)$

Oxygen: $\qquad\qquad -(\dot{M}_{O_2})_2 + (\dot{M}_{O_2})_3 = \mathcal{R}_{O_2} \qquad (7.25)$

In order to determine the *theoretical air* needed for *complete combustion*, we assume that all the fuel and all the oxygen that enter the combustion chamber are reacted, thus there is no fuel in Stream #3 and there is no oxygen in Stream #3. Under these circumstances Eqs. 7.24 and 7.25 reduce to

Fuel (*complete combustion*): $\quad -(\dot{M}_{C_mH_nO_p})_1 = \mathcal{R}_{C_mH_nO_p} \qquad (7.26)$

Oxygen (*complete combustion*): $\quad -(\dot{M}_{O_2})_2 = \mathcal{R}_{O_2} \qquad (7.27)$

At this point, we represent $\mathcal{R}_{O_2}$ in terms of $\mathcal{R}_{C_mH_nO_p}$ using Eq. 7.23, and this allows us to express the theoretical oxygen required for complete combustion as

$$(\dot{M}_{O_2})_{Theoretical} = (\dot{M}_{O_2})_2 = \frac{4m+n-2p}{4} (\dot{M}_{C_mH_nO_p})_1 \qquad (7.28)$$

If the fuel in Stream #1 contains N molecular species that can be represented as $C_mH_nO_p$, the theoretical oxygen required for complete combustion is given by

$$(\dot{M}_{O_2})_{Theoretical} = \sum_{i=1}^{i=N} \left[\frac{4m+n-2p}{4} \right]_i \left[(\dot{M}_{C_mH_nO_p})_1 \right]_i \qquad (7.29)$$

This analysis for the theoretical oxygen can be extended to other fuels that may contain sulfur provided that the molecules in the fuel can be characterized by $C_mH_nO_pS_q$. In this case, the *complete combustion* products would be CO_2, H_2O, and SO_2.

EXAMPLE 7.2   DETERMINATION OF THEORETICAL AIR

A fuel containing 60% CH_4, 15% C_2H_6, 5% CO, and 20% N_2 (all mole percent) is burned with air to produce a flue gas containing CO_2, H_2O, and N_2. The combustion process takes place in the unit illustrated in Figure 7.3, and we want to determine the molar flow rate of theoretical air needed for complete combustion. The solution to this problem is given by Eq. 7.29 that takes the form

$$(\dot{M}_{O_2})_{Theoretical} = \sum_{i=1}^{i=3} \left[\frac{4m+n-2p}{4}\right]\left[(\dot{M}_{C_mH_nO_p})_1\right]_i$$

$$= \left(\frac{4+4}{4}\right)(\dot{M}_{CH_4})_1 + \left(\frac{8+6}{4}\right)(\dot{M}_{C_2H_6})_1 + \left(\frac{4-2}{4}\right)(\dot{M}_{CO})_1$$

$$= 2(\dot{M}_{CH_4})_1 + \frac{7}{2}(\dot{M}_{C_2H_6})_1 + \frac{1}{2}(\dot{M}_{CO})_1 \qquad (1)$$

If we let $\dot{M}_1$ be the total molar flow rate of Stream #1, we can express Eq. 1 as

$$(\dot{M}_{O_2})_{Theoretical} = \left[2(0.60) + \frac{7}{2}(0.15) + \frac{1}{2}(0.05)\right]\dot{M}_1$$

$$= 1.75\,\dot{M}_1 \qquad (2)$$

This indicates that for every mole of feed we require 1.75 moles of oxygen for complete combustion. Taking the mole fraction of oxygen in air to be 0.21, we find the molar flow rate of air to be given by

$$(\dot{M}_{air})_{Theoretical} = \left(\frac{1}{0.21}\right)1.75\,\dot{M} = 8.33\,\dot{M}_1 \qquad (3)$$

The determination of the theoretical molar flow rate of air in Example 7.2 was relatively simple because the complete combustion process involved a feed stream that could be described in terms of a single molecular form given by $C_mH_nO_p$. In the next example, we consider a slightly more complex process in which the *excess air* is specified.

EXAMPLE 7.3   COMBUSTION OF RESIDUAL SYNTHESIS GAS

In Figure 7.3a, we have illustrated a synthesis gas (Stream #1) consisting of 0.4% methane, CH_4, 52.8% hydrogen, H_2, 38.3% carbon monoxide, CO, 5.5% carbon dioxide, CO_2, 0.1% oxygen, O_2, and 2.9% nitrogen, N_2. The synthesis gas reacts with air (Stream #2) that is supplied at a rate which provides 10% excess oxygen, and the composition of Stream #2 is assumed to be 79% nitrogen, N_2, and 21% oxygen, O_2.

In this example, we want to determine: (I) The *molar flow rate of air* relative to the molar flow rate of the synthesis gas, and (II) The *species molar flow rates* of the components of the flue gas relative to the molar flow rate of the synthesis gas. We assume complete combustion so that all hydrocarbon species in Stream #1 are converted to carbon dioxide, CO_2, and water, H_2O. In addition, we assume that the nitrogen is inert so that no NOX appears in the exit stream. This means that the molecular species in the flue gas are N_2, O_2, CO_2, and H_2O.

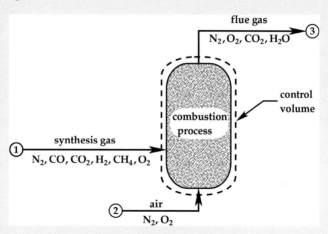

FIGURE 7.3a   Combustion of synthesis gas.

In this example, the development leading to Eq. 7.29 is applicable for the combustion of synthesis gas and that result takes the form

$$(\dot{M}_{O_2})_{Theoretical} = \sum_{i=1}^{i=3}\left[\frac{4m+n-2p}{4}\right]\left[(\dot{M}_{C_mH_nO_p})_1\right]_i$$

$$= \frac{4+4}{4}(\dot{M}_{CH_4})_1 + \frac{2}{4}(\dot{M}_{H_2})_1 + \frac{4-2}{4}(\dot{M}_{CO})_1 \qquad (1)$$

The species molar flow rates in Stream #1 can be expressed in terms of the total molar flow rate, $\dot{M}_1$ according to

$$(\dot{M}_{CH_4})_1 = 0.004\dot{M}_1, \quad (\dot{M}_{H_2})_1 = 0.528\dot{M}_1, \quad (\dot{M}_{CO})_1 = 0.383\dot{M}_1 \qquad (2)$$

and these results can be used in Eq. 1 to obtain the molar flow rate of theoretical air in terms of the molar flow rate of synthesis gas.

$$(\dot{M}_{O_2})_{Theoretical} = \left[\frac{4+4}{4}(0.004) + \frac{2}{4}(0.528) + \frac{4-2}{4}(0.383) \right] \dot{M}_1$$

$$= 0.4635 \dot{M}_1 \tag{3}$$

Taking into account that Stream #1 contains oxygen, the condition of 10% excess oxygen requires

$$(\dot{M}_{O_2})_1 + (\dot{M}_{O_2})_2 = (1.10)(\dot{M}_{O_2})_{Theoretical} \tag{4}$$

and this allows us to express the molar flow rate of oxygen in Stream #2 as

$$(\dot{M}_{O_2})_2 = (1.10)(\dot{M}_{O_2})_{Theoretical} - (y_{O_2})_1 \dot{M}_1 \tag{5}$$

Given that the mole fraction of oxygen, O_2, in the synthesis gas (Stream #1) is 0.001, we can use this result along with Eq. 3 to obtain

$$(\dot{M}_{O_2})_2 = 0.509 \, \dot{M}_1 \tag{6}$$

We are given that the mole fraction of oxygen, O_2, in the air (Stream #2) is 0.21 and this leads to the total molar flow rate for air given by

$$\dot{M}_2 = 2.42 \, \dot{M}_1 \tag{7}$$

Knowing the molar flow rate of the air stream relative to the molar flow rate of the synthesis gas that is required for complete combustion is crucial for the proper operation of the system. Knowing how much air is required to achieve complete combustion will generally be determined experimentally, thus the ratio 2.42 will be adjusted to achieve the desired operating condition.

In addition to knowing the required flow rate of air, we also need to know the molar flow rates of the species in the flue gas since this gas will often be discharged into the atmosphere. Determination of these molar flow rates requires the application of Axioms I and II. We begin with the atomic matrix

Molecular Species $\rightarrow$ CH_4 H_2 CO CO_2 H_2O O_2

$$\begin{array}{c} carbon \\ hydrogen \\ oxygen \end{array} \begin{bmatrix} 1 & 0 & 1 & 1 & 0 & 0 \\ 4 & 2 & 0 & 0 & 2 & 0 \\ 0 & 0 & 1 & 2 & 1 & 2 \end{bmatrix} \tag{8}$$

which leads to the following form of Axiom II:

$$\text{Axiom II:} \sum_{A=1}^{A=6} N_{JA} R_A = \begin{bmatrix} 1 & 0 & 1 & 1 & 0 & 0 \\ 4 & 2 & 0 & 0 & 2 & 0 \\ 0 & 0 & 1 & 2 & 1 & 2 \end{bmatrix} \begin{bmatrix} \mathcal{R}_{CH_4} \\ \mathcal{R}_{H_2} \\ \mathcal{R}_{CO} \\ \mathcal{R}_{CO_2} \\ \mathcal{R}_{H_2O} \\ \mathcal{R}_{O_2} \end{bmatrix} = \begin{bmatrix} 0 \\ 0 \\ 0 \end{bmatrix} \quad (9)$$

This axiom can be expressed in *row reduced echelon form* according to

$$\begin{bmatrix} 1 & 0 & 0 & -1 & -1 & -2 \\ 0 & 1 & 0 & 2 & 3 & 4 \\ 0 & 0 & 1 & 2 & 1 & 2 \end{bmatrix} \begin{bmatrix} \mathcal{R}_{CH_4} \\ \mathcal{R}_{H_2} \\ \mathcal{R}_{CO} \\ \mathcal{R}_{CO_2} \\ \mathcal{R}_{H_2O} \\ \mathcal{R}_{O_2} \end{bmatrix} = \begin{bmatrix} 0 \\ 0 \\ 0 \end{bmatrix} \quad (10)$$

The development in Sec. 6.4 indicates that the solution for the three *non-pivot* global net rates of production is given by

$$\begin{aligned} \mathcal{R}_{CH_4} &= \mathcal{R}_{CO_2} + \mathcal{R}_{H_2O} + 2\,\mathcal{R}_{O_2} \\ \mathcal{R}_{H_2} &= -2\,\mathcal{R}_{CO_2} - 3\,\mathcal{R}_{H_2O} - 4\,\mathcal{R}_{O_2} \\ \mathcal{R}_{CO} &= -2\,\mathcal{R}_{CO_2} - \mathcal{R}_{H_2O} - 2\,\mathcal{R}_{O_2} \end{aligned} \quad (11)$$

This completes our analysis of Axiom II, and we can move on to the steady-state form of Axiom I that is given by

$$\text{Axiom I:} \qquad \int_A c_A \mathbf{v}_A \cdot \mathbf{n}\, dA = \mathcal{R}_A, \quad A = 1, 2, \ldots, N \quad (12)$$

Application of this result to the control volume illustrated in Figure 7.3a leads to the general expression

$$-(\dot{M}_A)_1 - (\dot{M}_A)_2 + (\dot{M}_A)_3 = \mathcal{R}_A, \quad A = 1, 2, \ldots, 7 \quad (13)$$

while the individual species balances take the forms given by

CH_4:
$$-(\dot{M}_{CH_4})_1 = -0.004\,\dot{M}_1 = \mathcal{R}_{CH_4} \quad (14)$$

H_2: $-(\dot{M}_{H_2})_1 = -0.528\dot{M}_1 = \mathcal{R}_{H_2}$ (15)

CO: $-(\dot{M}_{CO})_1 = -0.383\dot{M}_1 = \mathcal{R}_{CO}$ (16)

CO_2: $-(\dot{M}_{CO_2})_1 + (\dot{M}_{CO_2})_3 = \mathcal{R}_{CO_2}$ (17)

N_2: $-(\dot{M}_{N_2})_1 - (\dot{M}_{N_2})_2 + (\dot{M}_{N_2})_3 = \mathcal{R}_{N_2}$ (18)

O_2: $-(\dot{M}_{O_2})_1 - (\dot{M}_{O_2})_2 + (\dot{M}_{O_2})_3 = \mathcal{R}_{O_2}$ (19)

H_2O: $(\dot{M}_{H_2O})_3 = \mathcal{R}_{H_2O}$ (20)

At this point, we need to simplify the balance equations for carbon dioxide, nitrogen, and oxygen. Beginning with carbon dioxide, we obtain

CO_2: $(\dot{M}_{CO_2})_3 = \mathcal{R}_{CO_2} + 0.055\dot{M}_1$ (21)

Moving on to the nitrogen balance given by Eq. 18, we make use of the conditions

N_2: $(\dot{M}_{N_2})_1 = 0.029\,\dot{M}_1, \quad (\dot{M}_{N_2})_2 = 0.79\,\dot{M}_2, \quad \mathcal{R}_{N_2} = 0$ (22)

to express the molar flow rate in the flue gas as

N_2: $(\dot{M}_{N_2})_3 = (\dot{M}_{N_2})_1 + (\dot{M}_{N_2})_2 = 0.029\,\dot{M}_1 + 0.79\,\dot{M}_2$ (23)

Finally, the use of Eq. 7 leads to the following result for the molar flow rate of nitrogen:

N_2: $(\dot{M}_{N_2})_3 = 1.94\,\dot{M}_1$ (24)

Turning our attention to the oxygen balance, we note that the mole fractions in Streams #1 and #2 are specified, thus we can represent Eq. 19 in the form

O_2: $(\dot{M}_{O_2})_3 = \mathcal{R}_{O_2} + 0.001\,\dot{M}_1 + 0.21\,\dot{M}_2$ (25)

Use of Eq. 7 provides the simplified version of the oxygen balance given by

O_2: $(\dot{M}_{O_2})_3 = \mathcal{R}_{O_2} + 0.509\dot{M}_1$ (26)

In order to determine the molar flow rates of oxygen, carbon dioxide, and water in the flue gas, we need to determine the global net rates of production of O_2, CO_2, and H_2O. From Axiom II, in the form given by Eq. 11, and with the use of Eqs. 14, 15, and 16, we can express the global net rates of production of the three pivot species as

$$-0.004\dot{M}_1 = \mathcal{R}_{CO_2} + \mathcal{R}_{H_2O} + 2\mathcal{R}_{O_2}$$
$$-0.528\dot{M}_1 = -2\,\mathcal{R}_{CO_2} - 3\,\mathcal{R}_{H_2O} - 4\,\mathcal{R}_{O_2} \qquad (27)$$
$$-0.383\dot{M}_1 = -2\,\mathcal{R}_{CO_2} - \mathcal{R}_{H_2O} - 2\,\mathcal{R}_{O_2}$$

The solution of these three equations is given by

$$\mathcal{R}_{CO_2} = 0.387\dot{M}_1, \quad \mathcal{R}_{H_2O} = 0.536\dot{M}_1, \quad \mathcal{R}_{O_2} = -0.463\dot{M}_1 \qquad (28)$$

and these results can be used with Eqs. 20, 22, and 26 to provide the species molar flow rates for the flue gas, Stream #3.

$$(\dot{M}_{N_2})_3 = 1.94\,\dot{M}_1, \qquad (\dot{M}_{O_2})_3 = 0.046\dot{M}_1$$
$$(\dot{M}_{CO_2})_3 = 0.442\dot{M}_1, \qquad (\dot{M}_{H_2O})_3 = 0.536\dot{M}_1 \qquad (29)$$

One should keep in mind that the solution for the species molar flow rates in the flue gas should be based on *three steps*. Application of Axiom I and Axiom II represent two of those steps while the third step, a degree-of-freedom analysis, was omitted.

7.3   RECYCLE SYSTEMS

In the previous section, we studied systems in which there were incomplete chemical reactions, such as combustion reactions, that produced both carbon dioxide and carbon monoxide. We must always consider the consequences of a desirable, but incomplete, chemical reaction. Do we simply discard the unused reactants? And if we do, where do we discard them? Can we afford to release carbon monoxide to the atmosphere without recovering the energy available in the oxidation of CO to produce CO_2? What is the impact on the environment of carbon monoxide? Even if we can achieve complete combustion of the carbon monoxide, CO, what do we do with the carbon dioxide, CO_2?

Incomplete chemical reactions demand the use of *recycle streams* because we cannot afford to release the unused reactants into our ecological system, both for environmental and economic reasons. A typical reaction might involve combining two species, *A* and *B*, to form a desirable product *C*. The product stream from the reactor will contain all three molecular species and we must separate the product C from this mixture and recycle the reactants as indicted in Figure 7.4. In a problem of this type, one would want to know the composition of the *product stream* and the magnitude of the *recycle stream*. The molar flow rate of the recycle stream depends on the degree of completion of the reaction in the catalytic reactor and the degree of separation achieved by the separator. The actual *design* of these units is the subject of subsequent courses on reactor design and mass transfer, and we will introduce students to a part of the reactor design process in Chapter 9. For the present, we will concern ourselves only with the *analysis* of systems for which the operating characteristics are known or can be determined from the information that is given.

The analysis of systems containing recycle streams is more complex than the systems we have studied previously, because recycle streams create *loops* in the flow of information. In a typical recycle configuration, a stream generated far downstream in the process is brought back to the front end of the process and mixed with an incoming feed. This is indicated in Figure 7.4 where we have illustrated a unit called a *mixer* that combines the feed stream with the recycle stream.

In addition to *mixers*, systems with recycle streams often contain *splitters* that produce recycle streams and purge streams. In Figure 7.5, we have illustrated a unit called an "ammonia converter" in which the feed consists of a mixture of nitrogen and hydrogen containing a small amount of argon. The ammonia produced in the reactor is removed as a liquid in a condenser, and the unconverted gas is recycled to the reactor through a *mixer*. In order to avoid the buildup of argon in the system, a purge stream is required and this gives rise to a *splitter* as illustrated in Figure 7.5. If the argon is not removed from the system by the splitter, the concentration of argon will *increase* and the efficiency of the reactor will *decrease*. The characteristics of splitters are discussed in detail in Sec. 7.3.1.

There are several methods for analyzing recycle systems. One approach, which we discuss first, is an extension of the method used previously. A set of control volumes is constructed and Axioms I and II are applied to those control volumes. Because more than one control volume is required, the most appropriate choice of control volumes is not always obvious. In previous chapters, we have discussed a set of rules or guidelines to be used in the construction of control volumes. Here we repeat those rules with the addition of one new rule that is important for the analysis of systems with recycle streams.

Rule I. Construct a *primary cut* where information is *required*.
Rule II. Construct a *primary cut* where information is *given*.
Rule III. Join these cuts with a surface located where $\mathbf{v}_A \cdot \mathbf{n}$ is *known*.

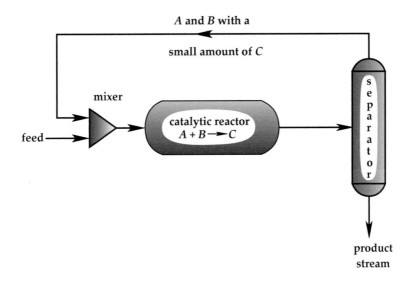

FIGURE 7.4   Catalytic reactor with recycle.

Rule IV. When joining the primary cuts to form control volumes, minimize the number of new or *secondary cuts* since these introduce information that is neither given nor required.

Rule V. Be sure that the surface specified by Rule III encloses regions in which volumetric information is either *given* or *required*.

Rule VI. When joining the primary cuts to form control volumes, minimize the number of *redundant cuts* since they provide no new information.

If one follows these rules carefully and pays careful attention to the associated degree-of-freedom analysis, the solution of many problems is either easy or impossible.

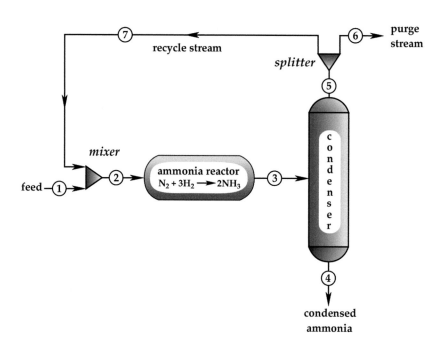

FIGURE 7.5   Recycle stream in an ammonia converter.

7.3.1 MIXERS AND SPLITTERS

Mixers and splitters are a natural part of recycle systems, and one must pay careful attention to their properties. In Figure 7.6, we have illustrated a mixer in which S streams are joined to form a single stream, and we have illustrated a splitter in which a single stream is split into S streams. Both accumulation and chemical reaction can be neglected in mixers and splitters since these devices consist only of tubes joined in some convenient manner.

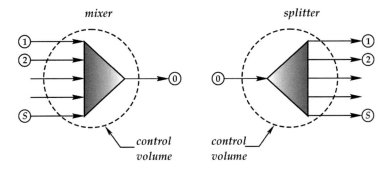

FIGURE 7.6   Mixer and splitter.

This means that they can be analyzed in terms of the steady form of Eq. 7.1 that simplifies to

$$\int_A c_A \mathbf{v}_A \cdot \mathbf{n}\, dA = 0, \quad A = 1, 2, ..., N \tag{7.30}$$

For the mixer shown in Figure 7.6, this result can be expressed as

Mixer (*species balances*): $-\displaystyle\sum_{i=1}^{i=S} (x_A)_i \dot{M}_i + (x_A)_o \dot{M}_o = 0, \quad A = 1, 2, ..., N$ (7.31)

This result also applies to the splitter shown in Figure 7.6, and for that case the macroscopic balance takes the form

Splitter (*species balances*): $-(x_A)_o \dot{M}_o + \displaystyle\sum_{i=1}^{i=S} (x_A)_i \dot{M}_i = 0, \quad A = 1, 2, ..., N$ (7.32)

The *physics* of a splitter require that the compositions in all the outgoing streams be equal to those in the incoming stream, and we express this idea as

Splitter (*physics*): $(x_A)_i = (x_A)_o, \quad i = 1, 2, ..., S, \quad A = 1, 2, ..., N$ (7.33)

In addition to understanding the physics of a splitter, we must understand how this constraint on the mole fractions is influenced by the constraints that we apply in terms of our degree of freedom analysis. All streams *cut by a control surface* are required to satisfy the constraint on the mole fractions given by

$$(x_A)_i + (x_B)_i + (x_C)_i + \cdots + (x_N)_i = 1, \quad i = 0, 1, 2, ..., S \tag{7.34}$$

This means that only $N-1$ mole fractions can be specified in the outgoing streams of a splitter. To illustrate how these constraints influence our description of a splitter, we consider Stream #2 of the splitter illustrated in Figure 7.6. For that stream, Eq. 7.33 provides the following $N-1$ equations:

$$(x_A)_2 = (x_A)_o \tag{7.35a}$$

$$(x_B)_2 = (x_B)_o \tag{7.35b}$$

$$(x_C)_2 = (x_C)_o \tag{7.35c}$$

$$\cdots \cdots \cdots$$

$$(x_{N-1})_2 = (x_{N-1})_o \tag{7.35n}$$

When we impose Eq. 7.34 for *both* Stream #2 *and* Stream #0, we obtain

$$1 - (x_N)_2 = 1 - (x_N)_o \tag{7.36}$$

This leads to the result that the two mole fractions of the Nth component must be equal.

$$(x_N)_2 = (x_N)_o \tag{7.37}$$

This indicates that we should impose Eq. 7.33 on only $N-1$ of the components so that our degree of freedom representation of the splitter takes the form

Splitter (*degree of freedom/mole fractions*):

$$(x_A)_i = (x_A)_o, \quad i = 1, 2, \ldots, S, \quad A = 1, 2, \ldots, N-1 \tag{7.38}$$

For many situations, a splitter may be *enclosed in a control volume*, as illustrated in Figure 7.6. In those situations, we will impose the macroscopic species balance given by Eq. 7.32, and we must again be careful to understand how this affects our degree of freedom analysis. In particular, we would like to prove that the splitter condition indicated by Eq. 7.38 need only be applied to $S-1$ streams when we make use of Eq. 7.32. To see that this is true, we first note that Eq. 7.32 can be summed over all N species to obtain the total molar balance given by

$$-\dot{M}_o + \left(\dot{M}_1 + \dot{M}_2 + \dot{M}_3 + \cdots + \dot{M}_S \right) = 0 \tag{7.39}$$

Next, we rearrange Eq. 7.32 in the form

$$-(x_A)_o \, \dot{M}_o + \sum_{i=1}^{i=S-1} (x_A)_i \, \dot{M}_i + (x_A)_S \, \dot{M}_S = 0, \quad A = 1, 2, \ldots, N \tag{7.40}$$

and then apply the constraints indicated by Eq. 7.38 for only $S-1$ of the streams leaving the splitter in order to obtain

$$-(x_A)_o \dot{M}_o + (x_A)_o \sum_{i=1}^{i=S-1} \dot{M}_i + (x_A)_S \dot{M}_S = 0, \quad A = 1, 2, \ldots, N \tag{7.41}$$

This result can be used with the total molar balance given by Eq. 7.39 to arrive at the condition

$$-(x_A)_o \, \dot{M}_S + (x_A)_S \, \dot{M}_S = 0, \quad A = 1, 2, \ldots, N \tag{7.42}$$

which obviously leads to

$$(x_A)_S = (x_A)_o, \quad A = 1, 2, ..., N \tag{7.43}$$

Here, we see that we have *derived*, independently, one of the conditions implied by Eq. 7.38 and this means that the splitter constraints indicated by Eq. 7.38 can be expressed as follows:

Splitter (*degree of freedom/mole fractions/species balances*):

$$(x_A)_i = (x_A)_o, \quad i = 1, 2, ..., S-1, \ A = 1, 2, ..., N-1 \tag{7.44}$$

Often, it is more convenient to work with species molar flow rates that are related to mole fractions by

$$x_A = \dot{M}_A / \dot{M} = \dot{M}_A \Big/ \sum_{B=1}^{B=N} \dot{M}_B \tag{7.45}$$

Use of this result in Eq. 7.44 provides the alternative representation for a splitter given by

$$(\dot{M}_A)_i \Big/ \sum_{B=1}^{B=N} (\dot{M}_B)_i = (\dot{M}_A)_o \Big/ \sum_{B=1}^{B=N} (\dot{M}_B)_o, \tag{7.46}$$
$$i = 1, 2, ..., S-1, \quad A = 1, 2, ..., N-1$$

To summarize, we note that the *physical conditions* associated with a splitter are given by Eq. 7.33. When we take into account the constraint on mole fractions given by Eq. 7.34, we must simplify Eq. 7.33 to the form given by Eq. 7.38 in order to be consistent with our degree of freedom analysis. When we further take into account the species mole balances that may be imposed on a control volume around the splitter, we must simplify Eq. 7.38 to the form given by Eq. 7.44 in order to be consistent with our degree of freedom analysis.

EXAMPLE 7.4   SPLITTER CALCULATION

In this example, we consider the splitter illustrated in Figure 7.4a in which Stream #1 is split into Streams #2, #3, and #4, each of which contain the three species entering the splitter. We will use $(\dot{M}_A)_j$, $(\dot{M}_B)_j$, and $(\dot{M}_C)_j$ to represent the molar flow rates of species A, B, and C in the jth stream and we will use $\dot{M}_1$ to represent the total molar flow rate entering Stream #1. The degree-of-freedom analysis given in Table 7.4a indicates that we have five degrees of freedom, and there are several ways in which the splitter problem can be solved. In this particular example, we consider the case in which five molar flow rates are specified according to:

$$(\dot{M}_A)_1 = 10 \text{ mol/h} \tag{1a}$$

$$(\dot{M}_B)_1 = 25 \text{ mol/h} \tag{1b}$$

$$(\dot{M}_C)_1 = 65 \text{ mol/h} \tag{1c}$$

$$(\dot{M}_B)_3 = 4 \text{ mol/h} \tag{1d}$$

$$(\dot{M}_B)_4 = 2 \text{ mol/h} \tag{1e}$$

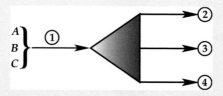

FIGURE 7.4a   Splitter producing three streams.

TABLE 7.4a
Degrees-of-freedom: Four Streams and Three Components

Stream Variables	
compositions (3 species)	$N = 3$
flow rates (4 streams)	$M = 4$
Generic Degrees of Freedom (A)	12
Number of Independent Balance Equations	
balance equations (3 species)	$N = 3$
Number of Constraints for Compositions (4 streams)	$M = 4$
Number of Constraints for Reactions (zero)	0
Generic Specifications and Constraints (B)	7
Degrees of Freedom (A − B)	5

The development given in Sec. 4.6 can be used to construct the general relations given by

$$(\dot{M}_A)_i = (x_A)_i \, \dot{M}_i, \quad A = 1,2,3, \quad i = 1,2,3,4 \tag{2a}$$

$$\dot{M}_i = (\dot{M}_A)_i + (\dot{M}_B)_i + (\dot{M}_C)_i, \quad i = 1,2,3,4 \tag{2b}$$

and from the data given in Eqs. 1a, 1b, and 1c we have

$$(x_A)_1 = 0.1, \quad (x_B)_1 = 0.25 \quad (x_C)_1 = 0.65 \tag{3}$$

This indicates that the conditions for Stream #1 are completely specified, and on the basis of Eq. 7.33 all the mole fractions in the other streams are determined.

Eq. 2a can be expressed in the form

$$\dot{M}_i = \frac{(\dot{M}_A)_i}{(x_A)_i}, \quad A=1,2,3, \quad i=1,2,3,4 \qquad (4)$$

and this can be used for Stream #3 to obtain

$$\dot{M}_3 = \frac{(\dot{M}_B)_3}{(x_B)_3} = \frac{(\dot{M}_B)_3}{(x_B)_1} = \frac{4 \text{ mol/h}}{0.25} = 16 \text{ mol/h} \qquad (5a)$$

Given the total molar flow rate in Stream #3, we can use Eq. 2a to obtain

$$(\dot{M}_A)_3 = (x_A)_3 \dot{M}_3 = (x_A)_1 \dot{M}_3 = (0.1)(16 \text{ mol/h}) = 1.6 \text{ mol/h} \qquad (5b)$$

$$(\dot{M}_C)_3 = (x_C)_3 \dot{M}_3 = (x_C)_1 \dot{M}_3 = (0.65)(16 \text{ mol/h}) = 10.4 \text{ mol/h} \qquad (5c)$$

indicating that all the molar flow rates in Stream #3 are determined.

Directing our attention to Stream #4, we repeat the analysis represented by Eqs. 5 to obtain

$$\dot{M}_4 = \frac{(\dot{M}_B)_4}{(x_B)_4} = \frac{(\dot{M}_B)_4}{(x_B)_1} = \frac{2 \text{ mol/h}}{0.25} = 8 \text{ mol/h} \qquad (6a)$$

$$(\dot{M}_A)_4 = (x_A)_4 \dot{M}_4 = (x_A)_1 \dot{M}_4 = (0.1)(8 \text{ mol/h}) = 0.8 \text{ mol/h} \qquad (6b)$$

$$(\dot{M}_C)_4 = (x_C)_4 \dot{M}_4 = (x_C)_1 \dot{M}_4 = (0.65)(8 \text{ mol/h}) = 5.2 \text{ mol/h} \qquad (6c)$$

and we are ready to move on to determine all the molar flow rates in Stream #2. This requires the use of Eq. 7.32 in terms of the species molar flow rates given by

$$(\dot{M}_A)_1 = (\dot{M}_A)_2 + (\dot{M}_A)_3 + (\dot{M}_A)_4 \qquad (7a)$$

$$(\dot{M}_B)_1 = (\dot{M}_B)_2 + (\dot{M}_B)_3 + (\dot{M}_B)_4 \qquad (7b)$$

$$(\dot{M}_C)_1 = (\dot{M}_C)_2 + (\dot{M}_C)_3 + (\dot{M}_C)_4 \qquad (7c)$$

These results can be used to determine the species molar flow rates leading to

$$(\dot{M}_A)_2 = 7.6 \text{ mol/h}, \quad (\dot{M}_B)_2 = 19 \text{ mol/h}, \quad (\dot{M}_C)_2 = 49.4 \text{ mol/h} \qquad (8)$$

Finally, we see that the total molar flow rate in Stream #2 is given by

$$\dot{M}_3 = 76 \text{ mol/h} \qquad (9)$$

Note that if all the species molar flow rates are specified for a single stream, as they are by Eqs. 1a, 1b, and 1c, then the additional specifications must be in the other streams. The above example illustrates why this is the case. Once all the species molar flow rates for a given stream are specified, the mole fractions for the species in that stream are determined, and from the splitter physics the mole fractions for all the other streams are known. If all the mole fractions are known, then a single molar flow rate for a given species (or the overall molar flow rate) determines the other molar flow rates for that stream.

The remaining four ways in which this splitter problem can be solved are left as exercises for the students. These exercises are essential to gain a comprehensive understanding of the behavior of splitters.

7.3.2 RECYCLE AND PURGE STREAMS

In this section, we analyze systems with recycle and purge streams. We begin with a system analogous to the one shown in Figure 7.4 in which there is a *mixer.* We then move on to a more complicated system analogous to the one shown in Figure 7.5 in which there is a *splitter.*

EXAMPLE 7.5   PYROLYSIS OF DICHLOROETHANE WITH RECYCLE

To illustrate a simple recycle system, we consider the pyrolysis of dichloro-ethane, $C_2H_4Cl_2$, to produce vinyl chloride, C_2H_3Cl, and hydrochloric acid, HCl. The pyrolysis reaction is not complete, and experimental measurements indicate that the conversion for a particular reactor is given by

$$C = \text{Conversion of } C_2H_4Cl_2 = \frac{-\mathcal{R}_{C_2H_4Cl_2}}{(\dot{M}_{C_2H_4Cl_2})_2} = 0.30 \qquad (1)$$

The unreacted dichloroethane is separated from the reaction products and recy-cled back to the reactor for the production of more vinyl chloride as indicated in Figure 7.5a. The composition of the feed Stream #1 is 98% dichloroethane, $C_2H_4Cl_2$, and 2% ethane, C_2H_6, on a molar basis. We assume that the separation column produces a *sharp separation* meaning that all the dichloroethane leaves in the bottom Stream #5, and the remaining components (vinyl chloride, hydro-chloric acid, and ethane) leave through the distillate Stream #4. Our objective in this example is to determine the recycle flow rate in Stream #5.

Our analysis of this process is based on Axioms I and II as given by Eqs. 7.4 and 7.5, and we begin with the steady form of the macroscopic mole bal-ance to obtain

Axiom I: $\displaystyle\int_A c_A \mathbf{v}_A \cdot \mathbf{n}\, dA = \mathcal{R}_A, \quad A \Rightarrow C_2H_6, \text{HCl}, C_2H_3Cl, C_2H_4Cl_2 \qquad (2)$

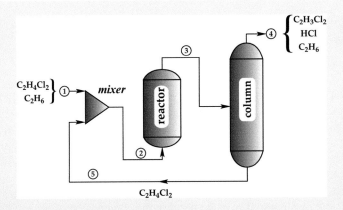

FIGURE 7.5a Reactor-separator with recycle.

The global rates of production within any control volume are constrained by Eq. 7.4 that takes the form

Axiom II:
$$\sum_{A=1}^{A=N} N_{JA} \mathcal{R}_A = 0, \quad J \Rightarrow C, H, Cl \tag{3}$$

For the case under consideration, the atomic matrix is given by

$$\text{Molecular Species} \rightarrow C_2H_6 \quad HCl \quad C_2H_3Cl \quad C_2H_4Cl_2$$

$$\begin{array}{r}
carbon \\
hydrogen \\
chlorine
\end{array}
\begin{bmatrix}
2 & 0 & 2 & 2 \\
6 & 1 & 3 & 4 \\
0 & 1 & 1 & 2
\end{bmatrix} \tag{4}$$

and Eq. 3 takes the form

$$\begin{bmatrix}
2 & 0 & 2 & 2 \\
6 & 1 & 3 & 4 \\
0 & 1 & 1 & 2
\end{bmatrix}
\begin{bmatrix}
\mathcal{R}_{C_2H_6} \\
\mathcal{R}_{HCl} \\
\mathcal{R}_{C_2H_3Cl} \\
\mathcal{R}_{C_2H_4Cl_2}
\end{bmatrix}
=
\begin{bmatrix}
0 \\
0 \\
0
\end{bmatrix} \tag{5}$$

Representing $[N_{JA}]$ in *row reduced echelon form* leads to

$$\begin{bmatrix}
1 & 0 & 0 & 0 \\
0 & 1 & 0 & 1 \\
0 & 0 & 1 & 1
\end{bmatrix}
\begin{bmatrix}
\mathcal{R}_{C_2H_6} \\
\mathcal{R}_{HCl} \\
\mathcal{R}_{C_2H_3Cl} \\
\mathcal{R}_{C_2H_4Cl_2}
\end{bmatrix}
=
\begin{bmatrix}
0 \\
0 \\
0
\end{bmatrix} \tag{6}$$

and application of the global pivot theorem (see Sec. 6.4) provides the following constraints on the global net rates of production:

Axiom II: $\quad \mathcal{R}_{C_2H_6} = 0, \quad \mathcal{R}_{HCl} = -\mathcal{R}_{C_2H_4Cl_2}, \quad \mathcal{R}_{C_2H_3Cl} = -\mathcal{R}_{C_2H_4Cl_2}$ (7)

Here we see that ethane acts as an inert, a conclusion that might have been extracted by intuition but has been made rigorous on the basis of Axiom II. The constraints given by Eq. 7 apply to each control volume that we construct for the system illustrated in Figure 7.5a, and we are now ready to construct those control volumes making use of the rules listed above. On the basis of those rules, we make the following five primary cuts:

   I. A cut of Stream #1 is made because the composition of Stream #1 is given. **NOTE**: "The composition of the feed Stream #1 is 98% dichloroethane and 2% ethane on a molar basis."

   II. A cut of Stream #5 is made because information about the composition is given for that stream. **NOTE**: "We assume that the separation column produces a *sharp separation* meaning that all the dichloroethane leaves in the bottom Stream #5, and the remaining components (ethane, hydrochloric acid and vinyl chloride) leave through the distillate Stream #4."

   III. A cut of Stream #4 is made because information about the composition is given for that stream. **NOTE**: "We assume that the separation column produces a *sharp separation* meaning that all the dichloroethane leaves in the bottom Stream #5, and the remaining components (ethane, hydrochloric acid and vinyl chloride) leave through the distillate Stream #4."

   IV. A cut of Stream #2 is required because information about the global net rate of production is given in terms of conditions in Stream #2. **NOTE**: The statement about the conversion requires that

$$\mathcal{R}_{C_2H_4Cl_2} = -C\,(\dot{M}_{C_2H_4Cl_2})_2$$

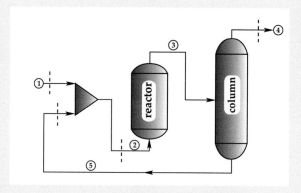

FIGURE 7.5b   Cuts for the construction of control volumes.

V. At least one control volume must enclose the reactor since information about the global net rate of production is given. **NOTE:**

$$\mathcal{R}_{C_2H_4Cl_2} = -C\,(\dot{M}_{C_2H_4Cl_2})_2$$

The cuts based on **Rules I** through **V** are indicated in Figure 7.5b. Two control volumes can be created that satisfy these rules, and these are illustrated in Figure 7.5c where we note that there is a single redundant cut of Stream #1. Given this choice of control volumes, our next step is to perform a degree-of-freedom analysis as indicated in Table 7.5a. There we see that we have a solvable problem.

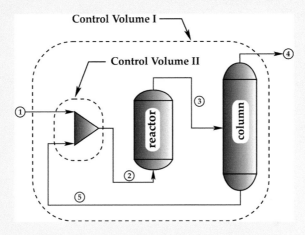

FIGURE 7.5c   Control volumes for vinyl chloride production unit.

Here, we note that the concentrations of HCl and C_2H_3Cl are zero in Stream #2 and this information has been ignored in the degree-of-freedom analysis. The explanation for this requires a degree-of-freedom analysis of the mixer that is left as an exercise for the student. Since, there is one degree of freedom, we will need to work in terms of dimensionless molar flow rates. For the control volumes illustrated in Figure 7.5c, we can express Eq. 2 as

Axiom I: $$\sum_{exits}(\dot{M}_A)_{exit} = \sum_{entrances}(\dot{M}_A)_{entrance} + \mathcal{R}_A,\quad A=1,2,3,4 \qquad (8)$$

We can utilize Eq. 2 of Example 7.4 to express the molar flow rates according to

At Entrances & Exits: $$\dot{M}_A = x_A\,\dot{M},\quad A=1,2,3,4 \qquad (9a)$$

TABLE 7.5a

Degrees-of-freedom for Production of Vinyl Chloride from Dichloroethane

<u>Stream & Net Rate of Production Variables</u>

compositions (4 species & 4 streams)	$N \times M = 4 \times 4 = 16$
flow rates (4 streams)	$M = 4$
net rates of production (4 species)	$N = 4$
Generic Degrees of Freedom (**A**)	**24**
<u>Number of Independent Balance Equations</u>	
mass/mole balance equations	$4 \times 2 = 8$
(4 species & two control volumes)	
<u>Number of Constraints for Compositions</u> (4 streams)	$M = 4$
<u>Number of Constraints for Reactions</u> (3 atomic species)	$T = 3$
Generic Specifications and Constraints (**B**)	**15**
<u>Specified Stream Variables</u>	
compositions (Streams #1, 4, & 5)	$3 + 1 + 3 = 7$
flow rates	0
<u>Constraints for Compositions</u>	0
<u>Auxiliary Constraints</u> (conversion)	1
Particular Specifications and Constraints (**C**)	**8**
Degrees of Freedom (**A** – **B** – **C**)	**1**

At Entrances & Exits: $\dot{M} = \dot{M}_A + \dot{M}_B + \dot{M}_C + \dot{M}_D$ (9b)

In many cases, it will be convenient to use Eq. 9a in the form

At Entrances & Exits: $x_A = \dfrac{\dot{M}_A}{\dot{M}}, \quad A = 1,2,3,4$ (9c)

Use of Eq. 8 allows us to express the species mole balances for Control Volume I in the form

<u>Control Volume I</u>

C_2H_6: $-(\dot{M}_{C_2H_6})_1 + (\dot{M}_{C_2H_6})_4 = \mathcal{R}_{C_2H_6}$ (10a)

$C_2H_4Cl_2$: $-(\dot{M}_{C_2H_4Cl_2})_1 + (\dot{M}_{C_2H_4Cl_2})_4 = \mathcal{R}_{C_2H_4Cl_2}$ (10b)

HCl: $-(\dot{M}_{HCl})_1 + (\dot{M}_{HCl})_4 = \mathcal{R}_{HCl}$ (10c)

C_2H_3Cl: $-(\dot{M}_{C_2H_3Cl})_1 + (\dot{M}_{C_2H_3Cl})_4 = \mathcal{R}_{C_2H_3Cl}$ (10d)

Since there are no chemical reactions taking place in the mixer illustrated in Figure 7.5c, the mole balances for Control Volume II take the simple forms given by

Control Volume II

C_2H_6: $-(\dot{M}_{C_2H_6})_1 - (\dot{M}_{C_2H_6})_5 + (\dot{M}_{C_2H_6})_2 = 0$ (11a)

$C_2H_4Cl_2$: $-(\dot{M}_{C_2H_4Cl_2})_1 - (\dot{M}_{C_2H_4Cl_2})_5 + (\dot{M}_{C_2H_4Cl_2})_2 = 0$ (11b)

HCl: $-(\dot{M}_{HCl})_1 - (\dot{M}_{HCl})_5 + (\dot{M}_{HCl})_2 = 0$ (11c)

C_2H_3Cl: $-(\dot{M}_{C_2H_3Cl})_1 - (\dot{M}_{C_2H_3Cl})_5 + (\dot{M}_{C_2H_3Cl})_2 = 0$ (11d)

We begin our analysis with Control Volume I and simplify the mole balances in terms of conditions on the molar flow rates. Some of those molar flow rates are zero and when we make use of this information the mole balances for Control Volume, I take the form

Control Volume I (*constraints on molar flow rates*)

C_2H_6: $-(\dot{M}_{C_2H_6})_1 + (\dot{M}_{C_2H_6})_4 = \mathcal{R}_{C_2H_6}$ (12a)

$C_2H_4Cl_2$: $-(\dot{M}_{C_2H_4Cl_2})_1 = \mathcal{R}_{C_2H_4Cl_2}$ (12b)

HCl: $(\dot{M}_{HCl})_4 = \mathcal{R}_{HCl}$ (12c)

C_2H_3Cl: $(\dot{M}_{C_2H_3Cl})_4 = \mathcal{R}_{C_2H_3Cl}$ (12d)

We now make use of the results from Axiom II given by Eq. 7 and impose those constraints on the global rates of production to obtain

Control Volume I (*constraints on global net rates of production*)

C_2H_6: $-(\dot{M}_{C_2H_6})_1 + (\dot{M}_{C_2H_6})_4 = 0$ (13a)

$C_2H_4Cl_2$: $-(\dot{M}_{C_2H_4Cl_2})_1 = \mathcal{R}_{C_2H_4Cl_2}$ (13b)

HCl: $(\dot{M}_{HCl})_4 = -\mathcal{R}_{C_2H_4Cl_2}$ (13c)

C_2H_3Cl: $(\dot{M}_{C_2H_3Cl})_4 = -\mathcal{R}_{C_2H_4Cl_2}$ (13d)

At this point, we can use Eq. 1 to express the global net rate of production in terms of the conversion leading to

Control Volume I (*global net rates of production in terms of the conversion*)

C_2H_6:
$$-(\dot{M}_{C_2H_6})_1 + (\dot{M}_{C_2H_6})_4 = 0 \tag{14a}$$

$C_2H_4Cl_2$:
$$(\dot{M}_{C_2H_4Cl_2})_1 = C\,(\dot{M}_{C_2H_4Cl_2})_2 \tag{14b}$$

HCl:
$$(\dot{M}_{HCl})_4 = C\,(\dot{M}_{C_2H_4Cl_2})_2 \tag{14c}$$

C_2H_3Cl:
$$(\dot{M}_{C_2H_3Cl})_4 = C\,(\dot{M}_{C_2H_4Cl_2})_2 \tag{14d}$$

Moving on to Control Volume II, we note that all global rates of production are zero and that only ethane and dichloroethane are present in this control volume. This leads to the two non-trivial species mole balances given by

Control Volume II (*constraints on molar flow rates*)

C_2H_6:
$$-(\dot{M}_{C_2H_6})_1 + (\dot{M}_{C_2H_6})_2 = 0 \tag{15a}$$

$C_2H_4Cl_2$:
$$-(\dot{M}_{C_2H_4Cl_2})_1 - (\dot{M}_{C_2H_4Cl_2})_5 + (\dot{M}_{C_2H_4Cl_2})_2 = 0 \tag{15b}$$

At this point, we use Eq. 14b to obtain

$$(\dot{M}_{C_2H_4Cl_2})_5 = \frac{1-C}{C}\,(\dot{M}_{C_2H_4Cl_2})_1 \tag{16}$$

Since, the molar flow rate of dichloroethane entering the system is not specified, it is convenient to work in terms of dimensionless molar flow rates, and this leads to

$C_2H_4Cl_2$:
$$(\dot{\mathcal{M}}_{C_2H_4Cl_2})_5 = \frac{(\dot{M}_{C_2H_4Cl_2})_5}{(\dot{M}_{C_2H_4Cl_2})_1} = \frac{1-C}{C} \tag{17a}$$

C_2H_3Cl:
$$(\dot{\mathcal{M}}_{C_2H_3Cl})_4 = \frac{(\dot{M}_{C_2H_3Cl})_4}{(\dot{M}_{C_2H_4Cl_2})_1} = 1.0 \tag{17b}$$

C_2H_6:
$$(\dot{\mathcal{M}}_{C_2H_6})_4 = \frac{(\dot{M}_{C_2H_6})_4}{(\dot{M}_{C_2H_4Cl_2})_1} = \frac{0.02}{0.98} \tag{17c}$$

HCl:
$$(\dot{\mathcal{M}}_{HCl})_4 = \frac{(\dot{M}_{HCl})_4}{(\dot{M}_{C_2H_4Cl_2})_1} = 1.00 \tag{17d}$$

In addition, the molar flow rates of the four species can be expressed in terms of the total molar flow rate entering the system, and this leads to

$C_2H_4Cl_2$:
$$(\dot{M}_{C_2H_4Cl_2})_5 = 2.2867\dot{M}_1 \tag{18a}$$

C_2H_3Cl:
$$(\dot{M}_{C_2H_3Cl})_4 = 0.98\dot{M}_1 \tag{18b}$$

C_2H_6:
$$(\dot{M}_{C_2H_6})_4 = 0.02\dot{M}_1 \tag{18c}$$

HCl:
$$(\dot{M}_{HCl})_4 = 0.98\dot{M}_1 \tag{18d}$$

These results can be used to determine the total molar flow rates in Streams #4 and #5

$$\dot{M}_5 = 2.2867\dot{M}_1, \quad \dot{M}_4 = 1.98\dot{M}_1 \tag{19}$$

along with the composition in Stream #4 that is given by

$$(x_{C_2H_3Cl})_4 = 0.495, \quad (x_{C_2H_6})_4 = 0.01, \quad (x_{HCl})_4 = 0.495 \tag{20}$$

A direct solution of the recycle problem given in Example 7.5 was possible because of the simplicity of the process, both in terms of the chemical reaction and in terms of the structure of the process. When purge streams are required, as illustrated in Figure 7.5, the analysis becomes more complex.

EXAMPLE 7.6   PRODUCTION OF ETHYLENE OXIDE WITH RECYCLE AND PURGE

In Figure 7.6a, we have illustrated a process in which ethylene oxide, C_2H_4O, is produced by the oxidation of ethylene, C_2H_4, over a catalyst containing silver. In a side reaction, ethylene is oxidized to carbon dioxide, CO_2, and water, H_2O. The feed stream, Stream #1, consists of Ethylene, C_2H_4, and air, N_2 and O_2, which is combined with Stream #5 that contains the unreacted ethylene, C_2H_4, carbon dioxide, CO_2, and nitrogen, N_2. The mole fraction of ethylene in Stream #2 entering the reactor must be maintained at 0.05 for satisfactory catalyst operation. The conversion of ethylene in the reactor is optimized to be 70% ($C = 0.70$) and the yield of ethylene oxide is 50% ($Y = 0.50$). All of the oxygen in the feed reacts, thus there is no oxygen in Stream #3. The reactor effluent is sent to an absorber where all the ethylene oxide is absorbed in the water entering in Stream #8. Water is fed to the absorber in a manner such that $(\dot{M}_{H_2O})_8 = 100\,(\dot{M}_{C_2H_4O})_7$ and we assume that all the water leaves the absorber in Stream #7. A portion of Stream #4 is recycled in Stream #5 and a portion is purged in Stream #6. In this example, we want to determine the fraction of $\dot{M}_4$ that needs to be purged in order that 100 mol/h of ethylene oxide are produced by the reactor.

We begin this example by constructing the equations containing information that is either given or required. To be specific, we want to determine the parameter α defined by

$$\dot{M}_6 = \alpha\,\dot{M}_4 \tag{1}$$

when 100 mol/h of ethylene oxide are produced by the reactor. We represent this information as

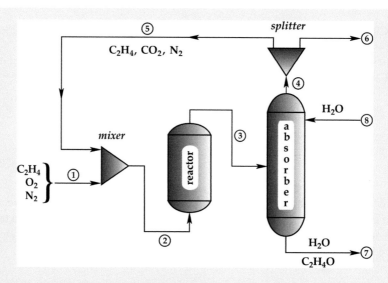

FIGURE 7.6a   Production of ethylene oxide.

$$\mathcal{R}_{C_2H_4O} = \beta, \quad \beta = 100 \text{ mol/h} \tag{2}$$

We are also given information about the conversion and yield (see Eqs. 7.8 and 7.13) that we express as

$$C = \text{Conversion of ethylene} = \frac{-\mathcal{R}_{C_2H_4}}{(\dot{M}_{C_2H_4})_2} = 0.70 \tag{3}$$

$$Y = \text{Yield of } (C_2H_4O/C_2H_4) = \frac{\mathcal{R}_{C_2H_4O}}{-\mathcal{R}_{C_2H_4}} = 0.50 \tag{4}$$

Other information concerning this process becomes clear when we consider the problem of constructing control volumes for the application of Axiom I. On the basis of the rules for constructing control volumes given at the beginning of this section, we make the following primary cuts of the streams indicated in Figure 7.6a:

   I. A cut of Stream #1 is made because information concerning the composition of Stream #1 is given. **NOTE:** "The feed stream, Stream #1, consists of ethylene (C_2H_4) and air (N_2 and O_2)."

   II. A cut of Stream #2 is made because a constraint on the composition is given. **NOTE:** "…the mole fraction of ethylene in Stream #2 leaving the mixer must be maintained at 0.05 for satisfactory catalyst operation."

   III. One might make a cut of Stream #3 because a constraint on the composition is given. **NOTE:** "All of the oxygen in the feed reacts, thus there is no oxygen in Stream #3." However, there are other interpretations of the original statement. For example, one could say that since all of the oxygen in the feed reacts, there is no oxygen in Stream #4 and in Stream #7. In addition, one could say that the molar flow rate of oxygen in Stream #2 is equal to the molar rate of consumption of oxygen in the reactor, i.e., $(\dot{M}_{O_2})_2 = -\mathcal{R}_{O_2}$. If we choose the second of these three possibilities, there is no need to cut Stream #3, thus we do not make a cut of Stream #3.

IV. The reactor is enclosed in a control volume because volumetric information about the reactor is given. **NOTE**: "The conversion of ethylene in the reactor is optimized to be 70% $(C = 0.70)$ and the yield of ethylene oxide is 50% $(Y = 0.50)$."

V. Cuts of Stream #4 and Stream #6 are made because the operating characteristics of the splitter are required. **NOTE**: "In this example, we want to determine the fraction of $\dot{M}_4$ that needs to be purged in order that 100 mol/h of ethylene oxide are produced by the reactor."

VI. A cut of Stream #7 is made because information is required. **NOTE**: "In this example, we want to determine the fraction of $\dot{M}_4$ that needs to be purged in order that 100 mol/h of ethylene oxide are produced by the reactor."

VII. A cut of Stream #8 is made because information is given. **NOTE**: "Water is fed to the absorber such that $(\dot{M}_{H_2O})_8 = 100\,(\dot{M}_{C_2H_4O})_7$."

The information identified in **II** and **IV** can be expressed as

$$(x_{C_2H_4})_2 = \eta, \quad \eta = 0.05 \tag{5}$$

$$(\dot{M}_{H_2O})_8 = \psi\,(\dot{M}_{C_2H_4O})_7, \quad \psi = 100 \tag{6}$$

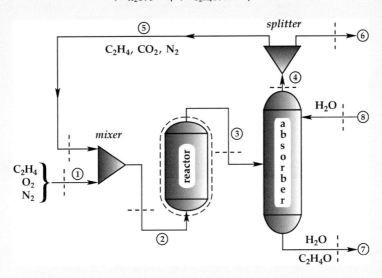

FIGURE 7.6b Cuts for the construction of control volumes.

The cuts and enclosures based on observations **I** through **VII** are illustrated in Figure 7.6b where we see one redundant cut of Stream #2 and an unnecessary cut of Stream #3. We can eliminate the latter with the control volumes illustrated in Figure 7.6c. There we see that we have avoided a cut of Stream #3 by enclosing the reactor and the absorber in Control Volume II; however, we have created redundant cuts of Stream #2, Stream #4, and Stream #5.

We begin the analysis of this process with Eq. 7.4 that takes the form

Axiom II:
$$\sum_{A=1}^{A=6} N_{JA}\, \mathcal{R}_A = 0, \quad J \Rightarrow C,H,O,N \tag{7}$$

and for the case under consideration, the atomic matrix is given by

$$
\begin{array}{c}
\text{Molecular Species} \rightarrow \\
\\
\begin{array}{r}
\textit{carbon} \\
\textit{hydrogen} \\
\textit{oxygen} \\
\textit{nitrogen}
\end{array}
\end{array}
\begin{array}{cccccc}
C_2H_4 & O_2 & CO_2 & H_2O & C_2H_4O & N_2 \\
\left[\begin{array}{cccccc}
2 & 0 & 1 & 0 & 2 & 0 \\
4 & 0 & 0 & 2 & 4 & 0 \\
0 & 2 & 2 & 1 & 1 & 0 \\
0 & 0 & 0 & 0 & 0 & 2
\end{array}\right]
\end{array} \tag{8}
$$

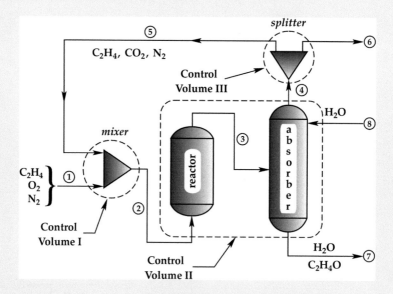

FIGURE 7.6c   Control volumes for the ethylene oxide production unit.

Given this form of the atomic matrix, we can express Eq. 7 as

$$
\begin{bmatrix}
2 & 0 & 1 & 0 & 2 & 0 \\
4 & 0 & 0 & 2 & 4 & 0 \\
0 & 2 & 2 & 1 & 1 & 0 \\
0 & 0 & 0 & 0 & 0 & 2
\end{bmatrix}
\begin{bmatrix}
\mathcal{R}_{C_2H_4} \\
\mathcal{R}_{O_2} \\
\mathcal{R}_{CO_2} \\
\mathcal{R}_{H_2O} \\
\mathcal{R}_{C_2H_4O} \\
\mathcal{R}_{N_2}
\end{bmatrix}
=
\begin{bmatrix}
0 \\
0 \\
0 \\
0
\end{bmatrix}
\tag{9}
$$

In *row reduced echelon form* Eq. 9 can be expressed as

$$
\begin{bmatrix}
1 & 0 & 0 & \dfrac{1}{2} & 1 & 0 \\
0 & 1 & 0 & \dfrac{3}{2} & \dfrac{1}{2} & 0 \\
0 & 0 & 1 & -1 & 0 & 0 \\
0 & 0 & 0 & 0 & 0 & 1
\end{bmatrix}
\begin{bmatrix}
\mathcal{R}_{C_2H_4} \\
\mathcal{R}_{O_2} \\
\mathcal{R}_{CO_2} \\
\mathcal{R}_{H_2O} \\
\mathcal{R}_{C_2H_4O} \\
\mathcal{R}_{N_2}
\end{bmatrix}
=
\begin{bmatrix}
0 \\
0 \\
0 \\
0
\end{bmatrix}
\tag{10}
$$

and the stoichiometric constraints on the global rates of production are given by application of the *global pivot theorem* (see Sec. 6.4)

$$
\mathcal{R}_{C_2H_4} = -\frac{1}{2}\mathcal{R}_{H_2O} - \mathcal{R}_{C_2H_4O}
\tag{11a}
$$

$$
\mathcal{R}_{O_2} = -\frac{3}{2}\mathcal{R}_{H_2O} - \frac{1}{2}\mathcal{R}_{C_2H_4O}
\tag{11b}
$$

$$
\mathcal{R}_{CO_2} = \mathcal{R}_{H_2O}
\tag{11c}
$$

$$
\mathcal{R}_{N_2} = 0
\tag{11d}
$$

DEGREE OF FREEDOM ANALYSIS

To gain some insight into a strategy for solving this problem, we need a degree-of-freedom analysis for the control volumes illustrated in Figure 7.6c. In this case, we will use molar flow rates and global rates of production as our independent variables. Of the eight streams illustrated in Figure 7.6c, only seven are cut by the surface of a control volume. Keeping in mind that we are dealing with six molecular species, we see that we have 42 stream variables in terms of species molar flow rates. In addition, we have six global rates of production so that the total number of *generic variables* is 48. The three control volumes illustrated in Figure 7.6c, coupled with the six molecular species, gives rise to 18 *generic constraints* associated with Axiom I. In addition, Axiom II provides four stoichiometric conditions (see Eq. 11), and the physics of the splitter provides $A - 1 = 5$ constraints on the composition (see Eq. 7.44) that we list here as

Splitter condition: $\qquad\qquad (x_A)_6 = (x_A)_4, \quad A = 1, 2, \ldots, 5$ $\qquad\qquad$ (12)

This provides us with 27 *generic constraints* and we can move on to the matter of *particular constraints*.

Control Volume I

We begin our exploration of the *particular constraints* with Control Volume I and direct our attention to Stream #1. The problem statement indicates that only ethylene, C_2H_4, oxygen, O_2, and nitrogen, N_2, are present in that stream, thus we have

Stream #1: $\qquad\qquad (\dot{M}_{C_2H_4O})_1 = (\dot{M}_{H_2O})_1 = (\dot{M}_{CO_2})_1 = 0$ $\qquad\qquad$ (13)

In addition, we are given that the oxygen and nitrogen in Stream #1 are supplied by air, thus we have a constraint on the composition given by

Stream #1: $\qquad\qquad (\dot{M}_{O_2})_1 = \varphi(\dot{M}_{N_2})_1, \quad \varphi = (21/79)$ $\qquad\qquad$ (14)

Moving on to the other stream entering Control Volume I, we draw upon information about the splitter to conclude that Stream #5 contains no ethylene oxide, C_2H_4O, no water, H_2O, and no oxygen, O_2, thus we have

Stream #5: $\qquad\qquad (\dot{M}_{C_2H_4O})_5 = (\dot{M}_{H_2O})_5 = (\dot{M}_{O_2})_5 = 0$ $\qquad\qquad$ (15)

It is obvious that there is no ethylene oxide, C_2H_4O, or water, H_2O, in Stream #2. However, we must be careful not to impose such a condition as a *particular constraint*, since this condition is "obvious" on the basis of the species mole balances and the conditions given by Eqs. 13 and 15. At this point, we conclude that there are seven particular constraints associated with Control Volume I and we list this result as

Control Volume I: $\qquad\qquad$ particular constraints $= 7$ $\qquad\qquad$ (16)

Moving on to Control Volume II, we note that the problem statement contained the comment "...the mole fraction of ethylene in Stream #2 entering the reactor must be maintained at 0.05 for satisfactory catalyst operation." This provided the basis for Eq. 5 and we list this particular constraint as

Stream #2: $\qquad\qquad (x_{C_2H_4})_2 = \eta, \quad \eta = 0.05$ $\qquad\qquad$ (17)

There are no particular constraints imposed on Stream #4, thus we move on to Stream #7 and Stream #8 for which the following particular constraints are imposed:

Stream #7:
$$(\dot{M}_{C_2H_4})_7 = (\dot{M}_{O_2})_7 = (\dot{M}_{N_2})_7 = (\dot{M}_{CO_2})_7 = 0 \tag{18}$$

Stream #8:
$$(\dot{M}_{C_2H_4O})_8 = (\dot{M}_{C_2H_4})_8 = (\dot{M}_{O_2})_8$$
$$= (\dot{M}_{N_2})_8 = (\dot{M}_{CO_2})_8 = 0 \tag{19}$$

In addition, we have a condition involving both Stream #7 and Stream #8 that was given earlier by Eq. 6 and repeated here as

Streams #7 and #8:
$$(\dot{M}_{H_2O})_8 = \psi\,(\dot{M}_{C_2H_4O})_7, \quad \psi = 100 \tag{20}$$

The requirement that 100 mol/h of ethylene oxide are produced by the reactor can be stated as

Reactor:
$$\mathcal{R}_{C_2H_4O} = \beta, \quad \beta = 100 \text{ mol/h} \tag{21}$$

while the information concerning the conversion and yield are repeated here as

Reactor:
$$C = \text{Conversion of ethylene} = \frac{-\mathcal{R}_{C_2H_4}}{(\dot{M}_{C_2H_4})_2} = 0.70 \tag{22}$$

Reactor:
$$Y = \text{Yield of } (C_2H_4O/C_2H_4) = \frac{\mathcal{R}_{C_2H_4O}}{-\mathcal{R}_{C_2H_4}} = 0.50 \tag{23}$$

Here, we conclude that there are 14 particular constraints associated with Control Volume II and we list this result as

Control Volume II: particular constraints $= 14$ (24)

Our third control volume encloses the splitter as indicated in Figure 7.6c. The problem statement indicates that "All of the oxygen in the feed reacts, … all the ethylene oxide is absorbed, … all the water leaves the absorber" thus it is clear Stream #4 is constrained by

Stream #4:
$$(\dot{M}_{C_2H_4O})_4 = (\dot{M}_{H_2O})_4 = (\dot{M}_{O_2})_4 = 0 \tag{25}$$

However, the splitter condition given by Eq. 12 along with the particular constraint given by Eq. 15 makes these three conditions *redundant*. Thus, there are no new constraints associated with Control Volume III.

Control Volume III: particular constraints $= 0$ (26)

The total number of particular constraints is $7+14 = 21$, thus the total number of constraints is given by

Total constraints: $18 + 27 = 48$ (27)

This means that we have zero degrees of freedom and the problem has a solution.

SOLUTION PROCEDURE

We can use Eqs. 3, 4, and 11 to express all the global rates of production in terms of the conversion (C), the yield (Y), and the molar flow rate of ethylene into the reactor, $(\dot{M}_{C_2H_4})_2$. These global rates of production are given by

$$\mathcal{R}_{C_2H_4} = -C \, (\dot{M}_{C_2H_4})_2 \tag{28a}$$

$$\mathcal{R}_{C_2H_4O} = Y C \, (\dot{M}_{C_2H_4})_2 \tag{28b}$$

$$\mathcal{R}_{H_2O} = 2C (1-Y)(\dot{M}_{C_2H_4})_2 \tag{28c}$$

$$\mathcal{R}_{CO_2} = 2C (1-Y)(\dot{M}_{C_2H_4})_2 \tag{28d}$$

$$\mathcal{R}_{O_2} = C(\frac{5}{2}Y - 3) \, (\dot{M}_{C_2H_4})_2 \tag{28e}$$

$$\mathcal{R}_{N_2} = 0 \tag{28f}$$

and this completes our application of Axiom II. Directing our attention to Axiom I, we begin our analysis of the control volumes illustrated in Figure 7.6c with the steady form of Eq. 7.1 given by

Axiom I: $\qquad \displaystyle\int_A c_A \mathbf{v}_A \cdot \mathbf{n} \, dA = \mathcal{R}_A, \quad A = 1, 2, \ldots, 6 \tag{29}$

We will apply this result to Control Volumes I, II, and III, and then begin the algebraic process of determining the fraction of $\dot{M}_4$ that needs to be purged in order that 100 mol/h of ethylene oxide are produced by the reactor.

Control Volume I
Since there are no chemical reactions in the mixer illustrated in Figure 7.6c, we can make use of Eq. 29 along with Eqs. 13 and 15 to obtain the following mole balance equations:

C_2H_4: $\qquad -(\dot{M}_{C_2H_4})_1 - (\dot{M}_{C_2H_4})_5 + (\dot{M}_{C_2H_4})_2 = 0 \tag{30a}$

CO_2: $\qquad -(\dot{M}_{CO_2})_5 + (\dot{M}_{CO_2})_2 = 0 \tag{30b}$

O_2: $\qquad -(\dot{M}_{O_2})_1 + (\dot{M}_{O_2})_2 = 0 \tag{30c}$

N_2: $\qquad -(\dot{M}_{N_2})_1 - (\dot{M}_{N_2})_5 + (\dot{M}_{N_2})_2 = 0 \tag{30d}$

The single constraint on these molar flow rates is given by Eq. 14 that we repeat here as

$$(\dot{M}_{O_2})_1 = \varphi(\dot{M}_{N_2})_1, \quad \varphi = (21/79) \tag{31}$$

We cannot make any significant progress with Eq. 30, thus we move on to the second control volume.

Control Volume II
In this case, we make use of Eqs. 28 and 29 to obtain the six species mole balances. These six equations are simplified by Eq. 18, Eq. 19, and Eq. 25 along with the observation that there is no ethylene oxide (C_2H_4O) or water (H_2O) in Stream #2. The six balance equations are given by

C_2H_4:
$$-(\dot{M}_{C_2H_4})_2 + (\dot{M}_{C_2H_4})_4 = -C\,(\dot{M}_{C_2H_4})_2 \tag{32a}$$

C_2H_4O:
$$(\dot{M}_{C_2H_4O})_7 = YC\,(\dot{M}_{C_2H_4})_2 \tag{32b}$$

H_2O:
$$-(\dot{M}_{H_2O})_8 + (\dot{M}_{H_2O})_7 = 2C\,(1-Y)(\dot{M}_{C_2H_4})_2 \tag{32c}$$

CO_2:
$$-(\dot{M}_{CO_2})_2 + (\dot{M}_{CO_2})_4 = 2C\,(1-Y)(\dot{M}_{C_2H_4})_2 \tag{32d}$$

O_2:
$$-(\dot{M}_{O_2})_2 = C\,(\frac{5}{2}Y - 3)\,(\dot{M}_{C_2H_4})_2 \tag{32e}$$

N_2:
$$-(\dot{M}_{N_2})_2 + (\dot{M}_{N_2})_4 = 0 \tag{32f}$$

Eqs. 21, 22, and 23 provide the constraint

$$(\dot{M}_{C_2H_4})_2 = \beta/CY \tag{33}$$

indicating that the molar flow rate of ethylene entering the reactor is specified. In addition, we can use Eq. 32d along with Eq. 32f and Eq. 17 to determine $(\dot{M}_{CO_2})_4 + (\dot{M}_{N_2})_4$. A summary of these results is given by

$$(\dot{M}_{C_2H_4})_2 = \beta/CY \tag{34a}$$

$$(\dot{M}_{C_2H_4})_4 = \beta(1-C)/CY \tag{34b}$$

$$(\dot{M}_{C_2H_4O})_7 = \beta \tag{34c}$$

$$(\dot{M}_{O_2})_2 = \beta(3 - \frac{5}{2}Y)/Y \tag{34d}$$

$$(\dot{M}_{H_2O})_7 = \beta\left[\psi + 2(1-Y)/Y\right] \tag{34e}$$

$$(\dot{M}_{H_2O})_8 = \beta\psi \tag{34f}$$

$$(\dot{M}_{CO_2})_4 + (\dot{M}_{N_2})_4 = \frac{\beta}{CY\eta}\left[1 - \eta\left(1 + C - \frac{1}{2}CY\right)\right] \tag{34g}$$

At this point, we need information about Stream #4 and this leads us to the splitter contained in Control Volume III.

Control Volume III

As indicated in Figure 7.6c, Stream #4 contains only ethylene, C_2H_4, carbon dioxide, CO_2, and nitrogen, N_2, and the appropriate mole balances are given by

C_2H_4:
$$-(\dot{M}_{C_2H_4})_4 + (\dot{M}_{C_2H_4})_5 + (\dot{M}_{C_2H_4})_6 = 0 \tag{35a}$$

CO_2:
$$-(\dot{M}_{CO_2})_4 + (\dot{M}_{CO_2})_5 + (\dot{M}_{CO_2})_6 = 0 \tag{35b}$$

N_2:
$$-(\dot{M}_{N_2})_4 + (\dot{M}_{N_2})_5 + (\dot{M}_{N_2})_6 = 0 \tag{35c}$$

On the basis of Eq. 12, and the fact that the splitter streams contain only three species (see Figure 7.6c), we apply Eq. 7.44 to obtain

$$(x_{C_2H_4})_6 = (x_{C_2H_4})_4 \tag{36a}$$

$$(x_{CO_2})_6 = (x_{CO_2})_4 \tag{36b}$$

$$(x_{N_2})_6 = (x_{N_2})_4 \tag{36c}$$

We now make use of the alternate form of Eq. 7.44 given by Eq. 7.46 in order to represent Eq. 36 as

$$(\dot{M}_{C_2H_4})_6 = \left(\dot{M}_6/\dot{M}_4\right)(\dot{M}_{C_2H_4})_4 \tag{37a}$$

$$(\dot{M}_{CO_2})_6 = \left(\dot{M}_6/\dot{M}_4\right)(\dot{M}_{CO_2})_4 \tag{37b}$$

$$(\dot{M}_{N_2})_6 = \left(\dot{M}_6/\dot{M}_4\right)(\dot{M}_{N_2})_4 \tag{37c}$$

At this point, we return to Eq. 1 which indicates that α is the fraction of Stream #4 that needs to be purged, i.e.,

$$\alpha = \dot{M}_6/\dot{M}_4 \tag{38}$$

This allows us to express the molar flow rates in Stream #6 as

$$(\dot{M}_{C_2H_4})_6 = \alpha(\dot{M}_{C_2H_4})_4 \tag{39a}$$

$$(\dot{M}_{CO_2})_6 = \alpha(\dot{M}_{CO_2})_4 \tag{39b}$$

$$(\dot{M}_{N_2})_6 = \alpha(\dot{M}_{N_2})_4 \tag{39c}$$

Use of these results with Eq. 35 leads to

C_2H_4:
$$(\dot{M}_{C_2H_4})_5 = (1-\alpha)(\dot{M}_{C_2H_4})_4 \tag{40a}$$

CO_2:
$$(\dot{M}_{CO_2})_5 = (1-\alpha)(\dot{M}_{CO_2})_4 \tag{40b}$$

N_2:
$$(\dot{M}_{N_2})_5 = (1-\alpha)(\dot{M}_{N_2})_4 \tag{40c}$$

and we can use Eq. 34b to represent the molar flow rate of ethylene in Stream #5 as

$$(\dot{M}_{C_2H_4})_5 = \frac{\beta(1-C)(1-\alpha)}{CY} \qquad (41)$$

At this point, we need to return to Control Volume I in order to acquire additional information needed to determine the parameter, α. Beginning with Eq. 30a, we use Eqs. 34a and 41 to obtain

C_2H_4:
$$(\dot{M}_{C_2H_4})_1 = \frac{\beta}{CY}\left[1-(1-C)(1-\alpha)\right] \qquad (42a)$$

Moving on to Eq. 30b, we use Eq. 32d along with Eqs. 34a and 40b to obtain

CO_2:
$$(\dot{M}_{CO_2})_2 = \frac{2\beta(1-\alpha)(1-Y)}{\alpha Y} \qquad (42b)$$

Next, we use Eq. 30c along with Eq. 34d to obtain

O_2:
$$(\dot{M}_{O_2})_1 = (\dot{M}_{O_2})_2 = \beta(3-\tfrac{5}{2}Y)/Y \qquad (42c)$$

and finally, we make use of Eq. 30c along with Eq. 31 to express the molar flow rate of nitrogen entering the system as

N_2:
$$(\dot{M}_{N_2})_1 = \beta(3-\tfrac{5}{2}Y)/\varphi Y \qquad (42d)$$

At this point, we are ready to return to Eq. 34g and make use of Eqs. 32d and 33 to obtain

$$(\dot{M}_{CO_2})_4 = (\dot{M}_{CO_2})_2 + 2\beta(1-Y)/Y \qquad (43)$$

This result can be used with Eq. 44b to provide the molar flow rate of carbon dioxide (CO_2) entering the splitter

$$(\dot{M}_{CO_2})_4 = 2\beta(1-Y)/\alpha Y \qquad (44)$$

This result can be used with Eq. 34g to determine the molar flow rate of nitrogen (N_2) entering the splitter as

$$(\dot{M}_{N_2})_4 = \frac{\beta}{CY\eta}\left[1-\eta(1+C-\tfrac{1}{2}CY)\right] - \frac{2\beta(1-Y)}{\alpha Y} \qquad (45)$$

We now recall Eq. 30d along with Eq. 32f and combine that result with Eq. 40c to obtain

$$(\dot{M}_{N_2})_1 = \alpha \, (\dot{M}_{N_2})_4 \tag{46}$$

This result could also be obtained directly by enclosing the entire system in a control volume and noting that the rate at which nitrogen enters the system should be equal to the rate at which nitrogen leaves the system.

At this point, we can use Eq. 42d along with Eq. 45 and Eq. 46 to eliminate all the molar flow rates and obtain an expression in which α is the only unknown.

$$\beta(3 - \frac{5}{2}Y)/\varphi Y = \frac{\alpha \beta}{C Y \eta}\left[1 - \eta(1 + C - \frac{1}{2}CY)\right] - \frac{2\beta(1-Y)}{Y} \tag{47}$$

This result provides a solution for the fraction of Stream #4 that must be purged, i.e.,

$$\alpha = \frac{\left(3 - \frac{5}{2}Y\right)(C\eta/\varphi) + 2C(1-Y)\eta}{1 - \eta\left(1 + C - \frac{1}{2}CY\right)} \tag{48}$$

Given the following data:

$C = $ conversion $= 0.70$

$Y = $ yield $= 0.50$

$\eta = $ mole fraction of ethylene entering the reactor $= 0.05$

$\varphi = (\dot{M}_{O_2})_1/(\dot{M}_{N_2})_1 = 21/79 = 0.266$

we determine the fraction of Stream #4 that must be purged to be $\alpha = 0.287$

7.4 SEQUENTIAL ANALYSIS FOR RECYCLE SYSTEMS

In Examples 7.5 and 7.6, we saw how the presence of a recycle stream created a *loop* in the flow of information. The determination of the molar flow rate of dichloroethane entering the reactor shown in Figure 7.5a required information about the recycle stream, i.e., information generated by the column used to purify the output stream for the process. The determination of the molar flow rate of ethylene entering the reactor shown in Figure 7.6a required information about the recycle stream, i.e., information generated by the absorber used to separate the ethylene oxide from the output of the reactor. For the systems described in Examples 7.5 and 7.6, it was possible to solve the linear set of equations simultaneously as we have done in other problems. However, most chemical engineering systems are nonlinear as a result of thermodynamic conditions and chemical kinetic models for reaction rates (see Chapter 9). For such systems, it becomes difficult to solve the system of equations globally, and in this section, we examine an alternative approach.

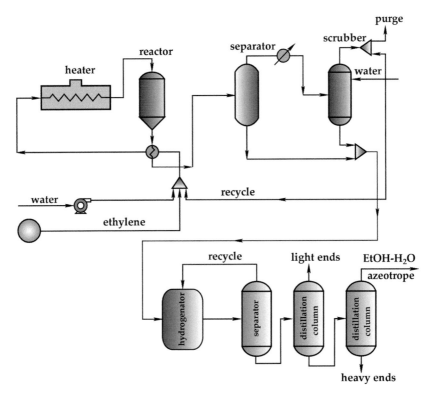

FIGURE 7.7 Flowsheet for the manufacture of ethyl alcohol from ethylene.

In Figure 7.7, we have illustrated a flowsheet for the manufacture of ethyl alcohol from ethylene that was presented earlier in Chapter 1. Here we see several units involving recycle and purge streams, and we need to think about what happens in those individual units in an operating chemical plant. *Each unit* is being monitored constantly on a day-to-day basis, and the data is interpreted in terms of macroscopic balances around *each unit*. If a mass or mole balance is not satisfied, there is a problem with *the unit* and a solution needs to be found. The engineer or operator in charge of a *specific unit* will be thinking about a single control volume around that unit, and this motivates the sequential problem-solving approach that we describe in this section. The sequential approach naturally leads to many redundant cuts; however, the advantage of this approach is that concepts associated with the analysis are simple and straightforward. We will use the pyrolysis of dichloroethane studied earlier in Example 7.5 to illustrate the sequential analysis of a simple recycle system.

EXAMPLE 7.7   SEQUENTIAL ANALYSIS OF
THE PYROLYSIS OF DICHLOROETHANE

In this example, we re-consider the pyrolysis of dichloroethane ($C_2H_4Cl_2$) to produce vinyl chloride (C_2H_3Cl) and hydrochloric acid (HCl) using a sequential analysis. The conversion of dichloroethane is given by

$$C = \text{Conversion of } C_2H_4Cl_2 = \frac{-\mathcal{R}_{C_2H_4Cl_2}}{(\dot{M}_{C_2H_4Cl_2})_2} = 0.30 \qquad (1)$$

and the unreacted dichloroethane is separated from the reaction products and recycled back to the reactor as indicated in Figure 7.7a.

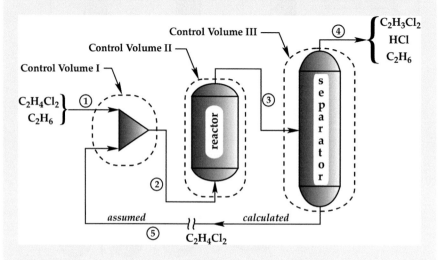

FIGURE 7.7a   Control volumes for sequential analysis of a recycle system.

The composition of Stream #1 is 98% dichloroethane, $C_2H_4Cl_2$, and 2% ethane, C_2H_6, on a molar basis, and the total molar flow rate is $\dot{M}_1$. We assume that the separation column produces a *sharp separation* meaning that all the dichloroethane leaves in Stream #5 and the remaining components (ethane, hydrochloric acid and vinyl chloride) leave in Stream #4. Our objective in this example is to determine the recycle flow rate in Stream #5 along with the composition and flow rate of Stream #4 in terms of the molar flow rate of Stream #1. Since our calculation is *sequential*, we require the *three* control volumes illustrated in Figure 7.7a instead of the *two* control volumes that were used in Example 7.5. In the sequential approach, we enclose each unit in a control volume and we assume that the *input conditions* to each control volume are known. For Control Volume I in Figure 7.7, this means that we must *assume* the molar flow rate of Stream #5. This process is referred to as *tearing the cycle* and Stream #5 is referred to as the *tear stream*.

Given the *input conditions* for the mixer, we can easily calculate the *output conditions*, i.e., the conditions associated with Stream #2. This means that we know the *input conditions* for the reactor and we can calculate the *output conditions* in Stream #3. Moving on to the separator, we use the conditions in Stream #3 to determine the conditions in Stream #4 and Stream #5. The calculated conditions in Stream #5 provide the *new assumed value* for the molar flow rate entering the mixer, and a sequential computational procedure can be repeated until a converged solution is obtained. For linear systems, convergence is assured; however, the matter is more complex for the non-linear system studied in Example 7.8.

Since the *input information* for each control volume is known, the structure for all the material balances has the form

$$(\dot{M}_A)_{out} = (\dot{M}_A)_{in} + \mathcal{R}_A, \quad A \Rightarrow C_2H_6, HCl, C_2H_3Cl, C_2H_4Cl_2 \quad (2)$$

Here, the *unknown quantities* are on the left-hand side and the *known quantities* are on the right-hand side. Directing our attention to the first control volume illustrated in Figure 7.7, we find

Control Volume I

C_2H_6: $\qquad\qquad\qquad (\dot{M}_{C_2H_6})_2 = (\dot{M}_{C_2H_6})_1 \qquad\qquad\qquad$ (3a)

$C_2H_4Cl_2$: $\qquad\qquad (\dot{M}_{C_2H_4Cl_2})_2 = (\dot{M}_{C_2H_4Cl_2})_1 + (\dot{M}_{C_2H_4Cl_2})_5^{(o)} \qquad$ (3b)

HCl: $\qquad\qquad\qquad (\dot{M}_{HCl})_2 = 0 \qquad\qquad\qquad$ (3c)

C_2H_3Cl: $\qquad\qquad (\dot{M}_{C_2H_3Cl})_2 = 0 \qquad\qquad\qquad$ (3d)

Here, we have used $(\dot{M}_{C_2H_4Cl_2})_5^{(o)}$ to identify the first *assumed value* for the molar flow rate of dichloroethane in the *tear stream*. Moving on to Control Volume II, we obtain

Control Volume II

C_2H_6: $\qquad\qquad (\dot{M}_{C_2H_6})_3 = (\dot{M}_{C_2H_6})_2 + \mathcal{R}_{C_2H_6} \qquad\qquad$ (4a)

$C_2H_4Cl_2$: $\qquad (\dot{M}_{C_2H_4Cl_2})_3 = (\dot{M}_{C_2H_4Cl_2})_2 + \mathcal{R}_{C_2H_4Cl_2} \qquad$ (4b)

HCl: $\qquad\qquad (\dot{M}_{HCl})_3 = (\dot{M}_{HCl})_2 + \mathcal{R}_{HCl} \qquad\qquad$ (4c)

C_2H_3Cl: $\qquad (\dot{M}_{C_2H_3Cl})_3 = (\dot{M}_{C_2H_3Cl})_2 + \mathcal{R}_{C_2H_3Cl} \qquad$ (4d)

The global net rates of production are constrained by Axiom II that provides the relations

Axiom II:
$$\mathcal{R}_{C_2H_6} = 0, \quad \mathcal{R}_{HCl} = -\mathcal{R}_{C_2H_4Cl_2}, \quad \mathcal{R}_{C_2H_3Cl} = -\mathcal{R}_{C_2H_4Cl_2} \tag{5}$$

which can be expressed in terms of the conversion (see Eq. 1) to obtain

$$\mathcal{R}_{C_2H_6} = 0, \quad \mathcal{R}_{HCl} = C\,(\dot{M}_{C_2H_4Cl_2})_2, \quad \mathcal{R}_{C_2H_3Cl} = C\,(\dot{M}_{C_2H_4Cl_2})_2 \tag{6}$$

These representations for the global net rates of production can be used with Eq. 4 to obtain

Control Volume II

C$_2$H$_6$:
$$(\dot{M}_{C_2H_6})_3 = (\dot{M}_{C_2H_6})_2 \tag{7a}$$

C$_2$H$_4$Cl$_2$:
$$(\dot{M}_{C_2H_4Cl_2})_3 = (\dot{M}_{C_2H_4Cl_2})_2(1-C) \tag{7b}$$

HCl:
$$(\dot{M}_{HCl})_3 = (\dot{M}_{HCl})_2 + C\,(\dot{M}_{C_2H_4Cl_2})_2 \tag{7c}$$

C$_2$H$_3$Cl:
$$(\dot{M}_{C_2H_3Cl})_3 = (\dot{M}_{C_2H_3Cl})_2 + C\,(\dot{M}_{C_2H_4Cl_2})_2 \tag{7d}$$

This completes the first two steps in the sequence, and we are ready to move on to Control Volume III shown in Figure 7.7. Application of Eq. 2 to Control Volume III provides the following mole balances:

Control Volume III

C$_2$H$_6$:
$$(\dot{M}_{C_2H_6})_4 = (\dot{M}_{C_2H_6})_3 \tag{8a}$$

C$_2$H$_4$Cl$_2$:
$$(\dot{M}_{C_2H_4Cl_2})_5^{(1)} = (\dot{M}_{C_2H_4Cl_2})_3 \tag{8b}$$

HCl:
$$(\dot{M}_{HCl})_4 = (\dot{M}_{HCl})_3 \tag{8c}$$

C$_2$H$_3$Cl:
$$(\dot{M}_{C_2H_3Cl})_4 = (\dot{M}_{C_2H_3Cl})_3 \tag{8d}$$

Here, we have indicated that the molar flow rate of dichloroethane, C$_2$H$_4$Cl$_2$, leaving the separator in Stream #5 is the *first approximation* based on the assumed value entering the mixer. This *assumed value* is given in Eq. 3b as $(\dot{M}_{C_2H_4Cl_2})_5^{(o)}$. Because Stream #5 contains only a single component, this problem is especially simple, and we only need to make use of Eqs. 3b, 7b, and 8b to obtain a relation between $(\dot{M}_{C_2H_4Cl_2})_5^{(1)}$ and $(\dot{M}_{C_2H_4Cl_2})_5^{(o)}$. This relation is given by

C$_2$H$_4$Cl$_2$:
$$(\dot{M}_{C_2H_4Cl_2})_5^{(1)} = (1-C)\left[(\dot{M}_{C_2H_4Cl_2})_1 + (\dot{M}_{C_2H_4Cl_2})_5^{(o)}\right] \tag{9}$$

Neither of the two molar flow rates on the right-hand side of this result are known; however, we can eliminate the molar flow rate of dichloroethane, $(\dot{M}_{C_2H_4Cl_2})_1$, by working in terms of a *dimensionless* molar flow rate defined by

$$\dot{\mathcal{M}}_5 = \frac{(\dot{M}_{C_2H_4Cl_2})_5}{(\dot{M}_{C_2H_4Cl_2})_1} \tag{10}$$

This allows us to express Eq. 9 as

$$\dot{\mathcal{M}}_5^{(1)} = (1-C)\left[1+\dot{\mathcal{M}}_5^{(o)}\right], \quad C=0.30 \tag{11}$$

and we can assume a value of $\dot{\mathcal{M}}_5^{(o)}$ in order to compute a value of $\dot{\mathcal{M}}_5^{(1)}$. On the basis of this computed value of $\dot{M}_5^{(1)}$, the analysis can be repeated to determine $\dot{\mathcal{M}}_5^{(2)}$ that is given by

$$\dot{\mathcal{M}}_5^{(2)} = (1-C)\left[1+\dot{\mathcal{M}}_5^{(1)}\right], \quad C=0.30 \tag{12}$$

This representation can be generalized to obtain the $(i+1)^{th}$ value that is given by

$$\dot{\mathcal{M}}_5^{(i+1)} = (1-C)\left[1+\dot{\mathcal{M}}_5^{(i)}\right], \quad C=0.30, \quad i=1,2,3,\ldots \tag{13}$$

This procedure is referred to as Picard's method[*], or as a *fixed-point iteration*, or as the *method of successive substitution* that is described in Sec. B.4 of Appendix B. Picard's method is often represented in the form

$$x_{i+1} = f(x_i), \quad i=1,2,3,\ldots \text{ etc.} \tag{14}$$

To illustrate how this iterative calculation is carried out, we assume that $\dot{\mathcal{M}}_5^{(o)} = 0$ to produce the values listed in Table 7.7a where we see a converged value given by $\dot{\mathcal{M}}_5 = 2.333$. One can avoid these detailed calculations by noting that for arbitrarily large values of i we arrive at the fixed-point condition given by

$$\dot{\mathcal{M}}_5^{(i+1)} = \dot{\mathcal{M}}_5^{(i)}, \quad i \to \infty \tag{15}$$

and the converged solution for the dimensionless molar flow rate is

$$\dot{\mathcal{M}}_5^{(\infty)} = \frac{1-C}{C} = 2.333 \tag{16}$$

---

[*] Bradie, B. 2006, *A Friendly Introduction to Numerical Analysis*, Pearson Prentice Hall, New Jersey.

This indicates that the molar flow rate of dichloroethane, $C_2H_4Cl_2$, in the recycle stream is

$$(\dot{M}_{C_2H_4Cl_2})_5 = 2.333\,(\dot{M}_{C_2H_4Cl_2})_1 \qquad (17)$$

In terms of the total molar flow rate in Stream #1, this takes the form

$$(\dot{M}_{C_2H_4Cl_2})_5 = 2.2867\,\dot{M}_1 \qquad (18)$$

which is exactly the answer obtained in Example 7.5.

TABLE 7.7a

Converging Values for Dimensionless Recycle Flow Rate (Picard's Method)

I	$\dot{\mathcal{M}}_5^{(i)}$	$\dot{\mathcal{M}}_5^{(i+1)}$
0	0.000	0.700
1	0.700	1.190
2	1.190	1.533
3	1.533	1.773
4	1.773	1.941
...	...	...
...	...	...
22	2.332	2.333
23	2.333	2.333

A variation on Picard's method is called Wegstein's method[†] and in terms of the nomenclature used in Eq. 14, this iterative procedure takes the form (see Sec. B.5 of Appendix B)

$$x_{i+1} = (1-q)\,f(x_i) + q\,x_i, \quad i = 1,2,3,\dots, \text{etc.} \qquad (19)$$

in which q is an adjustable parameter. When this adjustable parameter is equal to zero, $q = 0$, we obtain the *original* successive substitution scheme given by Eq. 14. When the adjustable parameter is greater than zero and less than one, $0 < q < 1$, we obtain a *damped* successive substitution process that improves stability for nonlinear systems. When the adjustable parameter is negative, $q < 0$, we obtain an *accelerated* successive substitution that may lead to an unstable procedure. For the problem under consideration in this example, Wegstein's method can be expressed as

$$\dot{\mathcal{M}}_5^{(i+1)} = (1-q)(1-C)\left[1 + \dot{\mathcal{M}}_5^{(i)}\right] + q\,\dot{\mathcal{M}}_5^{(i)}, \quad C = 0.30 \qquad (20)$$

---

† Wegstein, J.H. 1958, Accelerating convergences of iterative processes, Comm. ACM **1**, 9–13.

When the adjustable parameter is given by $q = -1.30$, we obtain the *accelerated convergence* illustrated in Table 7.7b. When confronted with nonlinear systems, the use of values of q near one may be necessary to obtain stable convergence.

TABLE 7.7b

Convergence for Dimensionless Recycle Flow Rate (Wegstein's Method)

I	q	$\dot{\mathcal{M}}_5^{(i)}$	$\dot{\mathcal{M}}_5^{(i+1)}$
0	−1.30	**0.000**	1.610
1	−1.30	1.610	2.109
2	−1.30	2.109	2.264
3	−1.30	2.264	2.312
4	−1.30	2.312	2.327
5	−1.30	2.327	2.331
6	−1.30	2.331	2.333
7	−1.30	**2.333**	**2.333**

In the next example, we return to a more complex problem associated with the production of ethylene oxide that was studied earlier in Example 7.6. In this analysis, we will find that a *damped* successive substitution process is necessary to obtain a converged solution.

EXAMPLE 7.8   SEQUENTIAL ANALYSIS OF THE PRODUCTION OF ETHYLENE OXIDE

In Figure 7.8a, we have illustrated a process in which ethylene oxide, C_2H_4O, is produced by the oxidation of ethylene, C_2H_4, over a catalyst containing silver. In a side reaction, ethylene is oxidized to carbon dioxide, CO_2, and water, H_2O. The feed stream, Stream #1, consists of ethylene, C_2H_4, and air, N_2 and O_2, which is combined with Stream #5 that contains the unreacted ethylene, C_2H_4, carbon dioxide, CO_2, and nitrogen, N_2. The mole fraction of ethylene in Stream #2 entering the reactor must be maintained at 0.05 for satisfactory catalyst operation. The conversion of ethylene in the reactor is optimized to be 70% ($C = 0.70$) and the yield of ethylene oxide is 50% ($Y = 0.50$). All of the oxygen in the feed reacts, thus there is no oxygen in Stream #3 and we have $(\dot{M}_{O_2})_3 = 0$. The reactor effluent is sent to an absorber where all the ethylene oxide is absorbed in the water entering in Stream #8. Water is fed to the absorber such that $(\dot{M}_{H_2O})_8 = 100 \, (\dot{M}_{C_2H_4O})_7$ and we assume that all the water leaves the absorber in Stream #7. A portion of Stream #4 is recycled in Stream #5 and a portion is purged in Stream #6. In this example, we want to determine the fraction of $\dot{M}_4$ that needs to be purged in order that 100 mol/h of ethylene oxide are produced by the reactor.

This problem statement is identical to that given by Example 7.6, and it is only the procedure for solving this problem that will be changed. In this case, we assume the values of the flow rates entering the mixer from the *tear stream*. We then update these assumed values on the basis of the sequential analysis of the four control volumes.

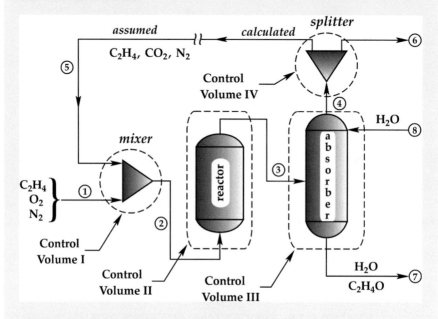

FIGURE 7.8a Control volumes for sequential analysis.

AVAILABLE DATA

The conversion and the yield are key parameters in this problem, and we define these quantities explicitly as

$$C = \text{Conversion of ethylene} = \frac{-\mathcal{R}_{C_2H_4}}{(\dot{M}_{C_2H_4})_2} = 0.70 \tag{1}$$

$$Y = \text{Yield of } (C_2H_4O/C_2H_4) = \frac{\mathcal{R}_{C_2H_4O}}{-\mathcal{R}_{C_2H_4}} = 0.50 \tag{2}$$

We are given that 100 mol/h of ethylene oxide are produced by the reactor and we express this condition as

Reactor: $\mathcal{R}_{C_2H_4O} = \beta, \quad \beta = 100 \text{ mol/h}$ (3)

The results expressed by Eqs. 1, 2, and 3 can be used to immediately deduce that

$$\mathcal{R}_{C_2H_4} = -\beta/Y \tag{4a}$$

$$(\dot{M}_{C_2H_4})_2 = \beta/CY \tag{4b}$$

In addition, we are given the following relations:

Streams #7 and #8: $\qquad (\dot{M}_{H_2O})_8 = \psi\,(\dot{M}_{C_2H_4O})_7, \quad \psi = 100 \tag{5a}$

Stream #1: $\qquad (\dot{M}_{O_2})_1 = \varphi(\dot{M}_{N_2})_1, \quad \varphi = (21/79) \tag{5b}$

Stream #2: $\qquad (x_{C_2H_4})_2 = \eta, \quad \eta = 0.05 \tag{5c}$

Stream #3: $\qquad (\dot{M}_{O_2})_3 = 0 \tag{5d}$

We can use Eq. 4b and Eq. 5c to provide

$$(x_{C_2H_4})_2 = \frac{(\dot{M}_{C_2H_4})_2}{\dot{M}_2} = \frac{\beta/CY}{\dot{M}_2} = \eta \tag{6a}$$

which indicates that the total molar flow rate entering the reactor is given by

$$\dot{M}_2 = \frac{\beta}{\eta\,CY} \tag{6b}$$

In Example 7.6, we carefully constructed control volumes that would minimize redundant cuts, while in Figure 7.8a we have simply enclosed each unit in a control volume and this creates four redundant cuts. As in Example 7.7, we will solve the material balances for each control volume in a *sequential manner*, and we will assume a value for any variable that is unknown.

STOICHIOMETRY

The details associated with Axiom II (see Eqs. 11 of Example 7.6) can be used along with the definitions of the conversion and yield to express the global rates of production in terms of the conversion and the yield. These results are given by

$$\mathcal{R}_{C_2H_4} = -\beta/Y \tag{7a}$$

$$\mathcal{R}_{CO_2} = 2(1-Y)\beta/Y \tag{7b}$$

$$\mathcal{R}_{N_2} = 0 \tag{7c}$$

$$\mathcal{R}_{O_2} = -(3 - \frac{5}{2}Y)\,\beta/Y \tag{7d}$$

$$\mathcal{R}_{C_2H_4O} = \beta \tag{7e}$$

$$\mathcal{R}_{H_2O} = 2(1-Y)\beta/Y \tag{7f}$$

MOLE BALANCES

We begin our analysis of this process with Axiom I for steady processes and fixed control volumes

Axiom I:
$$\int_A c_A \mathbf{v}_A \cdot \mathbf{n}\, dA = \mathcal{R}_A, \quad A = 1,2,3,\dots,6 \tag{8}$$

and we note that each mole balance can be expressed as

$$(\dot{M}_A)_{out} = (\dot{M}_A)_{in} + \mathcal{R}_A, \quad A = 1,2,3,\dots,6 \tag{9}$$

Application of Axiom I to the first control volume leads to

Control Volume I (Mixer)

C_2H_4:
$$(\dot{M}_{C_2H_4})_2 = (\dot{M}_{C_2H_4})_1 + (\dot{M}_{C_2H_4})_5 \tag{10a}$$

CO_2:
$$(\dot{M}_{CO_2})_2 = 0 + (\dot{M}_{CO_2})_5 \tag{10b}$$

N_2:
$$(\dot{M}_{N_2})_2 = (\dot{M}_{N_2})_1 + (\dot{M}_{N_2})_5 \tag{10c}$$

O_2:
$$(\dot{M}_{O_2})_2 = (\dot{M}_{O_2})_1 \tag{10d}$$

C_2H_4O:
$$(\dot{M}_{C_2H_4O})_2 = (\dot{M}_{C_2H_4O})_1 = 0 \tag{10e}$$

H_2O:
$$(\dot{M}_{H_2O})_2 = (\dot{M}_{H_2O})_1 = 0 \tag{10f}$$

Here, we have marked with a double asterisk (**) the three molecular species that are contained in the recycle stream. In order to determine the portion of Stream #4 that is recycled in Stream #5, we would normally require macroscopic balances for only those components that appear in the recycle stream. However, in this particular problem, the molar flow rates of oxygen and nitrogen entering the mixer are connected by Eq. 5b, thus we require the macroscopic balances for all four species that enter the mixer. This means that only Eqs. 10a, 10b, 10c, and 10d are required and we can move on to the remaining control volumes making use of only the mole balances associated with these four species.

Control Volume II (Reactor)
In this case, the four mole balances include a term representing the net global
rate of production owning to chemical reaction.

C_2H_4: $$(\dot{M}_{C_2H_4})_3 = (\dot{M}_{C_2H_4})_2 + \mathcal{R}_{C_2H_4} \tag{11a}$$

CO_2: $$(\dot{M}_{CO_2})_3 = (\dot{M}_{CO_2})_2 + \mathcal{R}_{CO_2} \tag{11b}$$

N_2: $$(\dot{M}_{N_2})_3 = (\dot{M}_{N_2})_2 \tag{11c}$$

O_2: $$(\dot{M}_{O_2})_3 = (\dot{M}_{O_2})_2 + \mathcal{R}_{O_2} \tag{11d}$$

Control Volume III (Absorber)
No chemical reaction occurs in this unit, thus the four balance equations are
given by

C_2H_4: $$(\dot{M}_{C_2H_4})_4 = (\dot{M}_{C_2H_4})_3 \tag{12a}$$

CO_2: $$(\dot{M}_{CO_2})_4 = (\dot{M}_{CO_2})_3 \tag{12b}$$

N_2: $$(\dot{M}_{N_2})_4 = (\dot{M}_{N_2})_3 \tag{12c}$$

O_2: $$(\dot{M}_{O_2})_4 = (\dot{M}_{O_2})_3 = 0 \tag{12d}$$

Here, we have made use of the information that all of the oxygen in the feed
reacts, and we have used the information that ethylene (C_2H_4), carbon dioxide
(CO_2), nitrogen (N_2) or oxygen, (O_2) do not appear in Stream #7 and Stream
#8.

Control Volume IV (Splitter)
This passive unit involves only three molecular species and the appropriate
mole balances are given by

C_2H_4: $$(\dot{M}_{C_2H_4})_5 + (\dot{M}_{C_2H_4})_6 = (\dot{M}_{C_2H_4})_4 \tag{13a}$$

CO_2: $$(\dot{M}_{CO_2})_5 + (\dot{M}_{CO_2})_6 = (\dot{M}_{CO_2})_4 \tag{13b}$$

N_2: $$(\dot{M}_{N_2})_5 + (\dot{M}_{N_2})_6 = (\dot{M}_{N_2})_4 \tag{13c}$$

Sequential Analysis for O_2
    At this point, it is convenient to direct our attention to the molar balances
for oxygen and carry out a sequential analysis to obtain

C.V. 1: $$(\dot{M}_{O_2})_2 = (\dot{M}_{O_2})_1 \tag{14a}$$

C.V. 2: $$(\dot{M}_{O_2})_3 = (\dot{M}_{O_2})_2 - (3 - \frac{5}{2}Y)\beta/Y \tag{14b}$$

C.V. 3: $\qquad$ $(\dot{M}_{O_2})_3 = 0$ $\qquad$ (14c)

These results allow us to determine the oxygen flow rate entering the mixer as

$$(\dot{M}_{O_2})_1 = (3 - \frac{5}{2}Y)\,\beta/Y \qquad (15)$$

Here, it is important to recall Eq. 5b and use that result to specify the molar flow rate of nitrogen (N_2) in Stream #1 as

$$(\dot{M}_{N_2})_1 = (3 - \frac{5}{2}Y)\,\beta/\varphi Y \qquad (16)$$

We are now ready to direct our attention to the molar flow rates of ethylene (C_2H_4), carbon dioxide (CO_2), and nitrogen (N_2) in the recycle stream and use these results to construct an iterative procedure that will allow us to calculate the fraction α.

ALGEBRA
We begin by noting that the splitter conditions can be expressed as

C_2H_4: $\qquad$ $(\dot{M}_{C_2H_4})_5 = (1-\alpha)\,(\dot{M}_{C_2H_4})_4$ $\qquad$ (17a)

CO_2: $\qquad$ $(\dot{M}_{CO_2})_5 = (1-\alpha)\,(\dot{M}_{CO_2})_4$ $\qquad$ (17b)

N_2: $\qquad$ $(\dot{M}_{N_2})_5 = (1-\alpha)\,(\dot{M}_{N_2})_4$ $\qquad$ (17c)

We will use these conditions in our analysis of ethylene (C_2H_4), carbon dioxide (CO_2), nitrogen (N_2). Directing our attention to the ethylene flow rate leaving the reactor, we note that the use of Eq. 4 in Eq. 11a leads to

$$(\dot{M}_{C_2H_4})_3 = \beta(1-C)/CY \qquad (18)$$

Moving from the reactor to the absorber, we use Eq. 12a with Eq. 18 to obtain

$$(\dot{M}_{C_2H_4})_4 = \beta(1-C)/CY \qquad (19)$$

When this result is used in the first splitter condition given by Eq. 17a we obtain

$$(\dot{M}_{C_2H_4})_5 = (1-\alpha)\,\beta\,(1-C)/CY \qquad (20)$$

Here, we must remember that α is the parameter that we want to determine by means of an iterative process. At this point, we move on to the mole balances for nitrogen (N_2) and make use of Eqs. 10c, 11c, 12c, 16, and 17c to obtain

$$(\dot{M}_{N_2})_5 = \frac{(1-\alpha)}{\alpha} \frac{\beta\left(3-\frac{5}{2}Y\right)}{\varphi Y} \tag{21}$$

Finally, we consider the mole balances for carbon dioxide (CO_2) and make use of Eqs. 7b, 10b, 11b, and 17b to obtain the following result for the molar flow rate of carbon dioxide in the recycle stream.

$$(\dot{M}_{CO_2})_5 = \frac{(1-\alpha)}{\alpha} \frac{2\beta(1-Y)}{Y} \tag{22}$$

At this point, we consider a total molar balance for the mixer

$$\dot{M}_1 + \dot{M}_5 = \dot{M}_2 \tag{23}$$

and note that the total molar flow rate of Stream #2 was given earlier by Eq. 6b. This leads to the mole balance around the mixer given by

Mixer: $$\dot{M}_1 = \beta/\eta\, CY - \dot{M}_5 \tag{24}$$

and we are ready to develop an iterative solution for the recycle flow and the parameter α.

At this point, we begin our analysis of the *tear stream*, i.e., Stream #5 that can be expressed as

$$\dot{M}_5 = (\dot{M}_{C_2H_4})_5 + (\dot{M}_{CO_2})_5 + (\dot{M}_{N_2})_5 \tag{25}$$

Given these values, we make use of Eq. 24 to express the molar flow rate entering the mixer in the form

$$\dot{M}_1 = \beta/\eta\, CY - \left[(\dot{M}_{C_2H_4})_5 + (\dot{M}_{CO_2})_5 + (\dot{M}_{N_2})_5\right] \tag{26}$$

Representing the total molar flow rate of Stream #1 in terms of the three species molar flow rates leads to

$$\dot{M}_1 = (\dot{M}_{C_2H_4})_1 + (\dot{M}_{O_2})_1 + (\dot{M}_{N_2})_1$$

in which the molar flow rates of oxygen, O_2, and nitrogen, N_2, are specified by Eqs. 15 and 16. Use of those results provides

$$\dot{M}_1 = (\dot{M}_{C_2H_4})_1 + (3 - \frac{5}{2}Y)\beta(1+\varphi)/\varphi Y \qquad (27)$$

and substitution of this result into Eq. 26 yields

$$(\dot{M}_{C_2H_4})_1 = \beta/\eta CY - (3-\frac{5}{2}Y)\beta(1+\varphi)/\varphi Y$$

$$- \left[(\dot{M}_{C_2H_4})_5 + (\dot{M}_{CO_2})_5 + (\dot{M}_{N_2})_5\right] \qquad (28)$$

Here we can use Eqs. 4b and 10a to obtain

$$\beta/CY = (\dot{M}_{C_2H_4})_1 + (\dot{M}_{C_2H_4})_5 \qquad (29)$$

which allows us to express Eq. 28 as

$$\beta/CY = \beta/\eta CY - (3-\frac{5}{2}Y)\beta(1+\varphi)/\varphi Y - \left[(\dot{M}_{CO_2})_5 + (\dot{M}_{N_2})_5\right] \qquad (30)$$

At this point, we return to Eqs. 20, 21, and 22 in order to express the molar flow rates of carbon dioxide, CO_2, and nitrogen, N_2, in the tear stream as

$$(\dot{M}_{CO_2})_5 + (\dot{M}_{N_2})_5 = \frac{(\dot{M}_{C_2H_4})_5\left[CY/\beta(1-C)\right]}{1-(\dot{M}_{C_2H_4})_5\left[CY/\beta(1-C)\right]}\left\{\frac{\beta\left(3-\frac{5}{2}Y\right)}{\varphi Y} + \frac{2\beta(1-Y)}{Y}\right\} \qquad (31)$$

Substitution of this result into Eq. 30 leads to an equation for the molar flow rate of ethylene, C_2H_4, in Stream 5. This result can be expressed in the compact form

$$\Omega - \left[\frac{(\dot{M}_{C_2H_4})_5}{\beta(1-C)/CY - (\dot{M}_{C_2H_4})_5}\right]\Lambda = 0 \qquad (32)$$

where the two parameters are given by

$$\Omega = \frac{1-\eta}{\eta CY} - \frac{\left(3-\frac{5}{2}Y\right)(1+\varphi)}{\varphi Y}, \quad \Lambda = \frac{\left(3-\frac{5}{2}Y\right)+2\varphi(1-Y)}{\varphi Y} \qquad (33)$$

At this point, we are ready to use a trial-and-error procedure to first solve for $(\dot{M}_{C_2H_4})_5$ and then solve for the parameter α.

PICARD'S METHOD

We begin by defining the dimensionless molar flow rate as

$$x = \frac{(\dot{M}_{C_2H_4})_5}{\beta} \tag{34}$$

so that the governing equation takes the form

$$H(x) = \Omega - \frac{\Lambda x}{(1-C)/CY - x} = 0 \tag{35}$$

In order to use Picard's method (see Sec. B.4 of Appendix B), we define a new function according to

Definition: $\qquad\qquad\qquad f(x) = x + H(x) \tag{36}$

and for any specific value of the dependent variable, x_i, we can *define* a new value, x_{i+1}, by

Definition: $\qquad\qquad x_{i+1} = f(x_i), \quad i = 1, 2, 3, \ldots, \infty \tag{37}$

To be explicit, we note that the new value of the dependent variable is given by

$$x_{i+1} = x_i + \left[\Omega - \frac{\Lambda x_i}{(1-C)/CY - x_i} \right], \quad i = 1, 2, 3, \ldots, \infty \tag{38}$$

In terms of the parameters for this particular problem

$$\Omega = 37.619, \quad \Lambda = 15.168, \quad C = 0.7, \quad Y = 0.5 \tag{39}$$

we have the following iterative scheme

$$x_{i+1} = x_i + \left[37.619 - \frac{15.168 \, x_i}{0.857 - x_i} \right], \quad i = 1, 2, 3, \ldots, \infty \tag{40}$$

By inspection, one can see that $x < 0.857$ thus we choose our first guess to be $x_o = 0.62$ and this leads to the results shown in Table 7.8a.

Clearly Picard's method *does not converge* for this case and we move on to Wegstein's method (see Sec. B.5 of Appendix B).

TABLE 7.8a

Iterative Values for Dimensionless Flow Rate (Picard's Method)

i	x_i	x_{i+1}
0	0.620	−1.424
1	−1.424	45.663
2	45.663	98.740
3	98.740	151.660
4	151.660	204.533
5	204.533	**257.383**
6	**257.383**	...
7	...	...

WEGSTEIN'S METHOD

In this case, we replace Eq. 37 with

Definition:
$$x_{i+1} = (1-q) f(x_i) + q x_i, \quad i = 1, 2, 3, \ldots, \infty \tag{41}$$

Next, we choose a value for the parameter, q, that provides for a severely damped convergence. The results are shown in Table 7.8b where we see that the iterative procedure converges rapidly to the value given by

$$x = (\dot{M}_{C_2H_4})_5 / \beta = 0.611 \tag{42}$$

Use of this result in Eq. 20 allows us to determine the fraction of Stream #4 that is purged and this fraction is given by

$$\alpha = 1 - \frac{(\dot{M}_{C_2H_4})_5 (CY)}{\beta(1-C)} = 0.287 \tag{43}$$

which is identical to that obtained in Example 7.6.

TABLE 7.8b

Iterative Values for Dimensionless Flow Rate (Wegstein's Method)

i	q	x_i	x_{i+1}
0	0.995	0.750	0.407
1	0.995	0.407	0.526
2	0.995	0.526	0.594
3	0.995	0.594	0.611
4	0.995	0.611	0.611
5	0.995	**0.611**	**0.611**

7.5 PROBLEMS

Note: Problems marked with the symbol ⌨ will be difficult to solve without the use of computer software.

SECTION 7.1

7.1. In the production of formaldehyde, CH_2O, by catalytic oxidation of methanol, CH_3OH, an equimolar mixture of methanol and air (21% oxygen and 79% nitrogen) is sent to a catalytic reactor. The reaction is catalyzed by finely divided silver supported on alumina as suggested in Figure 7-1 where we have indicated that carbon dioxide, CO_2, is produced as an undesirable product. The conversion for methanol, CH_3OH, is given by

$$C = \text{Conversion of } CH_3OH = \frac{-\mathcal{R}_{CH_3OH}}{\left(\dot{M}_{CH_3OH}\right)_1} = 0.20$$

and the selectivity for formaldehyde/carbon dioxide is given by

$$S = \text{Selectivity of } CH_2O/CO_2 = \frac{\mathcal{R}_{CH_2O}}{\mathcal{R}_{CO_2}} = 8.5$$

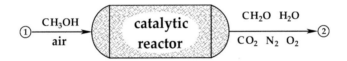

FIGURE 7-1 Production of formaldehyde.

In this problem, you are asked to determine the mole fraction of all components in the Stream #2 leaving the reactor.

7.2. Use the pivot theorem (Sec. 6.4) with Eq. 6 of Example 7.1 to develop a solution for $\mathcal{R}_{H_2}$ and $\mathcal{R}_{C_2H_6}$ using ethylene, C_2H_4, methane, CH_4, and propylene, C_3H_6, as the pivot species. Compare your solution with Eq. 7 and Eq. 8 of Example 7.1. In order to use ethylene, C_2H_4, ethane, C_2H_6, and propylene, C_3H_6, as the pivot species, one needs to use a column/row interchange (see Sec. 6.2.5) with Eq. 6 of Example 7.1. Carry out the appropriate column/row interchange and the necessary elementary row operations and use the pivot theorem to develop a solution for $\mathcal{R}_{H_2}$ and $\mathcal{R}_{CH_4}$.

7.3. Acetic anhydride can be made by direct reaction of ketene, CH_2CO, with acetic acid. Ketene, CH_2CO, is an important intermediary chemical used to produce acetic anhydride. The pyrolysis of acetone, CH_3COCH_3, in an externally heated empty tube, illustrated in Figure 7-3, produces ketene, CH_2CO, and methane, CH_4; however, some of the ketene reacts during the pyrolysis to form ethylene, C_2H_4, and carbon monoxide, CO. In turn, some of the ethylene is cracked to make coke, C, and hydrogen, H_2. At industrial reactor conditions, the yields for this set of reactions are given by

$$Y_{CH_2CO} = \text{Yield of } CH_2CO/CH_3COCH_3 = \frac{\mathcal{R}_{CH_2CO}}{-\mathcal{R}_{CH_3COCH_3}} = 0.95$$

$$Y_{C_2H_4} = \text{Yield of } C_2H_4/CH_3COCH_3 = \frac{\mathcal{R}_{C_2H_4}}{-\mathcal{R}_{CH_3COCH_3}} = 0.015$$

$$Y_{H_2} = \text{Yield of } H_2/CH_3COCH_3 = \frac{\mathcal{R}_{H_2}}{-\mathcal{R}_{CH_3COCH_3}} = 0.02$$

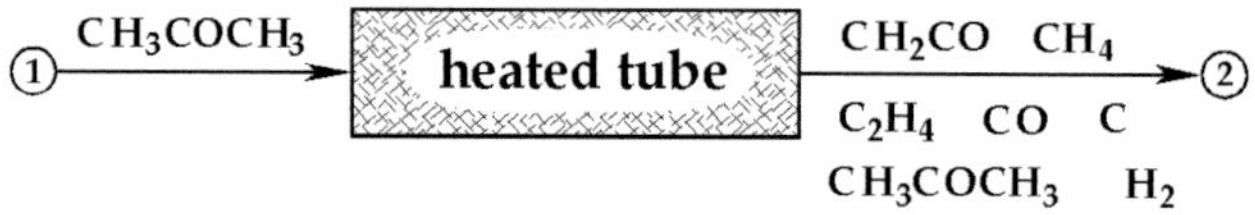

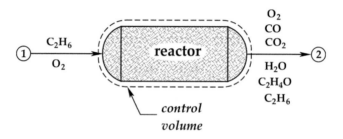

FIGURE 7-3   Pyrolysis of acetone.

Given the conversion of acetone

$$C = \text{Conversion of } CH_3COCH_3 = \frac{-\mathcal{R}_{CH_3COCH_3}}{(\dot{M}_{CH_3COCH_3})_{feed}} = 0.98$$

determine the mole fraction of all components in the product stream of the reactor.

7.4. Ethylene oxide can be produced by catalytic oxidation of ethane using pure oxygen. The stream leaving the reactor illustrated in Figure 7-4 contains non-reacted ethane and oxygen as well as ethylene oxide, carbon monoxide, carbon dioxide, and water. A gas stream of 10 kmol/min of ethane and oxygen is fed to the catalytic reactor with the mole fraction specified by

Stream #1: $y_{C_2H_6} = y_{O_2} = 0.50$ (1)

FIGURE 7-4   Production of ethylene oxide.

The reaction occurs over a platinum catalyst at a pressure of one atmosphere and a temperature of 250°C. Some of the mole fractions in the *exit stream* have been determined experimentally and the values are given by

Stream #2: $y_{C_2H_4O} = 0.287, \quad y_{C_2H_6} = 0.151, \quad y_{O_2} = 0.076$ (2)

In this problem, you are asked to determine the global rates of production of all the species participating in the catalytic oxidation reaction.

7.5💻. In a typical *experimental study*, such as that described in Problem 7.4, one would normally determine the complete composition of the entrance and exit streams. If these compositions were given by

Inlet Stream:
$$(y_{C_2H_6})_1 = 0.50, \quad (y_{O_2})_1 = 0.50, \quad (y_{H_2O})_1 = 0.0$$
$$(y_{CO})_1 = 0.0, \quad (y_{CO_2})_1 = 0.0, \quad (y_{C_2H_4O})_1 = 0.0$$

Outlet Stream:
$$(y_{C_2H_6})_2 = 0.14, \quad (y_{O_2})_2 = 0.08, \quad (y_{H_2O})_2 = 0.43$$
$$(y_{CO})_2 = 0.05, \quad (y_{CO_2})_2 = 0.03, \quad (y_{C_2H_4O})_2 = 0.27$$

how would you determine the six global rates of production associated with the partial oxidation of ethane? One should keep in mind that the experimental values of the mole fractions have been conditioned so that they sum to one.

7.6. Aniline is an important intermediate in the manufacture of dyes and rubber. A traditional process for the production of aniline consists of reducing nitrobenzene in the presence of iron and water at low pH. Nitrobenzene and water are fed in vapor form, at 250°C and atmospheric pressure, to a fixed-bed reactor containing the iron particles as illustrated in Figure 7-6. The solid iron oxide produced in the reaction remains in the reactor and it is later regenerated with hydrogen. The conversion is given by

$$C = \text{Conversion of } C_6H_5NO_2 = \frac{-\mathcal{R}_{C_6H_5NO_2}}{(\dot{M}_{C_6H_5NO_2})_1} = 0.80$$

and the feed consists of 100 kg/h of an equimolar gaseous mixture of nitrobenzene and water. Determine the mole fraction of all components leaving the reactor. If the reactor is initially charged with 2,000 kg of iron, Fe, use Eq. 7.5 to determine the time required to consume all the iron. Assume that the reactor operates at a steady state until the iron is depleted.

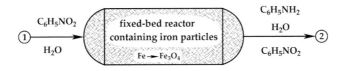

FIGURE 7-6   Production of aniline.

7.7. Vinyl chloride, CH_2CHCl, is produced in a fixed-bed catalytic reactor where a mixture of acetylene, C_2H_2, and hydrogen chloride, HCl, react in the presence of mercuric chloride supported on activated carbon. Assume that an equilibrium mixture leaves the reactor with the equilibrium relation given by

$$K_{eq} = \frac{y_A}{y_B \, y_C} = 300$$

Here, y_A, y_B and y_C are the mole fractions in the exit stream of vinyl chloride, CH_2CHCl, acetylene, C_2H_2, and hydrogen chloride, HCl, respectively. Determine the required excess of hydrogen chloride over the stoichiometric amount needed to achieve a conversion given by

$$C = \text{Conversion of } C_2H_2 = \frac{-\mathcal{R}_{C_2H_2}}{(\dot{M}_{C_2H_2})_{feed}} = 0.99$$

7.8. Carbon dioxide, CO_2, gas reacts over solid charcoal, C, to form carbon monoxide in the so-called Boudouard reaction illustrated in Figure 7-8. At 940 K, the equilibrium constant for the reaction of carbon dioxide with carbon to produce carbon monoxide is given by

$$K_{eq} = \frac{p_{CO}^2}{p_{CO_2}} = 1.2 \text{ atm} \qquad (1)$$

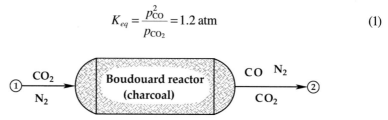

FIGURE 7-8   Production of carbon monoxide.

A mixture of carbon dioxide and nitrogen is fed to a fixed bed reactor filled with charcoal at 940 K. The mole fraction of carbon dioxide entering the reactor is 0.60. Assuming that the exit stream is in equilibrium with the solid charcoal, compute the mole fraction of all components in the exit stream.

7.9. Hydrogen cyanide (HCN) is made by reacting methane with anhydrous ammonia in the so-called Andrussov process illustrated in Figure 7-9. The equilibrium constant for this reaction at atmospheric pressure and 1,300 K is given by

$$K_{eq} = \frac{p_{HCN} \, p_{H_2}^3}{p_{CH_4} \, p_{NH_3}} = 380 \text{ atm}^2$$

A mixture of 50% by volume of methane and 50% anhydride ammonia is sent to a chemical reactor at atmospheric pressure and 1,300 K. Assuming the output of the reactor is in equilibrium, what would be the composition of the mixture of gases leaving the reactor, in mole fractions?

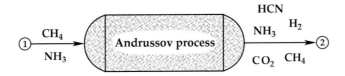

FIGURE 7-9   Hydrogen cyanide production.

7.10. Teflon, tetrafluoroethylene, C_2F_4, is made by pyrolysis of gaseous monocholorodifluoromethane, $CHClF_2$, in a reactor such as the one shown in Figure 7-2. The decomposition of $CHClF_2$ produces C_2F_4 and HCl in addition to the *undesirable* homologous polymer, $H(CF_2)_3Cl$. The conversion of gaseous monochlorodifluoromethane, $CHClF_2$, is given by

$$C = \text{Conversion of } (CHClF_2) = \frac{-\mathcal{R}_{CHClF_2}}{(\dot{M}_{CHClF_2})_{feed}} = 0.98$$

and the yield for Teflon takes the form

$$Y = \text{Yield of } (C_2F_4/CHClF_2) = \frac{\mathcal{R}_{C_2F_4}}{-\mathcal{R}_{CHClF_2}} = 0.475$$

For the first part of this problem, find a relation between the molar flow rate of the exit stream and the molar flow rate of the entrance stream in terms of the number of monomer units in the polymer species. For the second part, assume that $\alpha\, CHClF_2 \Rightarrow [H(CF_2)_3Cl]_\alpha$ where $\alpha = 10$ and assume that the input flow rate is $\dot{M}_1 = 100$ mol/s in order to determine the mole fractions of the four components in the product stream leaving the reactor.

7.11. In Example 7.1, numerical values for the conversion, selectivity and yield were given, in addition to the conditions for the inlet stream, i.e., $\dot{M} = 100$ mol/s and $x_{C_2H_6} = 1.0$. Indicate what other quantities had to be measured in order to determine the conversion, selectivity and yield.

7.12. In Example 7.1, the net rates of production for methane and hydrogen were determined to be $\mathcal{R}_{CH_4} = 1.0$ mol/s and $\mathcal{R}_{H_2} = 19.0$ mol/s. Determine the rates of production represented by $\mathcal{R}_{C_2H_6}$, $\mathcal{R}_{C_2H_4}$, and $\mathcal{R}_{C_3H_6}$.

SECTION 7.2

7.13. Show how to obtain the row reduced echelon form of $[N_{JA}]$ given in Eq. 7.18 from the form given in Eq. 7.16.

7.14. A stream of pure methane, CH_4, is partially burned with air in a furnace at a rate of 100 moles of methane per minute. The air is dry, the methane is in excess, and the nitrogen is inert in this particular process. The products of the reaction are illustrated in Figure 7-14. The exit gas contains a 1:1 ratio of H_2O:H_2 and a 10:1 ratio

FIGURE 7-14   Combustion of methane.

of $CO : CO_2$. Assuming that all of the oxygen and 94% of the methane are consumed by the reactions, determine the flow rate and composition of the exit gas stream.

7.15. Consider the special case in which two molecular species represented by $C_mH_nO_p$ and $C_qH_rO_s$ provide the fuel for *complete combustion* as illustrated in Figure 7-15. Assume that the molar flow rates of the two molecular species in the fuel are given in order to develop an expression for the theoretical air.

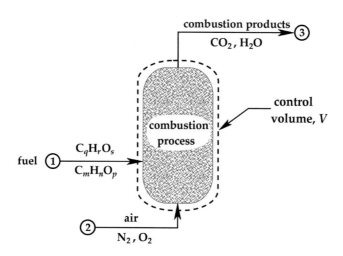

FIGURE 7-15   Combustion of two molecular species.

7.16. A flue gas (Stream #1) composed of carbon monoxide, carbon dioxide, and nitrogen can undergo reaction with "water gas" (Stream #2) and steam to produce the synthesis gas (Stream #3) for an ammonia converter. The carbon dioxide in the synthesis gas must be removed before the gas is used as feed for an ammonia converter. This process is illustrated in Figure 7-16, and the product gas in Stream #3 is required to contain hydrogen and nitrogen in a 3:1 molar ratio. In this problem, you are asked to determine the ratio of the molar flow rate of the flue gas to the molar flow rate of the water gas, i.e., $\dot{M}_1/\dot{M}_2$, that is required in order to meet the specification that $(y_{H_2})_3 = 3(y_{N_2})_3$.

Answer: $\dot{M}_1/\dot{M}_2 = 0.467$

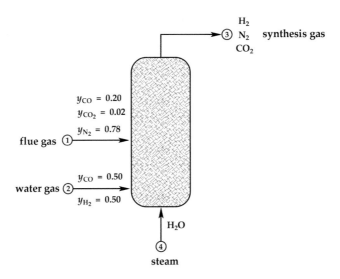

FIGURE 7-16 Synthesis gas reactor.

7.17. A process yields 10,000 ft³/day of a mixture of hydrogen chloride, HCl, and air. The volume fraction of hydrogen chloride, HCl, is 0.062, the temperature of the mixture is 550°F, and the total pressure is represented by 29.2 inches of mercury. Calculate the mass of limestone per day required to neutralize the hydrogen chloride, HCl, if the mass fraction of calcium carbonate, $CaCO_3$, in the limestone is 0.92. Determine the cubic feet of gas liberated per day at 70°F if the partial pressure of carbon dioxide, CO_2, is 1.8 inches of mercury. Assume that the reaction between HCl and $CaCO_3$ to form $CaCl_2$, CO_2, H_2O goes to completion.

7.18. Carbon is burned with air with all the carbon being oxidized to CO_2. Calculate the flue gas composition when the percent of excess air is 0, 50, and 100. The percent of excess air is defined as:

$$
\left\{ \begin{array}{l} \text{percent of} \\ \text{excess air} \end{array} \right\} = \frac{\left(\begin{array}{c} \text{molar flow} \\ \text{rate of oxygen} \\ \text{entering} \end{array} \right) - \left(\begin{array}{c} \text{molar rate of} \\ \text{consumption of oxygen} \\ \text{owing to reaction} \end{array} \right)}{\left(\begin{array}{c} \text{molar rate of} \\ \text{consumption of oxygen} \\ \text{owing to reaction} \end{array} \right)} \times 100
$$

Take the composition of air to be 79% nitrogen and 21% oxygen. Assume that no NOX is formed.

7.19. A waste gas from a petrochemical plant contains 5% HCN with the remainder being nitrogen. The waste gas is burned in a furnace with excess air to make sure the HCN is completely removed. The combustion process is illustrated in Figure 7-19. Assume that the percent excess air is 100% and determine the composition of the stream leaving the furnace. The percent of excess air is defined as:

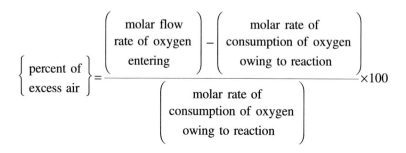

$$\left\{ \begin{array}{c} \text{percent of} \\ \text{excess air} \end{array} \right\} = \frac{\left(\begin{array}{c} \text{molar flow} \\ \text{rate of oxygen} \\ \text{entering} \end{array} \right) - \left(\begin{array}{c} \text{molar rate of} \\ \text{consumption of oxygen} \\ \text{owing to reaction} \end{array} \right)}{\left(\begin{array}{c} \text{molar rate of} \\ \text{consumption of oxygen} \\ \text{owing to reaction} \end{array} \right)} \times 100$$

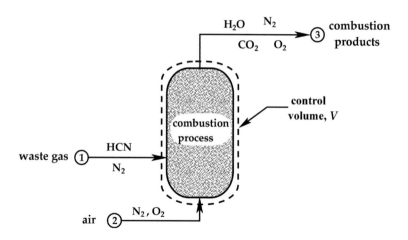

FIGURE 7-19   Combustion of hydrogen cyanide.

7.20. A fuel composed entirely of methane and nitrogen is burned with excess air. The dry flue gas composition in volume percent is: CO_2, 7.5%, O_2, 7%, and the remainder nitrogen. Determine the composition of the fuel gas and the percentage of excess air as defined in Problem 7.19.

7.21. In this problem, we consider the production of sulfuric acid illustrated in Figure 7-21. The mass flow rate of the dilute sulfuric acid stream is specified as $\dot{m}_1 = 100\,\text{lb}_m/\text{h}$ and we are asked to determine the mass flow rate of the pure sulfur trioxide stream, $\dot{m}_2$. As is often the custom with liquid systems the percentages given in Figure 7-21 refer to mass fractions, thus we desire to produce a final product in which the mass fraction of sulfuric acid is 0.98.

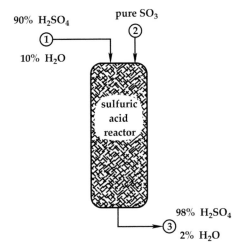

FIGURE 7-21   Sulfuric acid production.

7.22. In Example 7.3, use elementary row operations to obtain Eq. 10 from Eq. 9, and apply the pivot theorem to Eq. 10 to verify that Eqs. 11 are correct.

SECTION 7.3

7.23. Given $(x_A)_1$, $(x_A)_2$, and any three molar flow rates for the splitter illustrated in Figure 7.4, demonstrate that the compositions and total molar flow rates of all the streams are determined. For the three specified molar flow rates, use either species molar flow rates, or total molar flow rates, or a combination of both.

7.24. Given any three total molar flow rates and any two species molar flow rates for the splitter illustrated in Figure 7.4, demonstrate that the compositions and total molar flow rates of all the streams are determined. If the directions of the streams in Figure 7.4 are *reversed* we obtain a mixer as illustrated in Figure 7.6. Show that *six specifications* are needed to completely determine a mixer with three input streams, i.e., $S = 3$.

7.25. Given $(x_A)_1$, $(x_B)_1$, $\dot{M}_3/\dot{M}_2$, $\dot{M}_4/\dot{M}_2$, and any species molar flow rate for the splitter illustrated in Figure 7.4, demonstrate that the compositions and total molar flow rates of all the streams are determined.

7.26. Show how Eq. 6 is obtained from Eq. 5 in Example 7.5 using the concepts discussed in Sec. 6.2.5.

7.27. In the catalytic converter shown in Figure 7-27, a reaction of the form $A \rightarrow$ products takes place. The products are completely separated from the stream leaving the reactor, and pure A is recycled via stream #5. A certain fraction, φ, of species A that enters the reactor is converted to product, and we express this idea as

$$(x_A)_3 \dot{M}_3 = (1 - \varphi) \dot{M}_2$$

In this problem, you are asked to derive an expression for the ratio of molar flow rates, $\dot{M}_5/\dot{M}_1$, in terms of φ given that pure A enters the system in stream #1.

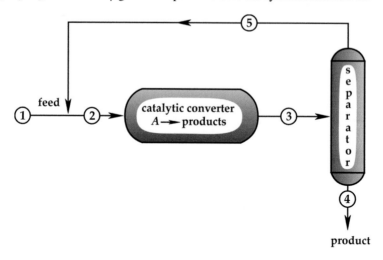

FIGURE 7-27   Catalytic converter.

7.28. A simple chemical reactor in which a reaction, $A \rightarrow$ products, is shown in Figure 7-28. The reaction occurs in the liquid phase and the feed stream is pure species A. The overall extent of reaction is designated by ξ where ξ is defined by the relation

$$(\omega_A)_4 = (1-\xi)(\omega_A)_2$$

Here, we see that $\xi = 0$ when no reaction takes place, and when $\xi = 1$ the reaction is complete and no species A leaves the reactor. We require that the mass fraction of species A entering the reactor be constrained by

$$(\omega_A)_2 = \varepsilon\,(\omega_A)_1$$

in which ε is some number less than one and greater than zero. Since the products of the reaction are not specified, assume that they can be lumped into a single "species" B. Under these conditions, the reaction can be expressed as $A \rightarrow B$ and the reaction rates for the two species system must conform to $r_A = -r_B$. The objective in this problem is to derive an expression for the ratio of mass flow rates $\dot{m}_5/\dot{m}_4$ in terms of ξ and ε.

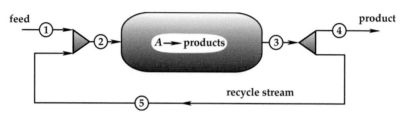

FIGURE 7-28   Chemical reactor with recycle stream.

7.29. Solve problem 7.28 using an iterative procedure (see Appendix B) for $\xi = 0.5$ and $\varepsilon = 0.3$ for a feed stream flow rate of $\dot{m}_1 = 100,000$ mol/h. Use a convergence criteria of 0.1 kmol/h for Stream #5.

7.30. Assume that the system described in Problem 7.28 contains N molecular species, thus species A represents the single reactant and there are $N-1$ product species. The reaction rates for the products can be expressed as

$$r_B = -r_A^{\text{I}}, \quad r_C = -r_A^{\text{II}}, \quad r_D = -r_A^{\text{III}}, \ldots, r_N = -r_A^{N-1}$$

where the overall mass rate of production for species A is given by

$$r_A = r_A^{\text{I}} + r_A^{\text{II}} + r_A^{\text{III}} + \cdots + r_A^{N-1}$$

Begin your analysis with the axioms for the mass of an N-component system and identify the conditions required in order that $N-1$ product species can be represented as a single species.

7.31. In the air dryer illustrated in Figure 7-31, part of the effluent air stream is to be recycled in an effort to control the inlet humidity. The solids entering the dryer (Stream #3) contain 20% water on a mass basis and the mass flow rate of the wet solids entering the dryer is 1,000 lb_m/h. The dried solids (stream #4) are to contain a maximum of 5% water on a mass basis. The partial pressure of water vapor in the fresh air entering the system (Stream #1) is equivalent to 10 mmHg and the partial pressure in the air leaving the dryer (Stream #5) must not exceed 200 mmHg. In this particular problem the flow rate of the recycle stream (stream #6) is to be regulated so that the partial pressure of water vapor in the air entering the dryer is equivalent to 50 mmHg. For this condition, calculate the total molar flow rate of fresh air entering the system (Stream #1) and the total molar flow rate of the recycle stream (Stream #6). Assume that the process operates at atmospheric pressure (760 mmHg).

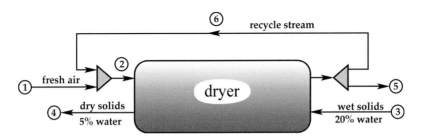

FIGURE 7-31 Air dryer with recycle stream.

7.32. Solve problem 7.31 using one of the iterative procedures described in Appendix B. Assume that the partial pressure of water in the fresh air stream (Stream #1) is 20 mmHg and that the maximum partial pressure in the stream leaving the unit (Stream #5), does not exceed 180 mmHg. If you use a spreadsheet or a MATLAB program to solve this problem, make sure all variables are conveniently labeled. Use a convergence criteria of 1 mmHg for Stream #6.

7.33. By manipulating the operating conditions (temperature, pressure and catalyst) in the reactor described in Example 7.5, the conversion can be increased from 0.30 to 0.47, i.e.,

$$C = \text{Conversion of } C_2H_4Cl_2 = \frac{-\mathcal{R}_{C_2H_4Cl_2}}{(\dot{M}_{C_2H_4Cl_2})_2} = 0.47$$

For this conversion, what are the changes in the molar flow rate of vinyl chloride in Stream #4 and the molar flow rate of dichloroethane in Stream #5?

7.34. For the conditions given in Example 7.5, determine the total molar flow rate and composition in Stream #3.

7.35. Metallic silver can be obtained from sulfide ores by roasting to sulfates, leaching with water, and precipitating the silver with copper. It is this latter process, involving the chemical reaction given by

$$Ag_2SO_4 + Cu \rightarrow 2Ag + CuSO_4$$

that we wish to consider here. In the system illustrated in Figure 7-35, the product leaving the second separator contains 90% (by mass) silver and 10% copper. The percent of excess copper in Stream #1 is defined by

$$
\left\{ \begin{array}{c} \text{percent of} \\ \text{excess copper} \end{array} \right\} = \frac{\left(\begin{array}{c} \text{molar flow rate} \\ \text{of copper entering} \\ \text{in the feed stream} \end{array} \right) - \left(\begin{array}{c} \text{molar rate of} \\ \text{consumption of} \\ \text{copper in the reactor} \end{array} \right)}{\left(\begin{array}{c} \text{molar rate of} \\ \text{consumption of} \\ \text{copper in the reactor} \end{array} \right)} \times 100
$$

and the conversion of silver sulfate is defined by

$$
C = \frac{\left\{ \begin{array}{c} \text{molar rate of consumption} \\ \text{of Ag}_2\text{SO}_4 \text{ in reactor} \end{array} \right\}}{\left\{ \begin{array}{c} \text{molar flow rate of Ag}_2\text{SO}_4 \\ \text{entering the reactor} \end{array} \right\}} = \frac{-\mathcal{R}_{Ag_2SO_4}}{(\dot{M}_{Ag_2SO_4})_3}
$$

For the conditions given, what is the percent of excess copper? If the conversion is given by $C = 0.75$, what is $\dot{m}_6/\dot{m}_5$?

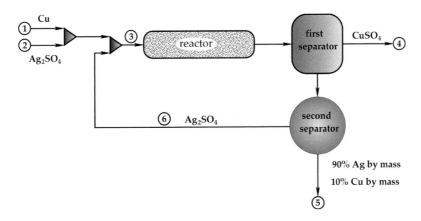

FIGURE 7-35   Metallic silver production.

7.36. In Figure 7-36 we, have illustrated a process in which $NaHCO_3$ is fed to a combined drying and calcining unit in which Na_2CO_3, H_2O, and CO_2 are produced.

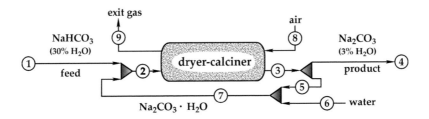

FIGURE 7-36   Dryer-calciner.

The partial pressure of water vapor in the entering air stream is equivalent to 12.7 mmHg and the system operates at one atmosphere (760 mmHg). The exit gas leaves at 300°F and a relative humidity of 5%. The $NaHCO_3$ is 70% solids and 30% water (mass basis) when fed to the system. The Na_2CO_3 leaves with a water content of 3% (mass basis). To improve the character of the solids entering the system, 50% (mass basis) of the dry material is moistened to produce $Na_2CO_3 \cdot H_2O$ and then recycled.

Calculate the following quantities per ton of Na_2CO_3 product.

a. mass of wet $NaHCO_3$ fed at 30% water.
b. mass of water fed to moisten recycle material.
c. cubic feet of dry air fed at 1 atm and 273 K.
d. total volume of exit gas at 1 atm and 300°F.

7.37. In Figure 7.5, we have illustrated an ammonia "converter" in which the unconverted gas is recycled to the reactor. In this problem, we consider a process in which the feed stream is a stoichiometric mixture of nitrogen and hydrogen containing

0.2% argon. In the reactor, 10% of the entering reactants (nitrogen and hydrogen) are converted to ammonia which is removed in a condenser. To be explicit, the conversion is given by

$$C = \text{Conversion of } N_2 = \frac{-\mathcal{R}_{N_2}}{(\dot{M}_{N_2})_1} = 0.10$$

The unconverted gas is recycled to the converter, and in order to avoid the buildup of argon in the system, a purge stream is incorporated in the recycle stream. In this problem, we want to determine the *fraction of recycle gas that must be purged* if the argon entering the reactor is limited to 0.5% on a molar basis.

SECTION 7.4

7.38. Solve Problem 7.37 using one of the following methods described in Appendix B.

  a. The bisection method
  b. The false position method
  c. Newton's method
  d. Picard's method
  e. Wegstein's method

8 Transient Material Balances

In Chapter 3, we studied *transient systems* that involved only a single molecular species. In this chapter, we extend our original study to include multicomponent, multiphase, reacting mixtures. First, we introduce the concept of a *perfectly mixed stirred tank* and study some simple mixing processes. This analysis of mixing is followed by a study of a *batch reactor* as an example of a transient process with chemical reaction. We continue our study of batch processes with an analysis of biomass production in a *chemostat* and then move on to the transient process of *batch distillation*.

Our analysis of transient systems begins with the *molar form* of Axiom I for species A given by

Axiom I:
$$\frac{d}{dt}\int_{V_a(t)} c_A \, dV + \int_{A_a(t)} c_A(\mathbf{v}_A - \mathbf{w}) \cdot \mathbf{n} \, dA = \int_{V_a(t)} R_A \, dV \qquad (8.1)$$

Here, one must remember that $V_a(t)$ represents an *arbitrary*, moving control volume having a surface $A_a(t)$ that moves with a speed of displacement given by $\mathbf{w} \cdot \mathbf{n}$. The second axiom requires that atomic species be conserved and is stated as

Axiom II:
$$\sum_{A=1}^{A=N} N_{JA} \, R_A = 0, \quad J = 1, 2, \ldots, T \qquad (8.2)$$

One can use this form to develop (see Sec. 6.1) a useful constraint on the net molar rates of production given by

$$\sum_{A=1}^{A=N} MW_A R_A = 0 \qquad (8.3)$$

The *mass form* of Axiom I will be useful in our analysis of biomass production in Sec. 8.4. This form is obtained from Eq. 8.1 by multiplying by the molecular mass, and the result is given by

$$\frac{d}{dt}\int_{V_a(t)} \rho_A \, dV + \int_{A_a(t)} \rho_A(\mathbf{v}_A - \mathbf{w}) \cdot \mathbf{n} \, dA = \int_{V_a(t)} r_A \, dV \qquad (8.4)$$

DOI: 10.1201/9781003283751-8

This form of Axiom I is used with a constraint on the species mass rates of production given by

$$\sum_{A=1}^{A=N} r_A = 0 \qquad (8.5)$$

We will use all of these forms in our study of transient systems.

8.1 PERFECTLY MIXED STIRRED TANK

In the chemical process industries, one encounters a system used for mixing which is referred to as a "completely mixed stirred tank" or a "perfect mixer." When used as a continuous reactor, such a system is often identified as a *continuous stirred tank reactor* or CSTR as an abbreviation. The essential characteristic of the perfectly mixed stirred tank is that the concentration in the tank is assumed to be uniform and equal to the effluent concentration even when the inlet conditions to the tank are changing with time. While this is impossible to achieve in any real system, it does provide an attractive model that represents an important limiting case for real stirred tank reactors and mixers.

As an example of a mixing process, we consider the system illustrated in Figure 8.1. The volumetric flow rate entering and leaving the system is assumed to be constant, thus the control volume is fixed in space. However, the concentration of the inlet stream is subject to changes, and we would like to know how the concentration in the tank responds to these changes. Since no chemical reaction is taking place, we can express Eq. 8.1 as

$$\frac{d}{dt} \int_{V(t)} c_A \, dV + \int_{A(t)} c_A \, (\mathbf{v}_A - \mathbf{w}) \cdot \mathbf{n} \, dA = 0 \qquad (8.6)$$

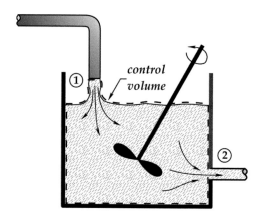

FIGURE 8.1   Perfectly mixed stirred tank.

While the gas-liquid interface may be moving normal to itself, it is reasonable to assume that there is no mass transfer of species A at that interface, thus $(\mathbf{v}_A - \mathbf{w}) \cdot \mathbf{n} = 0$ everywhere except at Streams #1 and #2. In addition, since the volumetric flow rates entering and leaving the tank are equal, it is reasonable to treat the control volume as a constant so that Eq. 8.6 simplifies to

$$\frac{d}{dt} \int_V c_A \, dV + \int_{A_e} c_A \, \mathbf{v}_A \cdot \mathbf{n} \, dA = 0 \qquad (8.7)$$

Here, A_e represents the area of *both* the entrance and the exit. Use of the traditional assumption for entrances and exits, $\mathbf{v}_A \cdot \mathbf{n} = \mathbf{v} \cdot \mathbf{n}$, along with the flat velocity profile assumption, allows us to write Eq. 8.7 as

$$\frac{d}{dt} \int_V c_A \, dV + \langle c_A \rangle_2 \, Q - \langle c_A \rangle_1 \, Q = 0 \qquad (8.8)$$

Here, $\langle c_A \rangle_1$ and $\langle c_A \rangle_2$ represent the *area-averaged* concentrations (see Sec. 4.5) at the entrance and exit respectively. The *volume-averaged* concentration is defined by

$$\langle c_A \rangle = \frac{1}{V} \int_V c_A \, dV \qquad (8.9)$$

and use of this definition in Eq. 8.8 leads to

$$\underbrace{V \frac{d\langle c_A \rangle}{dt}}_{\substack{\text{rate of accumulation} \\ \text{of species } A}} = \underbrace{\langle c_A \rangle_1 \, Q}_{\substack{\text{rate at which species } A \\ \text{enters the control volume}}} - \underbrace{\langle c_A \rangle_2 \, Q}_{\substack{\text{rate at which species } A \\ \text{leaves the control volume}}} \qquad (8.10)$$

Here, we have *two unknowns*, $\langle c_A \rangle$ and $\langle c_A \rangle_2$, and only a *single equation*; thus we need *more information* if we are to solve this problem. Under certain circumstances, the two concentrations are essentially equal and we express this limiting case as

$$\underbrace{\langle c_A \rangle}_{\substack{\text{volume average} \\ \text{concentration in} \\ \text{the tank}}} = \underbrace{\langle c_A \rangle_2}_{\substack{\text{area average} \\ \text{concentration} \\ \text{in the effluent}}} \qquad (8.11)$$

This allows us to write Eq. 8.10 in terms of the *single unknown*, $\langle c_A \rangle$, in order to obtain

$$V \frac{d\langle c_A \rangle}{dt} + \langle c_A \rangle Q = \langle c_A \rangle_1 \, Q \qquad (8.12)$$

One must be very careful to understand that Eq. 8.11 is an *approximation* based on the assumption that the difference between $\langle c_A \rangle_2$ and $\langle c_A \rangle$ is *small enough* so that it can be neglected[*].

It is convenient to divide Eq. 8.12 by the volumetric flow rate and express the result as

$$\tau \frac{d\langle c_A \rangle}{dt} + \langle c_A \rangle = \langle c_A \rangle_1 \tag{8.13}$$

Here τ represents the average residence time that is defined explicitly by

$$\left\{ \begin{array}{c} average \\ residence \\ time \end{array} \right\} = \frac{V}{Q} = \tau \tag{8.14}$$

In general, we are interested in processes for which the inlet concentration, $\langle c_A \rangle_1$, is a function of time, and a classic example is illustrated in Figure 8.2. There we have indicated that $\langle c_A \rangle_1$ undergoes a sudden change from c_A^o to c_A^l, and we want to determine how the concentration in the tank, $\langle c_A \rangle$, changes because of this change in the inlet concentration. The initial value problem associated with the sudden change in the inlet concentration is given by

$$\tau \frac{d\langle c_A \rangle}{dt} + \langle c_A \rangle = c_A^l, \quad t \geq 0 \tag{8.15a}$$

I.C. $$\langle c_A \rangle = c_A^o, \quad t = 0 \tag{8.15b}$$

Equations 8.15 are consistent with a situation in which the inlet concentration is originally fixed at c_A^o and then *instantaneously* switched from c_A^o to c_A^l at $t = 0$.

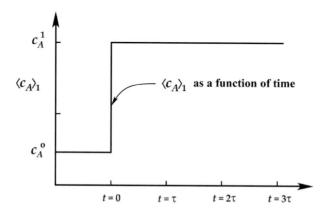

FIGURE 8.2 Inlet concentration as a function of time.

---

[*] Whitaker, S. 1988, Levels of simplification: The use of assumptions, restrictions, and constraints in engineering analysis, Chem. Eng. Ed. **22**, 104–108.

In order to solve Eq. 8.15a, we separate variables to obtain

$$\tau \frac{d\langle c_A \rangle}{\langle c_A \rangle - c_A^1} = -dt \tag{8.16}$$

The integrated form can be expressed as

$$\int_{\eta = c_A^o}^{\eta = \langle c_A \rangle} \frac{d\eta}{\eta - c_A^1} = -\tau^{-1} \int_{\xi = 0}^{\xi = t} d\xi \tag{8.17}$$

in which η and ξ are the dummy variables of integration. Carrying out the integration leads to

$$\ln \left[\frac{\langle c_A \rangle - c_A^1}{c_A^o - c_A^1} \right] = -t/\tau \tag{8.18}$$

which can be represented as

$$\langle c_A \rangle = c_A^1 + \left(c_A^o - c_A^1 \right) e^{-t/\tau} \tag{8.19}$$

This result is illustrated in Figure 8.3 where we see that a new steady-state condition is achieved for times on the order of *three to four residence times*. Even though *perfect mixing* can never be achieved in practice, and one can never change the inlet concentration *instantaneously* from one value to another, the results shown in Figure 8.3 can be used to provide an estimate of the *response time* of a mixer and this qualitative information is extremely useful. Experiments can be performed in systems similar to that shown in Figure 8.1 by suddenly changing the inlet concentration and continuously measuring the outlet concentration. If the results are in good agreement with the curve

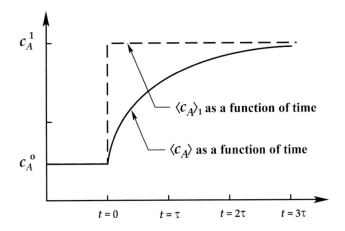

FIGURE 8.3   Response of a perfectly mixed stirred tank to a sudden change in the inlet concentration.

shown in Figure 8.3, one concludes that the system behaves as a perfectly mixed stirred tank *with respect to a passive mixing process*.

The solution for the mixing process described in the previous paragraphs was especially easy since the inlet concentration was a constant for all times greater than or equal to zero. The more general case would replace Eqs. 8.15 with

$$\tau \frac{d\langle c_A \rangle}{dt} + \langle c_A \rangle = f(t), \quad t \geq 0 \tag{8.20}$$

I.C. $$\langle c_A \rangle = c_A^\circ, \quad t = 0 \tag{8.21}$$

The solution to Eq. 8.20 can be obtained by means of a *transformation* known as the *integrating factor method* and this is left as an exercise for the student (see Problems 8.4 and 8.5).

8.2 BATCH REACTOR

In many chemical process industries, the *continuous* reactor is the most common type of chemical reactor. Petroleum refineries, for example, run day and night, and units are shut down on rare occasions. However, small-scale operations are a different matter and economic considerations often favor batch reactors for small-scale systems. The fermentation process that occurs during winemaking is an example of a batch reactor, and experimental studies of chemical reaction rates are often carried out in batch systems.

The analysis of a batch reactor, such as the liquid phase reactor shown in Figure 8.4, begins with the general form of the species mole balance

$$\frac{d}{dt} \int_{V_a(t)} c_A \, dV + \int_{A_a(t)} c_A \, (\mathbf{v}_A - \mathbf{w}) \cdot \mathbf{n} \, dA = \int_{V_a(t)} R_A \, dV \tag{8.22}$$

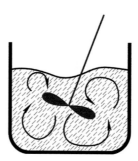

FIGURE 8.4   Batch reactor.

The batch reactor, by definition, has no entrances or exits, thus this result reduces to

$$\frac{d}{dt} \int_{V(t)} c_A \, dV = \int_{V(t)} R_A \, dV \tag{8.23}$$

Here, we have replaced $V_a(t)$ with $V(t)$ since the control volume is no longer *arbitrary* but is specified by the process under consideration. In terms of the volume averaged values of c_A and R_A, we can express our macroscopic balance as

$$\frac{d}{dt}\left[\langle c_A \rangle V(t)\right] = \langle R_A \rangle V(t) \tag{8.24}$$

In some batch reactors, the control volume is a function of time; however, in this development we assume that variations of the control volume are negligible so that Eq. 8.24 reduces to

$$\frac{d\langle c_A \rangle}{dt} = \langle R_A \rangle \tag{8.25}$$

The simplicity of this form of the *macroscopic species mole balance* makes the constant volume batch reactor an especially useful tool for studying the net rate of production of species A. Often it is important that the batch reactor be *perfectly mixed*; however, we will avoid imposing that condition for the present.

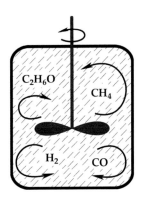

FIGURE 8.5   Constant volume batch reactor.

As an example, we consider the thermal decomposition of dimethyl ether in the constant volume batch reactor illustrated in Figure 8.5. The chemical species involved are the reactant C_2H_6O and the products CH_4, H_2, and CO. The visual representation of the atomic matrix is given by

$$\text{Molecular Species} \rightarrow CH_4 \quad H_2 \quad CO \quad C_2H_6O$$

$$\begin{array}{c} \textit{carbon} \\ \textit{hydrogen} \\ \textit{oxygen} \end{array} \begin{bmatrix} 1 & 0 & 1 & 2 \\ 4 & 2 & 0 & 6 \\ 0 & 0 & 1 & 1 \end{bmatrix} \tag{8.26}$$

and Axiom II takes the form

Axiom II:
$$
\begin{bmatrix} 1 & 0 & 1 & 2 \\ 4 & 2 & 0 & 6 \\ 0 & 0 & 1 & 1 \end{bmatrix}
\begin{bmatrix} R_{CH_4} \\ R_{H_2} \\ R_{CO} \\ R_{C_2H_6O} \end{bmatrix}
=
\begin{bmatrix} 0 \\ 0 \\ 0 \\ 0 \end{bmatrix}
\tag{8.27}
$$

Making use of the *row reduced echelon form* of the atomic matrix and applying the pivot theorem given in Sec. 6.4 leads to

$$
\begin{bmatrix} R_{CH_4} \\ R_{H_2} \\ R_{CO} \end{bmatrix}
=
\begin{bmatrix} -1 \\ -1 \\ -1 \end{bmatrix}
\begin{bmatrix} R_{C_2H_6O} \end{bmatrix}
\tag{8.28}
$$

in which C_2H_6O has been chosen as the *pivot species*. Hinshelwood and Asky[†] found that the reaction could be expressed as a first order decomposition providing a rate equation of the form

Chemical reaction rate equation: $R_{C_2H_6O} = -k \, c_{C_2H_6O}$ (8.29)

If we let dimethyl ether be species A, we can express the reaction rate equation as

$$
R_A = -k c_A
\tag{8.30}
$$

Use of this result in Eq. 8.25 leads to

$$
\frac{d\langle c_A \rangle}{dt} = -k \langle c_A \rangle
\tag{8.31}
$$

and we require only an initial condition to complete our description of this process. Given the following initial condition

I.C. $\langle c_A \rangle = c_A^o, \quad t = 0$ (8.32)

we find the solution for $\langle c_A \rangle$ to be

$$
\langle c_A \rangle = c_A^o \, e^{-kt}
\tag{8.33}
$$

This simple exponential decay is a classic feature of first order, irreversible processes. One can use this result along with experimental data from a batch reactor to determine the first order rate coefficient, k. This is often done by plotting the logarithm of $\langle c_A \rangle / c_A^o$ versus t, as illustrated in Figure 8.6, and noting that the slope is equal to $-k$.

[†] Hinshelwood, C.N. and Askey, P.J. 1927, Homogeneous reactions involving complex molecules: The kinetics of the decomposition of gaseous dimethyl ether, Proc. Roy. Soc. **A115**, 215–226.

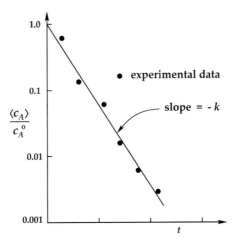

FIGURE 8.6   Batch reactor data, logarithmic scale.

If the rate coefficient in Eq. 8.33 is known, one can think of that result as a *design equation*. The idea here is that one of the key features of the design of a batch reactor is the specification of the *process time*. Under these circumstances, one is inclined to plot $\langle c_A \rangle / c_A^0$ as a function of time and this is done in Figure 8.7. The situation here is very similar to the mixing process described in the previous section. In that case, the *characteristic time* was the average residence time, V/Q, while in this case the *characteristic time* is the inverse of the rate coefficient, k^{-1}. When the rate coefficient is known one can quickly deduce that the *process time* is on the order of $3k^{-1}$ to $4k^{-1}$. Very few reactions are as simple as the first order irreversible reaction; however, it is a useful model for certain decomposition reactions.

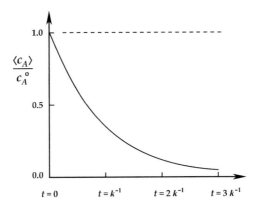

FIGURE 8.7   Concentration as a function of time, linear scale.

EXAMPLE 8.1   FIRST ORDER, REVERSIBLE REACTION IN A BATCH REACTOR

A variation of the first order *irreversible* reaction is the first order *reversible* reaction described by the following chemical kinetic schema:

Chemical kinetic schema: $A \underset{k_2}{\overset{k_1}{\rightleftharpoons}} B$ (1)

Here, k_1 is the forward reaction rate coefficient and k_2 is the reverse reaction rate coefficient. The *net rate of production* of species A can be modeled on the basis of the *picture* represented by Eq. 1 and this leads to a *chemical reaction rate equation* of the form

Chemical reaction rate equation: $R_A = -k_1 c_A + k_2 c_B$ (2)

Here, we remind the reader that in this text we use arrows to represent *pictures* and equal signs to represent *equations*.

Given an initial condition of the form

I.C. $c_A = c_A^o, \quad c_B = 0, \quad t = 0$ (3)

we want to derive an expression for the concentration of species A as a function of time for the batch reactor illustrated in Figure 8.1a.

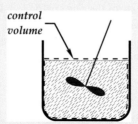

FIGURE 8.1a   Reversible reaction in a batch reactor.

We begin the analysis with the species mole balance for a fixed control volume

$$\frac{d}{dt} \int_V c_A \, dV = \int_V R_A \, dV$$ (4)

and express this result in terms of volume averaged quantities to obtain

$$\frac{d\langle c_A \rangle}{dt} = \langle R_A \rangle$$ (5)

The chemical kinetic rate equation given by Eq. 2 can now be used to write Eq. 5 in the form

$$\frac{d\langle c_A \rangle}{dt} = -k_1 \langle c_A \rangle + k_2 \langle c_B \rangle \tag{6}$$

In order to eliminate $\langle c_B \rangle$ from this result, we note that the development leading to Eq. 5 can be repeated for species B, and the use of $R_B = -R_A$ provides

$$\frac{d\langle c_B \rangle}{dt} = \langle R_B \rangle = -\langle R_A \rangle \tag{7}$$

From Eqs. 5 and 7, it is clear that

$$\frac{d\langle c_B \rangle}{dt} = -\frac{d\langle c_A \rangle}{dt} \tag{8}$$

indicating that the rate of *increase* of the concentration of species B is equal in magnitude to the rate of *decrease* of the concentration of species A. We can use Eq. 8 and the initial conditions to obtain

$$\langle c_B \rangle = -\left(\langle c_A \rangle - c_A^\circ \right) \tag{9}$$

This result allows us to eliminate $\langle c_B \rangle$ from Eq. 6 leading to

$$\frac{d\langle c_A \rangle}{dt} = -(k_1 + k_2)\langle c_A \rangle + k_2\, c_A^\circ \tag{10}$$

One can separate variables and form the indefinite integral to obtain

$$\frac{1}{(k_1 + k_2)} \ln\left[(k_1 + k_2)\langle c_A \rangle - k_2\, c_A^\circ \right] = -t + C_1 \tag{11}$$

where C_1 is the constant of integration. This constant can be determined by application of the initial condition which leads to

$$\ln\left[\left(\frac{k_1 + k_2}{k_1} \right) \frac{\langle c_A \rangle}{c_A^\circ} - \frac{k_2}{k_1} \right] = -(k_1 + k_2)t \tag{12}$$

An explicit expression for $\langle c_A \rangle$ can be extracted from Eq. 12 and the result is given by

$$\langle c_A \rangle = c_A^\circ \left[\frac{k_2}{k_1 + k_2} + \frac{k_1}{k_1 + k_2} e^{-(k_1 + k_2)t} \right] \tag{13}$$

It is always useful to examine any special case that can be extracted from a general result, and from Eq. 13 we can obtain the result for a first order, irreversible reaction by setting k_2 equal to zero. This leads to

$$\langle c_A \rangle = c_A^o e^{-k_1 t}, \quad k_2 = 0 \tag{14}$$

which was given earlier by Eq. 8.33.

Under *equilibrium conditions*, Eq. 2 reduces to

$$k_1 c_A = k_2 c_B, \quad \text{for} \quad R_A = 0 \tag{15}$$

and this can be expressed as

$$c_A = K_{eq} c_B, \quad \text{at equilibrium} \tag{16}$$

Here K_{eq} is the *equilibrium coefficient* defined by

$$K_{eq} = k_2 / k_1 \tag{17}$$

The general result expressed by Eq. 13 can also be written in terms of k_1 and K_{eq} to obtain

$$\langle c_A \rangle = c_A^o \left[\frac{K_{eq}}{1 + K_{eq}} + \frac{1}{1 + K_{eq}} e^{-k_1 (1 + K_{eq}) t} \right] \tag{18}$$

When $K_{eq} \ll 1$ we see that this result reduces to Eq. 14 as expected. In the design of a batch reactor for a reversible reaction, knowledge of the equilibrium coefficient (or equilibrium relation) is crucial since it immediately indicates the limiting concentration of the reactants and products.

8.3 DEFINITION OF REACTION RATE

If one assumes that the batch reactor shown in Figure 8.4 is a *perfectly mixed, constant volume* reactor, Eq. 8.25 takes the form

$$\frac{dc_A}{dt} = R_A, \quad \left\{ \begin{array}{l} \textit{perfectly mixed,} \\ \textit{constant volume} \\ \textit{batch reactor} \end{array} \right\} \tag{8.34}$$

Often there is a tendency to think of this result as defining the "reaction rate" (Dixon[‡]) and this is a perspective that one must avoid. Eq. 8.34 represents a special

---

[‡] Dixon, D.C. 1970, The definition of reaction rate, Chem. Engr. Sci. **25**, 337–338.

form of the *macroscopic mole balance for species A* and it does not represent a *definition* of R_A. In reality, Eq. 8.34 represents a very attractive special case that can be used with laboratory measurements to determine the species A *net molar rate of production*, R_A. Once R_A has been determined experimentally, one can search for chemical kinetic rate expressions such as that given by Eq. 8.30, and details of that search procedure are described in Chapter 9. If successful, the search provides both a satisfactory form of the rate expression and reliable values of the parameters that appear in the rate expression. To be convinced that Eq. 8.34 is not a definition of the reaction rate, one need only consider the perfectly mixed version of Eq. 8.24 which is given by

$$\frac{dc_A}{dt} + \frac{c_A}{V(t)}\frac{dV(t)}{dt} = R_A, \quad \left\{\begin{array}{c} \textit{perfectly mixed,} \\ \textit{batch reactor} \end{array}\right\} \tag{8.35}$$

Here, it should be clear that R_A is *not defined* by dc_A/dt; rather R_A is an *intrinsic property* of the system that represents the net molar rate of production of species A per unit volume. This is the sense in which the net rate of production was introduced in Chapter 4.

8.4 BIOMASS PRODUCTION

Biological compounds are produced by living cells, and the design and analysis of biological reactors requires both macroscopic balance analysis and kinetic studies of the complex reactions that occur within the cells. Given essential nutrients and a suitable temperature and pH, living cells will grow and divide to increase the cell mass. Cell mass production can be achieved in a chemostat where nutrients and oxygen are supplied as illustrated in Figure 8.8.

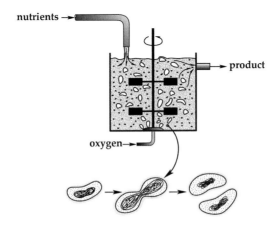

FIGURE 8.8 Cell growth in a chemostat.

Normally, the system is charged with cells, and a start-up period occurs during which the cells become accustomed to the nutrients supplied in the inlet stream. Oxygen and nutrients pass through the cell walls, and biological reactions within the cells lead to *cell growth* and the creation of new cells. In Figure 8.8 we have illustrated the process of cell division in which a single cell (called a *mother cell*) divides into two *daughter cells*. In Figure 8.9, we have identified species A and B as *substrates*, which is just another word for nutrients and oxygen. Species C represents *all the species* that leave the cell, while species D represents *all the species* that remain in the cell and create cell growth. The details of the enzyme-catalyzed reactions that occur within the cells are discussed in Sec. 9.2.

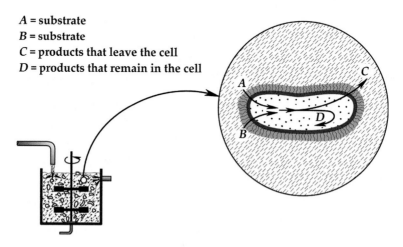

A = substrate
B = substrate
C = products that leave the cell
D = products that remain in the cell

FIGURE 8.9   Mass transfer and reaction in a cell.

To analyze cell growth in a chemostat, we need to know the rate at which species D is produced[§]. In reality, species D represents many chemical species which we identify explicitly as F, G, H, etc. The appropriate mass balances for these species are given by

$$\frac{d}{dt}\int_{V(t)} \rho_F \, dV + \int_{A(t)} \rho_F(\mathbf{v}_F - \mathbf{w}) \cdot \mathbf{n} \, dA = \int_{V(t)} r_F \, dV \qquad (8.36a)$$

$$\frac{d}{dt}\int_{V(t)} \rho_G \, dV + \int_{A(t)} \rho_G(\mathbf{v}_G - \mathbf{w}) \cdot \mathbf{n} \, dA = \int_{V(t)} r_G \, dV \qquad (8.36b)$$

$$\frac{d}{dt}\int_{V(t)} \rho_H \, dV + \int_{A(t)} \rho_H(\mathbf{v}_H - \mathbf{w}) \cdot \mathbf{n} \, dA = \int_{V(t)} r_H \, dV \qquad (8.36c)$$

$$\text{etc.} \qquad (8.36d)$$

---

§ Rodgers, A. and Gibon, Y. 2009, Enzyme Kinetics: Theory and Practice, Chapter 4 in *Plant Metabolic Networks*, edited by J. Schwender, Springer, New York.

Here, $V(t)$ represents the control volume illustrated in Figure 8.10 and $A(t)$ represents the surface area at which the speed of displacement is $\mathbf{w} \cdot \mathbf{n}$.

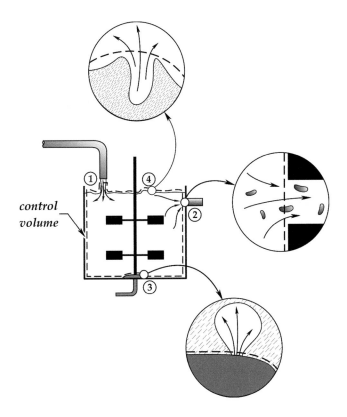

FIGURE 8.10 Control volume for chemostat.

In order to develop the macroscopic balance for the *total density of cellular material*, we simply add Eq. 8.36 to obtain

$$
\begin{aligned}
\frac{d}{dt} \int_{V(t)} \rho_D \, dV = &\int_{A(t)} \rho_F (\mathbf{v}_F - \mathbf{w}) \cdot \mathbf{n} \, dA + \int_{A(t)} \rho_G (\mathbf{v}_G - \mathbf{w}) \cdot \mathbf{n} \, dA \\
&+ \int_{A(t)} \rho_H (\mathbf{v}_H - \mathbf{w}) \cdot \mathbf{n} \, dA + \text{etc.} = \int_{V(t)} r_D \, dV
\end{aligned}
\tag{8.37}
$$

Here, we need to be very clear that ρ_D represents the *total density* of the cellular material and that this density is defined by

$$
\rho_D = \rho_F + \rho_G + \rho_H + \text{etc.}
\tag{8.38}
$$

In addition, we need to be very clear that r_D represents the *total mass rate of production* of cellular material, and that this mass rate of production is defined by

$$r_D = r_F + r_G + r_H + \text{etc.} \tag{8.39}$$

There are other molecular species in the system illustrated in Figure 8.9; however, we are interested in the rate of growth of cellular material, thus r_D is the quantity we wish to predict.

Returning to Eq. 8.37, we note that terms such as $(\mathbf{v}_G - \mathbf{w}) \cdot \mathbf{n}$ are *negligible* everywhere except at the *entrance* where cellular material may enter the chemostat, and at the *exit* where the product leaves the system. It is reasonable to assume that all the species associated with the cellular material move with the same velocity at the entrance and exit, and this allows us to express Eq. 8.37 as

$$\frac{d}{dt} \int_{V(t)} \rho_D \, dV + \int_{A_e} \rho_D \mathbf{v}_D \cdot \mathbf{n} \, dA = \int_{V(t)} r_D \, dV \tag{8.40}$$

where A_e represents the area of the entrance and the exit. It is important to understand that this result is based on the plausible assumption that all the velocities of the species remaining in the cell are the same

$$\mathbf{v}_F = \mathbf{v}_G = \mathbf{v}_H = \text{etc.} \tag{8.41}$$

and we have identified this *common velocity* by $\mathbf{v}_D$. For the typical chemostat, it is reasonable to ignore variations in the control volume and to assume that the velocities at the entrance and exit are constrained by

$$\mathbf{v}_D \cdot \mathbf{n} = \mathbf{v}_{H_2O} \cdot \mathbf{n}, \quad \text{at } A_e \tag{8.42}$$

so that Eq. 8.40 takes the form

$$\underbrace{V \frac{d \langle \rho_D \rangle}{dt}}_{\substack{\text{rate of accumulation of} \\ \text{cellular material} \\ \text{in the chemostat}}} + \underbrace{\langle \rho_D \rangle_2 Q_2}_{\substack{\text{rate at which} \\ \text{cellular material leaves} \\ \text{the chemostat}}}$$

$$\underbrace{- \quad \langle \rho_D \rangle_1 Q_1}_{\substack{\text{rate at which} \\ \text{cellular material enters} \\ \text{the chemostat}}} = \underbrace{\langle r_D \rangle V}_{\substack{\text{rate of production} \\ \text{of cellular material} \\ \text{in the chemostat}}} \tag{8.43}$$

It is the term on the right-hand side of this result that is important to us since it represents the *mass rate of production of cellular material* in the chemostat. Rather than work directly with this quantity, there is a tradition of using the *rate of production of cells* to describe the behavior of the chemostat. We define the *average mass of a cell* in the chemostat by

$$m_{cell} = \left\{ \begin{array}{c} \text{average mass} \\ \text{of a cell} \end{array} \right\} = \frac{\left\{ \begin{array}{c} \text{mass of cellular material} \\ \text{per unit volume} \end{array} \right\}}{\left\{ \begin{array}{c} \text{number of cells} \\ \text{per unit volume} \end{array} \right\}} \qquad (8.44)$$

and we represent the number of cells per unit volume by

$$\langle n \rangle = \left\{ \begin{array}{c} \text{number of cells} \\ \text{per unit volume} \end{array} \right\} \qquad (8.45)$$

This allows us to express the *mass of cellular material per unit volume* according to

$$\langle \rho_D \rangle = \langle n \rangle m_{cell} \qquad (8.46)$$

Given these definitions, we can divide Eq. 8.43 by the constant, m_{cell}, to obtain a macroscopic balance for the *number density* of cells that takes the form

$$V \frac{d\langle n \rangle}{dt} + \langle n \rangle_2 Q_2 - \langle n \rangle_1 Q_1 = \left(\langle r_D \rangle / m_{cell} \right) V \qquad (8.47)$$

Here, we have assumed that the average mass of a cell in the chemostat is *independent of time*, and this may not be correct for transient processes. In addition, Eq. 8.46 is based on the assumption that all of species D is contained within the cells. This is consistent with the cellular processes illustrated in Figure 8.9; however, that illustration *does not take into account* the process of cell death[**]. Because of cell death, Eq. 8.46 represents an over-estimate of the number of cells per unit volume.

Traditionally, one assumes that the volumetric flow rates entering and leaving the chemostat are equal so that Eq. 8.47 simplifies to

$$\frac{d\langle n \rangle}{dt} + \langle n \rangle_2 (Q/V) - \langle n \rangle_1 (Q/V) = \langle r_D \rangle / m_{cell} \qquad (8.48)$$

This represents a governing differential equation for cells per unit volume, $\langle n \rangle$; however, it is the cell concentration at the exit, $\langle n \rangle_2$, that we wish to predict, and this prediction is usually based on the assumption of a *perfectly mixed* system as described in Sec. 8.1. This assumption leads to $\langle n \rangle_2 = \langle n \rangle$ and it allows us to express Eq. 8.48 in the form

$$\underbrace{\frac{d\langle n \rangle}{dt}}_{accumulation} + \underbrace{\langle n \rangle (Q/V)}_{outflow} - \underbrace{\langle n \rangle_1 (Q/V)}_{inflow} = \underbrace{\left(\langle r_D \rangle / m_{cell} \right)}_{production} \qquad (8.49)$$

In previous sections of this chapter the quantity, V/Q, was identified as the *mean residence time* and denoted by τ. However, in the biochemical engineering literature,

---

[**] Bailey, J.E. and Ollis, D.F. 1986, Sec. 7.7, *Biochemical Engineering Fundamentals*, McGraw Hill Higher Education, 2nd Edition, New York.

the tradition is to identify Q/V as the *dilution rate* and denote it by D. Following this tradition, we express Eq. 8.49 in the form

$$\frac{d\langle n \rangle}{dt} + (\langle n \rangle - \langle n \rangle_1)D = \langle r_D \rangle / m_{cell} \tag{8.50}$$

where the term on the right-hand side should be interpreted as

$$\langle r_D \rangle / m_{cell} = \left\{ \begin{array}{c} number\ of\ cells \\ produced\ per\ unit \\ volume\ per\ unit\ time \end{array} \right\} \tag{8.51}$$

This rate of production, caused by biological reactions, is traditionally represented as

$$\langle r_D \rangle / m_{cell} = \mu \langle n \rangle \tag{8.52}$$

so that our governing differential equation for the cell concentration takes the form

$$\frac{d\langle n \rangle}{dt} + (\langle n \rangle - \langle n \rangle_1)D = \mu \langle n \rangle \tag{8.53}$$

The quantity, μ, is referred to as the *specific growth rate*, and if μ is known one can use this result to predict $\langle n \rangle$ as a function of time.

For many practical applications, there are no cells entering the chemostat, thus $\langle n \rangle_1$ is zero and we are dealing with what is called a *sterile feed*. For a sterile feed, the cell concentration is determined by the following governing equation and initial condition

$$\frac{d\langle n \rangle}{dt} + \langle n \rangle D = \mu \langle n \rangle \tag{8.54a}$$

I.C. $$\langle n \rangle = n_o, \quad t = 0 \tag{8.54b}$$

Here n_o represents the initial concentration of cells in the chemostat and this is usually referred to as the *inoculum*. If we treat the specific growth rate, μ, as a constant, the solution of the initial value problem for $\langle n \rangle$ is straightforward and is left as an exercise for the student.

The steady-state form of Eq. 8.54a is given by

$$(D - \mu)\langle n \rangle = 0 \tag{8.55}$$

and this indicates that the steady state can only exist when $\langle n \rangle = 0$ or when $D = \mu$. The first of these conditions is of no interest, while the second suggests that a steady-state chemostat might be rather rare since adjusting the *dilution rate*, $D = Q/V$, to be exactly equal to the *specific growth rate* might be very difficult if the specific growth rate were *constant*. However, a little thought indicates that the specific growth rate,

μ, must depend on the concentration of the nutrients entering the chemostat, thus μ can be controlled by adjusting the input conditions.

When the substrate B is present in excess, the rate of cell growth can be expressed in terms of the concentration of species A in the extracellular fluid, $\langle c_A \rangle$, a reference concentration, K_A, and other parameters according to

$$\mu = F\left(\langle c_A \rangle, K_A, \text{other parameters}\right) \qquad (8.56)$$

If a specific growth rate has the following characteristics

$$\mu = \begin{cases} 0 & \langle c_A \rangle \to 0 \\ \mu_{max} & \langle c_A \rangle \gg K_A \end{cases} \qquad (8.57)$$

it could be modeled by what is known as Monod's equation[††]

$$\text{Monod's equation: } \mu = \frac{\mu_{max} \langle c_A \rangle}{K_A + \langle c_A \rangle} \qquad (8.58)$$

The parameter K_A is sometimes referred to as the "half saturation" since it represents the concentration at which the growth rate is half the maximum growth rate, μ_{max}. It should be clear that there are many other functional representations that would satisfy Eq. 8.57; however, the form chosen by Monod has been used with reasonable success to correlate macroscopic experimental data[‡‡].

8.5   BATCH DISTILLATION

Distillation is a common method of separating the components of a solution. The degree of separation that can be achieved depends on the vapor-liquid equilibrium relation and the manner in which the distillation takes place. Salt and water are easily separated in solar ponds in a process that is analogous to batch distillation. In that case, the separation is essentially perfect since a negligible amount of salt is present in the vapor phase leaving the pond.

In this section, we wish to analyze the unit illustrated in Figure 8.11 which is sometimes referred to as a *simple still*. The process under consideration is obviously a transient one in which the unit is initially charged with M_o moles of a binary mixture containing species A and B. The initial mole fraction of species A is designated by x_A^o, and we will assume that the mole fraction of species A is small enough so that the ideal solution behavior discussed in Chapter 5 (see Eq. 5.30) provides an equilibrium relation of the form

$$y_A = \alpha_{AB}\, x_A \qquad (8.59)$$

[††] Monod, J. 1942, Recherche sur la Croissance des Cultures Bactériennes, Herman Editions, Paris.
[‡‡] Monod, J. 1949, The growth of bacterial cultures, Ann. Rev. Microbiol. **3**, 371–394.

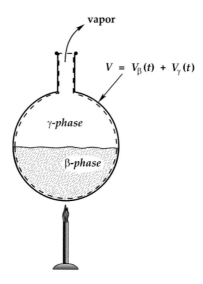

FIGURE 8.11   Batch distillation unit.

Here, y_A represents the mole fraction in the vapor phase and x_A represents the mole fraction in the liquid phase. In our analysis, we would like to predict the composition of the liquid during the course of the distillation process.

The control volume illustrated in Figure 8.11 is fixed in space and can be separated into the volume of the liquid (the β – phase) and the volume of the vapor (the γ – phase) according to

$$V = V_\beta(t) + V_\gamma(t) \tag{8.60}$$

Under normal circumstances, there will be no chemical reactions in a distillation process, and we can express the macroscopic mole balance for species A as

$$\frac{d}{dt}\int_V c_A \, dV + \int_A c_A \mathbf{v}_A \cdot \mathbf{n} \, dA = 0 \tag{8.61}$$

In addition to the mole balance for species A, we will need either the mole balance for species B or the total mole balance. The latter is more convenient in this particular case, and we express it as (see Sec. 4.4)

$$\frac{d}{dt}\int_V c \, dV + \int_A c \mathbf{v}^* \cdot \mathbf{n} \, dA = 0 \tag{8.62}$$

For the control volume shown in Figure 8.11, the molar flux is zero everywhere except at the exit of the unit and Eq. 8.61 takes the form

$$\frac{d}{dt}\left[\int_{V_\beta(t)} c_{A\beta} \, dV + \int_{V_\gamma(t)} c_{A\gamma} \, dV \right] + \int_{A_{exit}} c_{A\gamma} \mathbf{v}_{A\gamma} \cdot \mathbf{n} \, dA = 0 \tag{8.63}$$

Here, we have explicitly identified the control volume as consisting of the volume of the liquid (β – phase) and the volume of the vapor (γ – phase) At the exit of the control volume, we can ignore diffusive effects and replace $\mathbf{v}_{A\gamma} \cdot \mathbf{n}$ with $\mathbf{v}_\gamma \cdot \mathbf{n}$, and the concentration in both the liquid and vapor phases can be represented in terms of mole fractions so that Eq. 8.63 takes the form

$$\frac{d}{dt}\left[\int_{V_\beta(t)} x_A\, c_\beta\, dV + \int_{V_\gamma(t)} y_A\, c_\gamma\, dV\right] + \int_{A_{exit}} y_A\, c_\gamma \mathbf{v}_\gamma \cdot \mathbf{n}\, dA = 0 \tag{8.64}$$

If the total molar concentrations, c_β and c_γ, can be treated as constants, this result can be expressed as

$$\frac{d}{dt}\left(\langle x_A \rangle M_\beta\right) + \frac{d}{dt}\left(\langle y_A \rangle M_\gamma\right) + \int_{A_{exit}} y_A\, c_\gamma \mathbf{v}_\gamma \cdot \mathbf{n}\, dA = 0 \tag{8.65}$$

in which $\langle x_A \rangle$ and $\langle y_A \rangle$ are defined by

$$\langle x_A \rangle = \frac{1}{V_\beta(t)} \int_{V_\beta(t)} x_A\, dV, \quad \langle y_A \rangle = \frac{1}{V_\gamma(t)} \int_{V_\gamma(t)} y_A\, dV \tag{8.66}$$

In Eq. 8.65, we have used M_β and M_γ to represents the total number of moles in the liquid and vapor phases respectively. We can simplify Eq. 8.65 by imposing the restriction

$$\frac{d}{dt}\left(\langle y_A \rangle M_\gamma\right) \ll \frac{d}{dt}\left(\langle x_A \rangle M_\beta\right) \tag{8.67}$$

since c_γ is generally much, much less than c_β. Given this restriction, Eq. 8.65 takes the form

$$\frac{d}{dt}\left(\langle x_A \rangle M_\beta\right) + \int_{A_{exit}} y_A\, c_\gamma \mathbf{v}_\gamma \cdot \mathbf{n}\, dA = 0 \tag{8.68}$$

and we can express the flux at the exit in the traditional form to obtain

$$\frac{d}{dt}\left(\langle x_A \rangle M_\beta\right) + \langle y_A \rangle_{exit}\, c_\gamma\, Q_\gamma = 0 \tag{8.69}$$

This represents the governing equation for $\langle x_A \rangle$ and it is restricted to cases for which $c_\gamma \ll c_\beta$.

In addition to $\langle x_A \rangle$, there are other unknown terms in Eq. 8.69, and the *total mole balance* will provide information about one of these. Returning to Eq. 8.62, we apply that result to the control volume illustrated in Figure 8.11 to obtain

$$\frac{d}{dt}\left[\int_{V_\beta(t)} c_\beta\, dV + \int_{V_\gamma(t)} c_\gamma\, dV\right] + \int_{A_{exit}} c_\gamma \mathbf{v}_\gamma^* \cdot \mathbf{n}\, dA = 0 \tag{8.70}$$

At the exit of the control volume, we again ignore diffusive effects and replace $\mathbf{v}_\gamma^* \cdot \mathbf{n}$ with $\mathbf{v}_\gamma \cdot \mathbf{n}$ so that this result takes the form

$$\frac{d}{dt}\left(M_\beta + M_\gamma \right) + c_\gamma Q_\gamma = 0 \tag{8.71}$$

At this point, we again impose the restriction that $c_\gamma \ll c_\beta$ which allows us to simplify this result to the form

$$\frac{d M_\beta}{dt} + c_\gamma Q_\gamma = 0 \tag{8.72}$$

We can use this result to eliminate $c_\gamma Q_\gamma$ from Eq. 8.69 so that the mole balance for species A takes the form

$$\frac{d}{dt}\left(\langle x_A \rangle M_\beta \right) - \langle y_A \rangle_{exit} \frac{d M_\beta}{dt} = 0 \tag{8.73}$$

At this point, we have a single equation and three unknowns: $\langle x_A \rangle$, M_β and $\langle y_A \rangle_{exit}$, and our analysis has been only moderately restricted by the condition that $c_\gamma \ll c_\beta$. We have yet to make use of the equilibrium relation indicated by Eq. 8.59, and to be very precise in the next step in our analysis we repeat that equilibrium relation according to

Equilibrium relation: $y_A = \alpha_{AB}\, x_A$,   *at the vapor – liquid interface*   (8.74)

In our macroscopic balance analysis, we are confronted with the mole fractions indicated by $\langle x_A \rangle$ and $\langle y_A \rangle_{exit}$, and the values of these mole fractions *at the vapor-liquid interface* illustrated in Figure 8.11 are *not available to us*. Knowledge of x_A and y_A at the $\beta - \gamma$ interface can only be obtained by a detailed analysis of the *diffusive transport*[§§] that is responsible for the separation that occurs in batch distillation. In order to proceed with an *approximate solution* to the batch distillation process, we replace Eq. 8.74 with

Process equilibrium relation: $\langle y_A \rangle_{exit} = \alpha_{eff} \langle x_A \rangle$ (8.75)

Here we note that Eqs. 8.74 and 8.75 are analogous to Eqs. 5.49 and 5.50 if the approximation $\alpha_{eff} = \alpha_{AB}$ is valid. The *process equilibrium relation* suggested by Eq. 8.75 may be acceptable if the batch distillation process is *slow enough*, but we do not know what is meant by *slow enough* without a more detailed theoretical analysis or an experimental study in which theory can be compared with experiment.

Keeping in mind the uncertainty associated with Eq. 8.75, we use that result in Eq. 8.73 to obtain

$$M_\beta \frac{d\langle x_A \rangle}{dt} + (1 - \alpha_{eff})\langle x_A \rangle \frac{dM_\beta}{dt} = 0 \tag{8.76}$$

[§§] Bird, R.B., Stewart, W.E. and Lightfoot, E.N. 2002, *Transport Phenomena*, 2nd Edition, John Wiley & Sons, Inc., New York.

The initial conditions for the mole fraction, $\langle x_A \rangle$, and the number of moles in the still, $M_\beta(t)$, are given by

I.C.1 $\langle x_A \rangle = x_A^o, \quad t = 0$ (8.77)

I.C.2 $M_\beta = M_\beta^o, \quad t = 0$ (8.78)

At this point, we have a *single* differential equation and *two* unknowns, $\langle x_A \rangle$ and $M_\beta(t)$. Obviously, we cannot determine both of these quantities as a function of time unless some additional information is given. For example, if $M_\beta(t)$ were specified as a function of time we could use Eq. 8.76 to determine $\langle x_A \rangle$ as a function of time; however, without the knowledge of how $M_\beta(t)$ changes with time we can only determine $\langle x_A \rangle$ as a function of M_β. This represents a classic situation in many batch processes where one can only determine the *changes that take place between one state and another.* In this analysis, the state of the system is characterized by $\langle x_A \rangle$ and M_β.

Returning to Eq. 8.76, we divide by $\langle x_A \rangle M_\beta$ and multiply by dt in order to obtain

$$\frac{d\langle x_A \rangle}{\langle x_A \rangle} = -\left(1 - \alpha_{eff}\right) \frac{dM_\beta}{M_\beta}$$ (8.79)

Since α_{eff} will generally depend on the temperature, and the temperature at which the solution boils will depend on $\langle x_A \rangle$, we need to determine how α_{eff} depends upon $\langle x_A \rangle$ before the variables in Eq. 8.79 can be completely separated. Here, we will avoid this complication and treat α_{eff} as a constant so that Eq. 8.79 can be integrated leading to

$$\int_{\eta=x_A^o}^{\eta=\langle x_A \rangle} \frac{d\eta}{\eta} = -\left(1 - \alpha_{eff}\right) \int_{\xi=M_\beta^o}^{\xi=M_\beta(t)} \frac{d\xi}{\xi}$$ (8.80)

Evaluation of the integrals allows one to obtain a solution for $\langle x_A \rangle$ given by

$$\langle x_A \rangle = x_A^o \left[\frac{M_\beta(t)}{M_\beta^o} \right]^{\alpha_{eff} - 1}$$ (8.81)

Once again, we must remember that α_{eff} is a *process equilibrium relation* that will generally depend on the temperature which will change during the course of a batch distillation. Nevertheless, we can use Eq. 8.81 to provide a *qualitative indication* of how the mole fraction of the liquid phase changes during the course of a batch distillation process.

When α_{eff} is greater than one ($\alpha_{eff} > 1$) we can see from Eq. 8.75 that the vapor phase is *richer* in species A than the liquid, and Eq. 8.81 predicts a *decreasing* value of $\langle x_A \rangle$ as the number of moles of liquid decreases. For the case in which α_{eff} takes on a variety of values, we have indicated the normalized mole fraction $\langle x_A \rangle / x_A^o$ as a

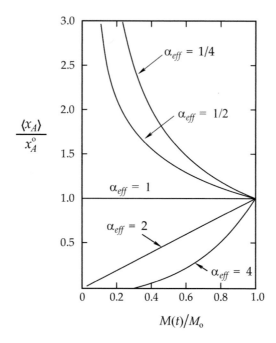

FIGURE 8.12 Composition of a binary system in a batch still.

function of $M_\beta(t)/M_\beta^o$ in Figure 8.12. There we can see that a significant separation takes place when α_{eff} is either large or small compared to one. The results presented in Figure 8.12 are certainly quite plausible; however, one must keep in mind that they are based on the *process equilibrium relation* represented by Eq. 8.75. Whenever one is confronted with an assumption of uncertain validity, experiments should be performed, or a more comprehensive theory should be developed, or both.

8.6 PROBLEMS

SECTION 8.1

8.1. A tank containing 200 gal of saturated salt solution (3 lb_m of salt per gallon) is to be diluted by the addition of brine containing 1 lb_m of salt per gallon. If this solution enters the tank at a rate of 4 gal/min and the mixture leaves the tank at the same rate, how long will it take for the concentration in the tank to be reduced to a concentration of 1.01 lb_m/gal?

8.2. A salt solution in a perfectly stirred tank is washed out with fresh water at a rate such that the average residence time, V/Q, is 10 minutes. Calculate the following:

  a. The time in minutes required to remove 99% of the salt originally present
  b. The percentage of the original salt removed after the addition of one full tank of fresh water.

8.3. Two reactants are added to a stirred tank reactor as illustrated in Figure 8-3. In the inlet stream containing reactant #2, there is also a miscible liquid catalyst which is added at a level that yields a concentration *in the reactor* of 0.002 mol/L. The volumetric flow rates of the two reactant streams are equal and the total volumetric rate entering and leaving the stirred tank is 15gal/min.

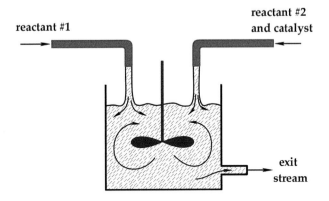

FIGURE 8-3   Catalyst mixing process.

It has been decided to change the type of catalyst and the change will be made by a substitution of the new catalyst for the old as the reactant and catalyst are continuously pumped into the 10,000-gal tank. The inlet concentration of the new catalyst is adjusted to provide a final concentration of 0.0030 mol/L when the mixing process is operating at steady state. Determine the time in minutes required for the concentration of the new catalyst to reach 0.0029 mol/L.

8.4. Three perfectly stirred tanks, each of 10,000-gal capacity, are arranged so that the effluent of the first is the feed to the second and the effluent of the second is the feed to the third. Initially, the concentration in each tank is c_o. Pure water is then fed to the first tank at the rate of 50 gal/min. You are asked to determine:

a. The time required to reduce the concentration in the first tank to 0.10 c_o
b. The concentrations in the second and third tanks at this time
c. A general equation for the concentration in the *nth* tank in the cascade system illustrated in Figure 8-4.

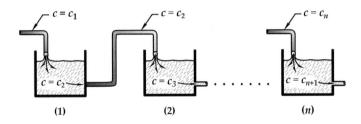

FIGURE 8-4   Sequence of stirred tanks.

In order to solve an ordinary differential equation of the form

$$\frac{d\langle c_A \rangle}{dt} + g(t)\langle c_A \rangle = f(t) \tag{1}$$

explore the possibility that this *complex equation* can be transformed to the *simple equation* given by

$$\frac{d}{dt}\left[a(t)\langle c_A \rangle\right] = b(t) \tag{2}$$

We refer to this as a simple equation because it can be integrated directly to obtain

$$a(t)\langle c_A \rangle = \left[a(t)\langle c_A \rangle\right]_{t=0} + \int_{\xi=0}^{\xi=t} b(\xi)d\xi \tag{3}$$

The new functions, $a(t)$ and $b(t)$ can be determined by noting that

$$\frac{1}{a(t)}\frac{da}{dt} = g(t), \quad b(t) = a(t)f(t) \tag{4}$$

Any initial condition for $a(t)$ will suffice since the solution for $\langle c_A \rangle$ does not depend on the initial condition for $a(t)$. Use of dummy variables of integration is essential in order to avoid confusion.

8.5. Develop a solution for Eqs. 8.20 and 8.21 when $f(t)$ is given by

$$f(t) = \begin{cases} c_A^o + \left(c_A^1 - c_A^o \right) t/\Delta t, & 0 < t < \Delta t \\ c_A^1, & t \geq \Delta t \end{cases} \tag{1}$$

This represents a process in which the original steady-state concentration is c_A^o and the final steady-state concentration is c_A^1. The *response time* for the mechanism that creates the change in the concentration of the incoming stream is Δt while the response time of the tank can be thought of as the *mean residence time*, τ. The time required to approach within 1% of the new steady state is designated as t_1, and we express this idea as

$$\langle c_A \rangle = c_A^1 + 0.01\left(c_A^o - c_A^1 \right), \quad t = t_1 \tag{2}$$

For $\Delta t = 0$, we know that $t_1/\tau = 4.6$ on the basis of Eq. 8.19. In this problem, we want to know what value of t_1/τ is required to approach within 1% of the new steady state when $\Delta t/\tau = 0.2$.

8.6. Volume 2 of the Guidelines for Incorporating Safety and Health into Engineering Curricula published by the Joint Council for Health, Safety, and Environmental

Education of Professionals (JCHSEEP) introduces, without derivation, the following equation for the determination of concentration of contaminants inside a room:

$$\frac{G - Q^* C}{G} = e^{-Q^* t/V}$$

where:

C = concentration of contaminant at time t
G = rate of generation of contaminant
Q = effective rate of ventilation
Q^* = Q/K
V = volume of room enclosure
K = design distribution ventilation constant
t = length of time to reach concentration C.

Making the appropriate assumptions, derive the above equation from the material
balance of contaminants. What would be a proper set of units for the variables contained in this equation?

8.7. Often, it is convenient to express the transient concentration in a perfectly mixed
stirred tank in terms of the dimensionless concentration defined by

$$C_A = \frac{\langle c_A \rangle - c_A^o}{c_A^1 - c_A^o}$$

Represent the solution given by Eq. 8.19 in terms of this dimensionless concentration.

8.8. A perfectly mixed stirred tank reactor is illustrated in Figure 8-8. A feed stream
of reactants enters at a volumetric flow rate of Q_o and the volumetric flow rate leaving the reactor is also Q_o. Under steady state conditions, the tank is half full and the
volume of the reacting mixture is V_o. At the time $t = 0$, an *inert species* is added to the
system at a concentration c_A^o and a volumetric flow rate Q_1. Unfortunately, someone
forgot to change a downstream valve setting and the volumetric flow rate leaving
the tank remains constant at the value Q_o. This means that the tank will overflow at

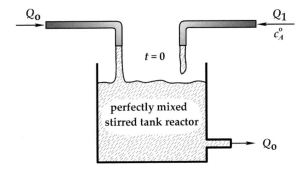

FIGURE 8-8   Accidental overflow from a stirred tank reactor.

$t = V_o/Q_1$. While species A is *inert* in terms of the reaction taking place in the tank, it is *mildly toxic* and you need to predict the concentration of species A in the fluid when the tank overflows. Derive a general expression for this concentration taking into account that the concentration of species A in the tank is zero at $t = 0$.

SECTION 8.2

8.9. Show how Eq. 8.27 can be used to derive Eq. 8.28.

8.10. A perfectly mixed batch reactor is used to carry out the reversible reaction described by

Chemical kinetic schema: $$A + E \underset{k_2}{\overset{k_1}{\rightleftharpoons}} B \tag{1}$$

The use of mass action kinetics provides a chemical reaction rate equation given by

Chemical reaction rate equation: $$R_A = \underbrace{-k_1 c_A c_E}_{\substack{\text{second order} \\ \text{reaction}}} + \underbrace{k_2 c_B}_{\substack{\text{first order} \\ \text{decomposition}}} \tag{2}$$

If species E is present in great excess, the concentration c_E will undergo negligible changes during the course of the process and we can define a pseudo first order rate coefficient by

$$k_1' = k_1 c_E \tag{3}$$

Given initial conditions of the form

I.C. $$c_A = c_A^o, \quad c_B = c_B^o, \quad t = 0 \tag{4}$$

use the pseudo first order rate coefficient to determine the concentration of species A and B as a function of time.

8.11. When molecular species A and B combine to form a product, one often adopts the chemical kinetic schema given by

$$A + B \xrightarrow{k} \text{products} \tag{1}$$

Use of mass action kinetics then leads to a chemical kinetic rate equation of the form

$$R_A = -k c_A c_B \tag{2}$$

One must always look upon such rate expressions as assumptions to be tested by experimental studies. For a homogeneous, liquid-phase reaction, this test can be carried out in a batch reactor which is subject to the initial conditions

I.C.1 $\qquad\qquad\qquad c_A = c_A^\circ, \quad t = 0$ $\qquad\qquad$ (3)

I.C.2 $\qquad\qquad\qquad c_B = c_B^\circ, \quad t = 0$ $\qquad\qquad$ (4)

Use the macroscopic mole balances for both species A and B, along with the stoichiometric constraint,

$$R_A = R_B \qquad\qquad (5)$$

in order to derive an expression for c_A as a function of time. In this case, one is forced to assume perfect mixing so that $\langle c_A c_B \rangle$ can be replaced by $\langle c_A \rangle\langle c_B \rangle$.

8.12. A batch reactor illustrated in Figure 8-12 is used to study the irreversible, decomposition reaction

$$A \xrightarrow{\ k\ } \text{products} \qquad\qquad (1)$$

The proposed chemical kinetic rate equation is

$$R_A = -k c_A \qquad\qquad (2)$$

and this decomposition reaction is catalyzed by sulfuric acid. To initiate the batch process, a small volume of catalyst is placed in the reactor as illustrated in Figure 8-12. At the time, $t = 0$, the solution of species A is added at a volumetric flow rate Q_o and a concentration c_A°. When the reactor is full, the stream of species A is shut off and the system proceeds in the normal manner for a batch reactor. During the start-up time, the volume of fluid in the reactor can be expressed as

$$V(t) = V_o + Q_o t \qquad\qquad (3)$$

and the final volume of the fluid is given by

$$V_1 = V_o + Q_o t_1 \qquad\qquad (4)$$

Here, t_1 is the start-up time. In this problem, you are asked to determine the concentration of species A during the start-up time and all subsequent times. The analysis for the start-up time can be simplified by means of the transformation

$$y(t) = \langle c_A \rangle V(t) \qquad\qquad (5)$$

and use of the initial condition

I.C. $\qquad\qquad\qquad y = 0, \quad t = 0$ $\qquad\qquad$ (6)

After you have determined $y(t)$, you can easily determine $\langle c_A \rangle$ during the start-up period. The concentration at $t = t_1$ then becomes the initial condition for the analysis of the system for all subsequent times. Often one simplifies the analysis of a batch reactor by assuming that the time required to fill the reactor is *negligible*. Use your solution to this problem to identify what is meant by "negligible" for this particular problem.

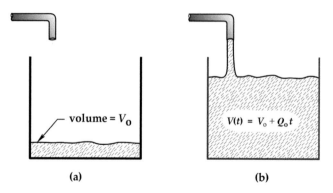

FIGURE 8-12   Batch reactor start-up process.

8.13. In the perfectly mixed continuous stirred tank reactor illustrated in Figure 8-13a, species A undergoes an irreversible reaction to form products according to

$$A \xrightarrow{k} products, \quad R_A = -k \, c_A \tag{1}$$

The original volume of fluid in the reactor is V_0 and the original volumetric flow rate *into and out of* the reactor is Q_0. The concentration of species A *entering* the reactor is fixed at c_A° and under steady state operating conditions the concentration at the exit (and therefore the concentration in the reactor) is $\langle c_A \rangle$.

Part (a). Determine the concentration $\langle c_A \rangle$ under steady state operating conditions.

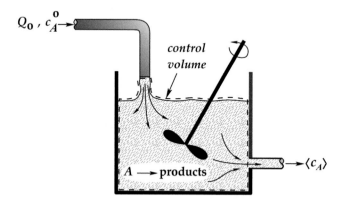

FIGURE 8-13a   Steady batch reactor.

Part (b). Because of changes in the downstream processing, it is necessary to reduce the concentration of species A leaving the reactor. This is to be accomplished by increasing the volume of the reactor by adding pure liquid to the reactor, as illustrated in Figure 8-13b, at a volumetric flow rate Q_1 until the desired reactor volume is achieved. During the transient period when the volume is given by $V(t) = V_0 + Q_1 t$,

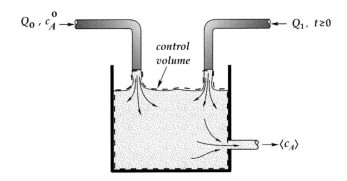

FIGURE 8-13b   Transient process in a perfectly mixed stirred tank reactor.

the exit flow rate is constant at Q_o. In this problem, you are asked to determine the exit concentration during this transient period.

SECTION 8.3

8.14. Consider the process studied in Example 8.1, subject to an initial condition of the form,

I.C. $$c_A = c_A^o, \quad c_B = c_B^o, \quad t = 0$$

and determine the concentration of species A as a function of time.

8.15. For the process studied in Example 8.1, assume that the equilibrium coefficient and the first order rate coefficient have the values

$$K_{eq} = 10^{-1}, \quad k_1 = 10 \text{ min}^{-1} \tag{1}$$

and determine the time, t_1 required for the concentration of species A to be given by

$$\langle c_A \rangle = c_A^o + 0.99 \left(\langle c_A \rangle_{eq} - c_A^o \right), \quad t = t_1 \tag{2}$$

Here $\langle c_A \rangle_{eq}$ represents the equilibrium concentration.

8.16. In this problem, we consider the heated, semi-batch reactor shown in Figure 8-16 where we have identified the vapor phase as the γ – phase and the liquid phase as the β – phase. This reactor has been designed to determine the chemical kinetics of the dehydration of t-butyl alcohol (species A) to produce isobutylene (species B) and water (species C). The system is initially charged with t-butyl alcohol; a catalyst is then added which causes the dehydration of the alcohol to form isobutylene and water. The isobutylene escapes through the top of the reactor while the water and t-butyl alcohol are condensed and remain in the reactor. If one measures the concentration of the t-butyl alcohol in the liquid phase, the rate of reaction can be determined and that is the objective of this particular experiment.

The analysis begins with the fixed control volume shown in Figure 8-16 and the general macroscopic balance given by

$$\frac{d}{dt}\int_V c_D \, dV + \int_A c_D \mathbf{v}_D \cdot \mathbf{n} \, dA = \int_V R_D \, dV, \quad D = A, B, C \tag{1}$$

The moles of species A (t-butyl alcohol) in the γ – phase can be neglected, thus the macroscopic balance for this species takes the form

t-butyl alcohol:
$$\frac{d}{dt}\int_{V_\beta(t)} c_{A\beta} \, dV = \int_{V_\beta(t)} R_{A\beta} \, dV \tag{2}$$

and in terms of average values for the concentration and the net rate of production of species A we have

t-butyl alcohol:
$$\frac{d}{dt}\Big[\langle c_{A\beta}\rangle V_\beta(t)\Big] = \langle R_{A\beta}\rangle V_\beta(t) \tag{3}$$

If we also assume that the moles of species B (isobutylene) and species C (water) are negligible in the γ – phase, the macroscopic balances for these species take the form

isobutylene:
$$\frac{d}{dt}\Big[\langle c_{B\beta}\rangle V_\beta(t)\Big] + \dot{M}_{B\gamma} = \langle R_{B\beta}\rangle V_\beta(t) \tag{4}$$

water:
$$\frac{d}{dt}\Big[\langle c_{C\beta}\rangle V_\beta(t)\Big] = \langle R_{C\beta}\rangle V_\beta(t) \tag{5}$$

This indicates that alcohol and water are retained in the system by the condenser while the isobutylene leaves the system at a molar rate given by $\dot{M}_{B\gamma}$. The initial conditions for the three molecular species are given by

I.C. $$\langle c_A\rangle = c_A^o, \quad t = 0 \tag{6a}$$

I.C. $$\langle c_B\rangle = 0, \quad t = 0 \tag{6b}$$

I.C. $$\langle c_C\rangle = 0, \quad t = 0 \tag{6c}$$

Since the molar rates of reaction are related by

$$R_{C\beta} = -R_{A\beta}, \quad \text{and} \quad R_{B\beta} = -R_{A\beta} \tag{7}$$

we need only be concerned with the rate of reaction of the t-butyl alcohol. If we treat the reactor as *perfectly mixed*, the mole balance for t-butyl alcohol can be expressed as

$$R_{A\beta} = \frac{dc_{A\beta}}{dt} + \frac{c_{A\beta}}{V_\beta(t)}\frac{dV_\beta(t)}{dt} \tag{8}$$

This indicates that we need to know both $c_{A\beta}$ and V_β as functions of time in order to obtain *experimental values* of $R_{A\beta}$. The volume of fluid in the reactor can be expressed as

$$V_\beta(t) = n_{A\beta}\overline{V}_A + n_{B\beta}\overline{V}_B + n_{C\beta}\overline{V}_C \qquad (9)$$

in which $n_{A\beta}$, $n_{B\beta}$ and $n_{C\beta}$ represent the moles of species A, B and C in the β − phase and $\overline{V}_A$, $\overline{V}_B$ and $\overline{V}_C$ represent the partial molar volumes.

To develop a useful expression for $R_{A\beta}$, *assume* that the liquid mixture is ideal so that the partial molar volumes are constant. In addition, *assume* that the moles of species B in the liquid phase are negligible. On the basis of these assumptions, show that Eq. 8 can be expressed as

$$R_{AB} = \frac{dc_{A\beta}}{dt}\left\{1 + \frac{c_{A\beta}(\overline{V}_A - \overline{V}_C)}{\left[1 - c_{A\beta}(\overline{V}_A - \overline{V}_C)\right]}\right\} \qquad (10)$$

This form is especially useful for the interpretation of *initial rate data*, i.e., experimental data can be used to determine both $c_{A\beta}$ and $dc_{A\beta}/dt$ at $t = 0$ and this provides an experimental determination of $R_{A\beta}$ for the initial conditions associated with the experiment.

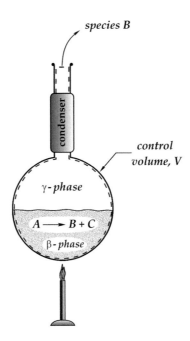

FIGURE 8-16 Semi-batch reactor for determination of chemical kinetics.

An alternate approach[***] to the determination of $R_{A\beta}$ is to measure the molar flow rate of species B that leaves the reactor in the γ – phase and relate that quantity to the rate of reaction.

Section 8.4

8.17. When Eqs. 8.41 and 8.42 are valid, Eq. 8.43 represents a valid result for the chemostat shown in Figure 8.10. One can divide this equation by a constant, m_{cell}, in order to obtain Eq. 8.47; however, the average mass of a cell, m_{cell}, in the chemostat may not be the average mass of a cell in the incoming stream. If m_{cell} is different than $(m_{cell})_1$, Eq. 8.47 is still correct, but our *interpretation* of $\langle n \rangle_1$ is not correct. Consider the case, $(m_{cell})_1 \neq m_{cell}$, and develop a new version of Eq. 8.47 in which the incoming flux of cells is interpreted properly in terms of the number of cells per unit volume in the incoming stream.

8.18. Develop a general solution for Eq. 8.54 and consider the behavior of the system for $D < \mu$, $D = \mu$ and $D > \mu$.

Section 8.5

8.19. Repeat the analysis of binary batch distillation when Raoult's law is applicable, i.e., the *process equilibrium relation* is given by

$$\langle y_A \rangle_{exit} = \frac{\alpha_{eff} \langle x_A \rangle}{1 + \langle x_A \rangle (\alpha_{eff} - 1)}$$

Here, α_{eff} is the effective relative volatility which is temperature dependent; however, in this problem, you may assume that α_{eff} is constant. Use your result to construct figures comparable to Figure 8.12 for $x_A^o = 0.1$ and $x_A^o = 0.5$ with values of α_{eff} given by 1/4, 1/2, 1, 2, and 4.

[***] Gates, B.C. and Sherman, J.D. 1975, Experiments in heterogeneous catalysis: Kinetics of alcohol dehydration reactions, Chem. Eng. Ed. Summer, 124–127.

9 Reaction Kinetics

In Chapter 6, we introduced stoichiometry as the concept that atomic species are neither created nor destroyed by chemical reactions. In Chapters 7 and 8, we applied Axioms I and II to the *analysis* of systems with reactors, separators, and recycle streams. The pivot theorem (see Sec. 6.4) is an essential part of the *analysis* of a chemical reactor. To *design* a chemical reactor[*], we need information about the rates of chemical reaction in terms of the concentration of the reacting species.

9.1 CHEMICAL KINETICS

In order to *predict* the concentration changes that occur in reactors, we need to make use of Axiom I (see Eq. 6.1) and Axiom II (see Eq. 6.2) in addition to *chemical reaction rate equations* that allow us to express the net rates of production, R_A, R_B, etc., in terms of the concentrations, c_A, c_B, etc. The subject of chemical kinetics brings us into contact with the *chemical kinetic schemas* that are used to illustrate reaction mechanisms. To be useful, these *schemas* must be translated to *equations* and we will illustrate how this is done in the following paragraphs.

Hydrogen bromide (HBr) reaction

As an example of both stoichiometry and chemical kinetics, we consider the reaction of hydrogen with bromine to produce hydrogen bromide. Keeping in mind the principle of stoichiometric skepticism (see Sec. 6.1.1) one could *assume* that the molecular species involved are H_2, Br_2, and HBr, and this idea is illustrated in Figure 9.1. There we have suggested that the reaction does not go to completion since both hydrogen and bromine appear in the product stream. Here it is important to note that the products of a chemical reaction are determined by *experiment*, and in this case, experimental data indicate that hydrogen bromide can be produced by reacting hydrogen and bromine.

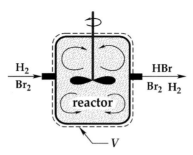

FIGURE 9.1   Production of hydrogen bromide.

---

[*] Whitaker, S. and Cassano, A.E. 1986, *Concepts and Design of Chemical Reactors*, Gordon and Breach Science Publishers, New York.

DOI: 10.1201/9781003283751-9   **323**

For the process illustrated in Figure 9.1, the visual representation of the atomic matrix takes the form

$$\text{Molecular species} \rightarrow \begin{array}{ccc} H_2 & Br_2 & HBr \end{array}$$

$$\begin{array}{c} hydrogen \\ bromine \end{array} \begin{bmatrix} 2 & 0 & 1 \\ 0 & 2 & 1 \end{bmatrix} \tag{9.1}$$

and the elements of this matrix can be expressed explicitly as

$$[N_{JA}] = \begin{bmatrix} 2 & 0 & 1 \\ 0 & 2 & 1 \end{bmatrix} \quad \text{or} \quad A = \begin{bmatrix} 2 & 0 & 1 \\ 0 & 2 & 1 \end{bmatrix} \tag{9.2}$$

The components of N_{JA} are used with Axiom II (see Eq. 6.20)

Axiom II:
$$\sum_{A=1}^{A=N} N_{JA} R_A = 0, \quad J = 1, 2, \ldots, T \tag{9.3}$$

in order to develop the stoichiometric relations between the three net rates of production represented by R_{H_2}, R_{Br_2}, and R_{HBr}. For the atomic matrix given by Eq. 9.2, we see that Axiom II provides

Axiom II:
$$\begin{bmatrix} 2 & 0 & 1 \\ 0 & 2 & 1 \end{bmatrix} \begin{bmatrix} R_{H_2} \\ R_{Br_2} \\ R_{HBr} \end{bmatrix} = \begin{bmatrix} 0 \\ 0 \end{bmatrix} \tag{9.4}$$

and the use of the *row reduced echelon form* of the atomic matrix leads to

$$\begin{bmatrix} 1 & 0 & 1/2 \\ 0 & 1 & 1/2 \end{bmatrix} \begin{bmatrix} R_{H_2} \\ R_{Br_2} \\ R_{HBr} \end{bmatrix} = \begin{bmatrix} 0 \\ 0 \end{bmatrix} \tag{9.5}$$

If HBr is chosen to be the single *pivot species* (see Example 6.1) we can express Axiom II in the form

Pivot theorem:
$$\begin{bmatrix} R_{H_2} \\ R_{Br_2} \end{bmatrix} = \underbrace{\begin{bmatrix} -1/2 \\ -1/2 \end{bmatrix}}_{pivot\ matrix} [R_{HBr}] \tag{9.6}$$

This matrix equation provides the following representations for the net rates of production of hydrogen and bromine

Local stoichiometry:
$$R_{H_2} = -\frac{1}{2} R_{HBr}, \quad R_{Br_2} = -\frac{1}{2} R_{HBr} \tag{9.7}$$

At this point, we wish to apply Axioms I and II to the control volume illustrated in Figure 9.1. The axioms are given by Eqs. 7.4 and 7.5 and repeated here as

Axiom I:
$$\frac{d}{dt} \int_V c_A \, dV + \int_A c_A \mathbf{v}_A \cdot \mathbf{n} \, dA = \mathcal{R}_A, \quad A = 1, 2, \ldots, N \tag{9.8}$$

Axiom II:
$$\sum_{A=N}^{A=1} N_{JA} \mathcal{R}_A = 0, \quad J = 1, 2, \ldots, T \tag{9.9}$$

For the particular case under consideration, Eq. 9.8 leads to

$$-Q\langle c_A \rangle_{entrance} + Q\langle c_A \rangle_{exit} = \mathcal{R}_A, \quad A \Rightarrow H_2, Br_2, HBr \tag{9.10}$$

while Eq. 9.9 takes the special form given by

Global stoichiometry:
$$\mathcal{R}_{H_2} = -\frac{1}{2}\mathcal{R}_{HBr}, \quad \mathcal{R}_{Br_2} = -\frac{1}{2}\mathcal{R}_{HBr} \tag{9.11}$$

Since there is no hydrogen bromide in the inlet stream illustrated in Figure 9.1, the steady-state macroscopic balance provides the following result for hydrogen bromide

$$Q\langle c_{HBr} \rangle_{exit} = \mathcal{R}_{HBr} \tag{9.12}$$

Here we see that measurements of the volumetric flow rate and the exit concentration of HBr provide an experimental determination of $\mathcal{R}_{HBr}$.

The concepts of local and global stoichiometry are illustrated in Figure 9.2 where we suggest that *activated* hydrogen atoms, $H\bullet$, and *activated* bromine atoms, $Br\bullet$, participate in the reaction at the *local level*, but are not detectable at the *macroscopic level*. What is not detectable at the macroscopic level is often neglected, and we have done this in our representation of the hydrogen bromide reaction given by Eq. 9.11. We represent this simplification as

$$\mathcal{R}_{HBr} \ggg \langle R_{H\bullet} \rangle \quad \mathcal{R}_{HBr} \ggg \langle \mathcal{R}_{Br\bullet} \rangle \tag{9.13}$$

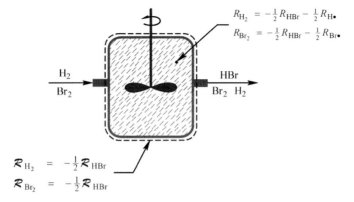

$$R_{H_2} = -\frac{1}{2}R_{HBr} - \frac{1}{2}R_{H\bullet}$$
$$R_{Br_2} = -\frac{1}{2}R_{HBr} - \frac{1}{2}R_{Br\bullet}$$

H_2
Br_2

HBr
$Br_2 \ H_2$

$$\mathcal{R}_{H_2} = -\frac{1}{2}\mathcal{R}_{HBr}$$
$$\mathcal{R}_{Br_2} = -\frac{1}{2}\mathcal{R}_{HBr}$$

FIGURE 9.2 Production of hydrogen bromide: Local and global stoichiometry.

and this is indicated by the global stoichiometry shown in Figure 9.2. At this point, we assume that the reactor is *perfectly mixed* and this provides the simplification

$$\underbrace{\langle c_A \rangle}_{\substack{\text{volume average} \\ \text{concentration in} \\ \text{the reactor}}} = \underbrace{\langle c_A \rangle_{exit}}_{\substack{\text{area average} \\ \text{concentration} \\ \text{in the exit}}} = \underbrace{c_A}_{\substack{\text{concentration} \\ \text{at a point in} \\ \text{the reactor}}}, \quad A = 1, 2, \dots, N \quad (9.14)$$

For the specific system illustrated in Figure 9.2, the assumption of perfect mixing leads to

$$\langle c_A \rangle = \langle c_A \rangle_{exit} = c_A, \quad A \Rightarrow H_2, Br_2, HBr \quad (9.15)$$

Given these simplifications we can discuss the process illustrated in Figure 9.1 in terms of local conditions for which the chemical kinetics may be illustrated by a schema of the form

Chemical kinetic schema: $\quad H_2 + Br_2 \rightarrow 2HBr \quad (9.16)$

This schema suggests that a molecule of hydrogen collides with a molecule of bromine to produce two molecules of hydrogen bromide as illustrated in Figure 9.3. The frequency of the collisions that cause the reaction depends on the product of the two concentrations, c_{H_2} and c_{Br_2}, and this leads to the kinetic model given by[†]

Mass action kinetics: $\quad R_{HBr} = 2\, k c_{H_2} c_{Br_2} \quad (9.17)$

Chemical kinetics are traditionally represented in *local form*, as indicated in Eq. 9.17, *even when* they are based on *macroscopic observations* as we have suggested in Figures 9.1 and 9.2.

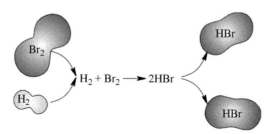

FIGURE 9.3 Molecular collision leading to a chemical reaction.

Experimental studies of the reaction of hydrogen and bromine to form hydrogen bromide were carried out by Bodenstein and Lind[‡] in a well-mixed batch reactor, and those experiments indicate that the net rate of production of hydrogen bromide can be expressed as

Experimental kinetics: $\quad R_{HBr} = \dfrac{k\, c_{H_2} \sqrt{c_{Br_2}}}{1 + k'\left(c_{HBr} / c_{Br_2}\right)} \quad (9.18)$

[†] Horn, F. and Jackson, R. 1972, General mass action kinetics, Arch. Rat. Mech. **47**, 81–116.
[‡] Bodenstein, M. and Lind, S.C. 1907, Geschwindigkeit der Bildung der Bromwasserstoffes aus sienen Elementen, Z. physik. Chem. **57**, 168–192.

This experimental result is certainly *not consistent* with the mass action kinetics in the form given by Eq. 9.17.

At this point, we need to summarize our results for the reaction of hydrogen with bromine to produce hydrogen bromide.

1. Eq. 9.16 is *not* an equation but it is a useful representation of what occurs at the macroscopic level.
2. Eq. 9.17 is an alternate method of describing what is represented by Eq. 9.16; however, it does not accurately represent the physical process.
3. Eq. 9.18 represents a reliable expression for the rate of reaction and can be used in the design of a chemical reactor.

As another example of an apparently simple reaction, we consider the gas-phase decomposition of azomethane, $(CH_3)_2N_2$, to produce ethane, C_2H_6, and nitrogen, N_2. This reaction is illustrated in Figure 9.4 where we have indicated that azomethane appears in both the input and the output streams. Here we have *not indicated* the presence of any *activated species* and we will return to this matter in subsequent paragraphs.

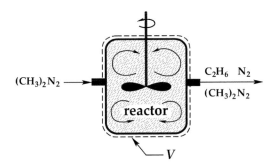

FIGURE 9.4   Decomposition of azomethane.

The visual representation of the atomic matrix for this system is given by

$$
\text{Molecular species} \quad \rightarrow \quad
\begin{array}{ccc}
C_2H_6 & N_2 & (CH_3)_2N_2
\end{array}
$$

$$
\begin{array}{l}
\textit{carbon} \\
\textit{nitrogen} \\
\textit{hydrogen}
\end{array}
\begin{bmatrix}
2 & 0 & 2 \\
0 & 2 & 2 \\
6 & 0 & 6
\end{bmatrix}
\tag{9.19}
$$

Use of this representation with Axiom II provides

$$
\text{Axiom II:} \qquad
\begin{bmatrix}
2 & 0 & 2 \\
0 & 2 & 2 \\
6 & 0 & 6
\end{bmatrix}
\begin{bmatrix}
R_{C_2H_6} \\
R_{N_2} \\
R_{(CH_3)_2N_2}
\end{bmatrix}
=
\begin{bmatrix}
0 \\
0 \\
0
\end{bmatrix}
\tag{9.20}
$$

This can be expressed in terms of the *row reduced echelon form* of the atomic matrix in order to obtain

$$
\begin{bmatrix} 1 & 0 & 1 \\ 0 & 1 & 1 \\ 0 & 0 & 0 \end{bmatrix}
\begin{bmatrix} R_{C_2H_6} \\ R_{N_2} \\ R_{(CH_3)_2N_2} \end{bmatrix}
=
\begin{bmatrix} 0 \\ 0 \\ 0 \end{bmatrix}
\tag{9.21}
$$

and a row-row partition of this matrix leads to

$$
\begin{bmatrix} 1 & 0 & 1 \\ 0 & 1 & 1 \end{bmatrix}
\begin{bmatrix} R_{C_2H_6} \\ R_{N_2} \\ R_{(CH_3)_2N_2} \end{bmatrix}
=
\begin{bmatrix} 0 \\ 0 \end{bmatrix}
\tag{9.22}
$$

Use of the pivot theorem (see Sec. 6.2.4) allows us to express the net rates of production for ethane and nitrogen in terms of azomethane according to

$$
\begin{bmatrix} R_{C_2H_6} \\ R_{N_2} \end{bmatrix}
=
\underbrace{\begin{bmatrix} -1 \\ -1 \end{bmatrix}}_{pivot\ matrix}
\begin{bmatrix} R_{(CH_3)_2N_2} \end{bmatrix}
\tag{9.23}
$$

and this result leads to the local stoichiometric relations given by

Local stoichiometry: $\qquad R_{C_2H_6} = - R_{(CH_3)_2N_2}, \quad R_{N_2} = - R_{(CH_3)_2N_2}$ $\qquad$ (9.24)

The result for global stoichiometry is based on Eq. 9.9 that leads to

Global stoichiometry: $\qquad \mathscr{R}_{C_2H_6} = -\mathscr{R}_{(CH_3)_2N_2}, \quad \mathscr{R}_{N_2} = -\mathscr{R}_{(CH_3)_2N_2}$ $\qquad$ (9.25)

The relation between local and global stoichiometry is illustrated in Figure 9.5. The fact that these two relations are identical in form is based on the *assumption* that only

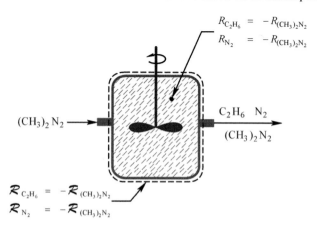

FIGURE 9.5 Local and global stoichiometry for decomposition of azomethane.

azomethane, ethane, and nitrogen are present in *significant amounts*. At this point
we accept Eqs. 9.24 and 9.25 as being valid; however, we note that the principle of
stoichiometric skepticism discussed in Sec. 6.1.1 should always be kept in mind.

As we did in the case of the hydrogen bromide reaction, we begin with the simplest possible chemical kinetic schema given by

Chemical kinetic schema: $(CH_3)_2N_2 \rightarrow C_2H_6 + N_2$ (9.26)

This schema suggests that a molecule of azomethane spontaneously decomposes
into a molecule of ethane and a molecule of nitrogen, and this decomposition is
illustrated in Figure 9.6. On the basis of the chemical kinetic schema indicated by
Eq. 9.26 and illustrated in Figure 9.6, the local rate equation for the production of
ethane takes the form (see footnote †)

Mass action kinetics: $R_{C_2H_6} = k\, c_{(CH_3)_2N_2}$ (9.27)

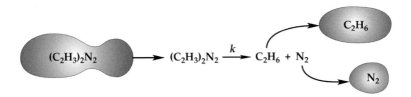

FIGURE 9.6   Spontaneous decomposition leading to products.

This result is not in agreement with experimental observations[§] which indicates that
the reaction is *first order* with respect to azomethane at high concentrations and *second order* at low concentrations. The experimental observations can be expressed as

Experimental kinetics: $R_{C_2H_6} = \dfrac{k[c_{(CH_3)_2N_2}]^2}{1 + k' c_{(CH_3)_2N_2}}$ (9.28)

in which k and k' are not to be confused with the analogous coefficients in Eq. 9.18.

The experimental results represented by Eq. 9.18 and Eq. 9.28 indicate that both
reaction processes are more complex than suggested by Figure 9.3 and suggested by
Figure 9.6. The fundamental difficulty results from the fact that mass action kinetics
provide a *picture of the process*, but in general they are unrelated to *the physics of
the process*.

Global observations cannot necessarily be used to correctly infer local processes,
and we need to explore the local processes more carefully if we are to correctly
predict the forms given by Eq. 9.18 and Eq. 9.28. In order to do so, we need to examine the classic mass action kinetics in more detail and this is done in subsequent
paragraphs.

[§] Ramsperger, H.C. 1927, Thermal decomposition of azomethane over a large range of pressures, J. Am.
Chem. Soc. **49**, 1495–1512.

9.1.1 ELEMENTARY STOICHIOMETRY

The concept of *stoichiometry* was introduced in Chapter 6, identified above by Eq. 9.3, and repeated here as

Axiom II:
$$\sum_{A=1}^{A=N} N_{JA} R_A = 0, \quad J = 1, 2, \ldots, T \tag{9.29}$$

If we consider a set of K elementary reactions *involving* the species indicated by $A = 1, 2, \ldots, N$, we encounter a set of *net rates of production* that are designated by $R_A^I, R_A^{II}, R_A^{III}, \ldots, R_A^K$. Associated with each *elementary reaction* is a condition of *elementary stoichiometry* that we express as

Elementary Stoichiometry
$$\sum_{A=1}^{A=N} N_{JA} R_A^k = 0, \quad J = 1, 2, \ldots, T, \quad k = I, II, \ldots, K \tag{9.30}$$

The sum of the K elementary net rates of production for species A is the total net rate of production for species A indicated by

$$\sum_{k=I}^{k=K} R_A^k = R_A \tag{9.31}$$

Since N_{JA} is independent of $k = I, II, \ldots, K$, we can sum Eq. 9.30 over all K reactions and interchange the order of summation to recover the stoichiometric condition given by

$$\sum_{k=I}^{k=K}\sum_{A=1}^{A=N} N_{JA} R_A^k = \sum_{A=1}^{A=N} N_{JA} \sum_{k=I}^{k=K} R_A^k = \sum_{A=1}^{A=N} N_{JA} R_A = 0, \quad J = 1, 2, \ldots, T \tag{9.32}$$

Clearly, when there is a single elementary reaction, the *elementary stoichiometry* is identical to the traditional stoichiometry introduced in Chapter 6.

9.1.2 MASS ACTION KINETICS AND ELEMENTARY STOICHIOMETRY

In this section, we want to summarize the concept of mass action kinetics and indicate how it can be connected to elementary stoichiometry. As an example, we consider a system in which there are four participating molecular species indicated by A, B, C, and D. The chemical kinetic schema for one possible reaction associated with these four molecular species is indicated by

Elementary chemical kinetic schema I: $\qquad \alpha A + \beta B \xrightarrow{\ k_I\ } \gamma C + \delta D \tag{9.33}$

Here, we have avoided the use of k_1, k_2, etc., to represent rate coefficients and instead we have employed a nomenclature that makes use of k_I, k_{II}, k_{III}, etc., to identify the chemical reaction rate coefficients. At this point, we need to *translate* this *picture* to an *equation* associated with mass action kinetics and then explore what can be extracted from this picture in terms of elementary stoichiometry. According to the rules of mass action kinetics, the chemical kinetic translation of Eq. 9.33 is given by

Elementary chemical reaction rate equation I: $R_A^I = -k_I c_A^\alpha c_B^\beta$ (9.34)

Here, we have used the first species in the chemical kinetic schema as the basis for the proposed rate equation, and this represents a reasonable convention but not a necessary one. One should remember that binary collisions dominate chemical reactions and that ternary collisions are rare. This means that we expect the sum of the integers α and β to be less than or equal to two. Often there is a second reaction involving species A, B, C, and D, and we express the second chemical kinetic schema as

Elementary chemical kinetic schema II: $\varepsilon B + \eta C \xrightarrow{k_{II}} \xi D$ (9.35)

This second chemical kinetic schema leads to a chemical reaction rate equation of the form

Elementary chemical reaction rate equation II: $R_B^{II} = -k_{II} c_B^\varepsilon c_C^\eta$ (9.36)

In general, we are interested in the *net rate of production* which is given by the sum of the elementary rates of production according to

$$R_A = R_A^I + R_A^{II}, \quad R_B = R_B^I + R_B^{II}$$
$$R_C = R_C^I + R_C^{II}, \quad R_D = R_D^I + R_D^{II}$$

(9.37)

At this point, we need *stoichiometric information* to develop useful chemical reaction rate equations. Since stoichiometry is associated with the conservation of *atomic species*, we need to be *very careful* when using a representation in which there are no identifiable atomic species. The translation associated with kinetic schema and elementary stoichiometry must be consistent with Axiom II. In terms of *stoichiometry*, we identify the meaning of Eq. 9.34 as follows:

α moles of species A react with β moles of species B to form γ moles of species C and δ moles of species D.

To make things very clear, we consider the *highly unlikely* prospect that 8 moles of species A react with 3 moles of species B. This would lead to the condition

$$\frac{R_A^I}{8} = \frac{R_B^I}{3}$$

(9.38)

and a little thought will indicate that the general *stoichiometric translation* of Eq. 9.34 is given by

Elementary stoichiometry I

$$\frac{R_A^I}{\alpha} = \frac{R_B^I}{\beta}, \quad \frac{R_A^I}{\alpha} = -\frac{R_C^I}{\gamma}, \quad \frac{R_A^I}{\alpha} = -\frac{R_D^I}{\delta}$$

(9.39)

This result is based on the assumption that species A, B, C, and D are all *unique* species. For example, if species C is actually identical to species A, the second of Eq. 9.39 takes the form

Unacceptable stoichiometry:
$$\frac{R_A^I}{\alpha} = -\frac{R_A^I}{\gamma} \qquad (9.40)$$

In this special case, it should be clear that Eq. 9.33 cannot be used as a *picture of the stoichiometry*. If species B, C, and D are all *unique* species, we can follow the same thought process that led to Eq. 9.38 to conclude that the stoichiometric translation of Eq. 9.35 is given by

Elementary stoichiometry II

$$\frac{R_B^{II}}{\varepsilon} = \frac{R_C^{II}}{\eta}, \quad \frac{R_B^{II}}{\varepsilon} = -\frac{R_D^{II}}{\xi}, \quad R_A^{II} = 0 \qquad (9.41)$$

Throughout our study of stoichiometry in Chapter 6, we used representations such as C_2H_5OH and $CH_3OC_2H_3$ to identify the atomic structure of various molecular species, and with those representations it was easy to keep track of atomic species. The representations given by Eq. 9.33 and Eq. 9.35 are less informative, and we need to proceed with greater care when the atomic structure is not given explicitly.

With the elementary stoichiometry now available in terms of Eq. 9.39 and Eq. 9.41, we can develop the local chemical reaction rate equations for species A and B based on Eqs. 9.34 and 9.36. This leads to

$$R_A = -k_1 c_A^\alpha c_B^\beta, \quad R_B = -\left(\beta/\alpha\right)k_1 c_A^\alpha c_B^\beta - k_{II} c_B^\varepsilon c_C^\eta \qquad (9.42)$$

and the rate equations for the other species can be constructed in the same manner.

9.1.3 Decomposition of Azomethane and Reactive Intermediates

We are now ready to return to the decomposition of azomethane to produce ethane and nitrogen. The rate equation given by Eq. 9.28 is based on the work of Ramsperger[**]and an explanation of that rate equation requires the existence of *reactive intermediates* (Herzfeld[††], Polanyi[‡‡]) or *Bodenstein products* (Frank-Kamenetsky[§§]). Most chemical reactions involve reactive intermediate species, and this idea is illustrated in Figure 9.7, where we have indicated the existence of an *activated form* of azomethane identified as $(CH_3)_2N_2\bullet$. This form exists in such small

[**] Ramsperger, H.C. 1927, Thermal decomposition of azomethane over a large range of pressures, J. Am. Chem. Soc. **49**, 1495–1512.

[††] Herzfeld, K.F. 1919, The theory of the reaction speeds in gases, Ann. Physic. **59**, 635–667.

[‡‡] Polanyi, M. 1920, Reaction isochore and reaction velocity from the standpoint of statistics, Z. Elektrochem. **26**, 49–54.

[§§] Frank-Kamenetsky, D.A. 1940, Conditions for the applicability of Bodenstein's method in chemical kinetics, J. Phys. Chem. (USSR) **14**, 695–702.

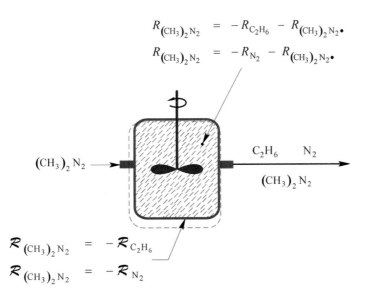

FIGURE 9.7 Local and global stoichiometry for decomposition of azomethane.

concentrations that it is difficult to detect in the exit stream and thus does not appear in the representation of the *global stoichiometry*. A key idea here is that the expression for a chemical reaction rate is based on experiments, and when a specific chemical species cannot be detected experimentally it often does not appear in the *first effort* to construct a chemical reaction rate expression. For simplicity we represent the molecular species suggested by Figure 9.7 as

$$A = (CH_3)_2N_2, \quad B = C_2H_6, \quad C = N_2, \quad A\bullet = (CH_3)_2N_2\bullet \quad (9.43)$$

in which $A\bullet$ represents the activated form of azomethane or the so-called *reactive intermediate*. On the basis of the analysis of Lindemann[***], we explore the following set of *elementary chemical kinetic schemas*:

Elementary chemical kinetic schema I: $\qquad 2A \xrightarrow{\ k_{\mathrm{I}}\ } A + A\bullet \qquad (9.44a)$

Elementary chemical kinetic schema II: $\qquad A\bullet \xrightarrow{\ k_{\mathrm{II}}\ } B + C \qquad (9.44b)$

Elementary chemical kinetic schema III: $\qquad A\bullet + A \xrightarrow{\ k_{\mathrm{III}}\ } 2A \qquad (9.44c)$

The schema represented by Eq. 9.44a is illustrated in Figure 9.8 where we see that a collision between two molecules of azomethane leads to the creation of the reactive intermediate denoted by $(CH_3)_2N_2\bullet$. Eq. 9.44a represents an example of the situation illustrated by Eqs. 9.39 and 9.41, and one must be careful in terms of the stoichiometric interpretation. In this case, we draw upon Figure 9.8 to conclude that the

---

[***] Lindemann, F.A. 1922, The radiation theory of chemical action, Trans. Faraday Soc. **17**, 598–606.

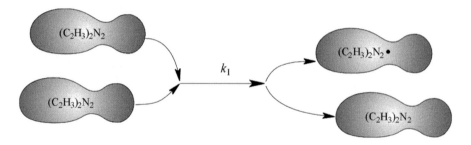

FIGURE 9.8   Creation of a reactive intermediate for the decomposition of azomethane.

stoichiometric schema associated with Eq. 9.44a is the activation of an azomethane molecule that we represent in the form

Stoichiometric schema: $(CH_3)_2N_2 \rightarrow (CH_3)_2N_2\bullet$ or $A \rightarrow A\bullet$ (9.45)

Given this stoichiometric schema for the first elementary step, we see that Eqs. 9.44 lead to the following four representations:

Elementary stoichiometric schema I: $A \rightarrow A\bullet$ (9.46a)

Elementary chemical kinetic schema I: $2A \xrightarrow{\ k_I\ } A + A\bullet$ (9.46b)

Elementary stoichiometry I: $R_A^I = -R_{A\bullet}^I$ (9.46c)

Elementary chemical reaction rate equation I: $R_A^I = -k_I\, c_A^2$ (9.46d)

The second elementary step involves the decomposition of the activated molecule to form ethane and nitrogen according to:

Elementary stoichiometric schema II: $A\bullet \rightarrow B + C$ (9.47a)

Elementary chemical kinetic schema II: $A\bullet \xrightarrow{\ k_{II}\ } B + C$ (9.47b)

Elementary stoichiometry II: $R_{A\bullet}^{II} = -R_B^{II}, \quad R_{A\bullet}^{II} = -R_C^{II}$ (9.47c)

Elementary chemical reaction rate equation II: $R_{A\bullet}^{II} = -k_{II}\, c_{A\bullet}$ (9.47d)

The final elementary step consists of the recombination of an activated molecule with azomethane to form two molecules of azomethane. This final step is described by the following representations:

Elementary stoichiometric schema III: $A\bullet \rightarrow A$ (9.48a)

Elementary chemical kinetic schema III: $A\bullet + A \xrightarrow{\ k_{III}\ } 2A$ (9.48b)

Elementary stoichiometry III: $R_{A\bullet}^{III} = -R_A^{III}$ (9.48c)

Elementary chemical reaction rate equation III: $\quad R_{A_\bullet}^{III} = -k_{III}\, c_A\, c_{A_\bullet}.$ (9.48d)

According to Eq. 9.32, the local net rates of production are given by

$$R_A = R_A^I + R_A^{II} + R_A^{III}$$ (9.49a)

$$R_{A_\bullet} = R_{A_\bullet}^I + R_{A_\bullet}^{II} + R_{A_\bullet}^{III}$$ (9.49b)

$$R_B = R_B^I + R_B^{II} + R_B^{III}$$ (9.49c)

We now have a complete description of the reaction process for the schemas represented by Eqs. 9.44, and from these results we would like to extract a representation for R_B in terms of c_A. The classic simplification of this algebraic problem is to assume that the net rate of production of the *reactive intermediate* or the *Bodenstein product* can be approximated by

Local reaction equilibrium: $\quad R_{A_\bullet} = 0$ (9.50)

This simplification is often referred to as the *steady-state assumption* or the *steady state hypothesis* or the *pseudo steady state hypothesis*. These are appropriate phrases when kinetic mechanisms are being studied by means of a batch reactor; however, the phrase *local reaction equilibrium* is preferred since it is not process-dependent. Use of Eq. 9.50 with Eq. 9.49b leads to

$$R_{A_\bullet}^I + R_{A_\bullet}^{II} + R_{A_\bullet}^{III} = k_1\, c_A^2 - k_{II}\, c_{A_\bullet} - k_{III}\, c_A\, c_{A_\bullet} = 0$$ (9.51)

and from this we determine the concentration of the reactive intermediate to be

$$c_{A_\bullet} = \frac{k_1\, c_A^2}{k_{II} + k_{III}\, c_A}$$ (9.52)

We now make use of Eq. 9.49c to express the net rate of production of ethane as

$$R_B = R_B^I + R_B^{II} + R_B^{III} = R_{C_2H_6}$$ (9.53)

and application of Eqs. 9.47c and 9.47d provides the chemical reaction rate equation given by

$$R_{C_2H_6} = k_{II}\, c_{A_\bullet}.$$ (9.54)

At this point, we use Eq. 9.52 in order to express the net rate of production of ethane as

$$R_{C_2H_6} = \frac{k_1\, k_{II}\, c_A^2}{k_{II} + k_{III}\, c_A}$$ (9.55)

in which c_A represents the concentration of azomethane, $(CH_3)_2N_2$. Here we can see that the two limiting rate expressions for high and low concentrations are given by

$$R_{C_2H_6} = \frac{k_I k_{II} c_A^2}{k_{II} + k_{III} c_A} = \begin{cases} \left(k_I k_{II}/k_{III} \right) c_A, & \text{high concentration} \\ k_I c_A^2, & \text{low concentration} \end{cases} \tag{9.56}$$

which is consistent with the experimental results of Ramsperger (see footnote §) illustrated by Eq. 9.28. We can be more precise about what is meant by high concentration and low concentration by expressing these ideas as

$$\begin{aligned} c_A \gg k_{II}/k_{III}, & \quad \text{high concentration} \\ c_A \ll k_{II}/k_{III}, & \quad \text{low concentration} \end{aligned} \tag{9.57}$$

Here, we see that the relatively simple process suggested by Eq. 9.26 is governed by the relatively complex rate equation indicated by Eq. 9.55. The analysis leading to this result is based on *three concepts*: (A) local and elementary stoichiometry, (B) mass action kinetics, and (C) the approximation of local reaction equilibrium. The simplifying assumptions associated with this development are discussed in the following paragraphs.

9.1.3.1 Assumptions and Consequences

A reasonable *assumption* concerning the continuous stirred tank reactor shown in Figure 9.5 is that only azomethane, ethane, and nitrogen participate in the reaction. This assumption, in turn, leads to the chemical kinetic schema illustrated both in Eq. 9.26 and in Figure 9.6. Experimental measurement of the concentrations in the inlet and outlet streams might confirm the assumption that only $(CH_3)_2N_2$, C_2H_6, and N_2 are present in the reactor. However, the experimental determination of the reaction rate is not in agreement with Eq. 9.27. In reality, our analysis is based on the *restriction* that no *significant* amount of *reactive intermediate* enters or exits the reactor, and we state this idea as

Restriction: $\qquad c_{A\cdot} \ll c_A, c_B, c_C, \begin{cases} \text{at the entrance and} \\ \text{the exit of the reactor} \end{cases}$ (9.58)

While the concentration of the reactive intermediate might be small compared to the other species, it is certainly not zero. If it were zero, Eq. 9.54 would indicate that the rate of production of ethane would be zero and that is not in agreement with experimental observation.

Given that $c_{A\cdot}$ is not zero, one can wonder about the assumption (see Eq. 9.51) that $R_{A\cdot}$ is zero. In reality, $R_{A\cdot}$ must be *small enough* so that it can be approximated by zero, and we need to know how small is *small enough*. To find out, we make use of Eq. 9.49b to determine that the net rate of production of A• is given by

$$R_{A\cdot} = k_I c_A^2 - \left(k_{II} + k_{III} c_A \right) c_{A\cdot}. \tag{9.59}$$

and we use this result to show that the concentration of the reactive intermediate takes the form

$$c_{A\bullet} = \frac{k_1 \, c_A^2}{\left(k_{II} + k_{III} \, c_A\right)} - \frac{R_{A\bullet}}{\left(k_{II} + k_{III} \, c_A\right)} \tag{9.60}$$

Use of this result in Eq. 9.54 leads to the net rate of production of C_2H_6 given by

$$R_{C_2H_6} = \frac{k_1 \, k_{II} \, c_A^2}{k_{II} + k_{III} \, c_A} - \frac{R_{A\bullet} \, k_{II}}{\left(k_{II} + k_{III} \, c_A\right)} \tag{9.61}$$

If the second term on the right-hand side of this result is negligible compared to the first term, we obtain the result given earlier by Eq. 9.55. This indicates that the *assumption* given by $R_{A\bullet} = 0$ is a reasonable substitute for the *restriction* given by

Restriction: $\qquad\qquad\qquad\qquad R_{A\bullet} \ll k_1 \, c_A^2 \qquad\qquad\qquad\qquad$ (9.62)

When this inequality is imposed on Eq. 9.61, we obtain the result given previously by Eq. 9.55 *provided that* we are willing to assume that *small causes produce small effects* (Birkhoff[†††]). Even though, Eqs. 9.50 and 9.62 lead to the same result, Eq. 9.62 should serve as a reminder that neglecting something that is *small* always requires the crucial assumption that small causes produce small effects.

One important part of this analysis is the fact that the assumption concerning $c_{A\bullet}$ at entrances and exits cannot be extended into the reactor where finite values of the concentration of the *reactive intermediate* control the rate of reaction. This is clearly indicated by Eq. 9.54. The situation we have encountered in this study occurs often in the transport and reaction of chemical species and can generalized as:

> Sometimes a small quantity, such as $R_{A\bullet}$ or $c_{A\bullet}$, can be ignored and set equal to zero for the purposes of analysis. Sometimes a small quantity cannot be ignored and setting it equal to zero represents a serious mistake.

Knowing when small causes produce small effects requires experience, intuition, experiment, and analysis. These are skills that are acquired steadily over time.

In this section, we have examined the concepts of global, local, and elementary stoichiometry, along with the concept of mass action kinetics. We have made use of *pictures* to describe both elementary stoichiometry and elementary chemical kinetics, and we have illustrated how these pictures are related to *equations*. The concept of *local reaction equilibrium*, also known as the *steady-state assumption* or the *steady state hypothesis* or the *pseudo steady-state hypothesis*, has been applied in order to develop a simplified rate expression for the production of ethane and nitrogen from azomethane. The resulting rate expression compares favorably with experimental observations.

[†††] Birkhoff, G. 1960, *Hydrodynamics: A Study in Logic, Fact, and Similitude*, Princeton University Press, Princeton, New Jersey.

9.2 MICHAELIS-MENTEN KINETICS

In Sec. 8.4, we presented a brief analysis of the cell growth phenomena that occurs in a chemostat. In addition, we presented the well-known Monod equation that has been used extensively to model *macroscopic* cell growth. In this section, we briefly explore an enzyme-catalyzed reaction that occurs in all cellular systems. Within a cell, such as the one illustrated in Figure 9.9, hundreds of reactions occur. To appreciate the complexity of cells, we note that a typical eukaryotic cell contains the following subcellular components: nucleolus, nucleus, ribosome, vesicle, rough endoplasmic reticulum, Golgi apparatus, Cytoskeleton, smooth endoplasmic reticulum, mitochondria, vacuole, cytoplasm, lysosome, and centrioles within centrosome[‡‡‡]. Obviously, the cell is a busy place, and much of that business is associated with the enzyme-catalyzed reactions that produce intracellular material represented by species D and extracellular material represented by species C. Species D provides the material that leads to cell growth as described in Sec. 8.4, while species C provides desirable products to be harvested by chemical engineers and others.

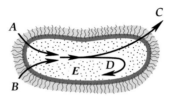

FIGURE 9.9 Transport and reaction in a cell.

9.2.1 CATALYSTS

A catalyst is an agent that causes an increase in the reaction rate without undergoing any permanent change, and the enzymes represented by species E in Figure 9.9 perform precisely that function in the production of intracellular and extracellular material. Enzymes are global proteins that *bind* substrates (reactants) in particular configurations that enhance the reaction rate. The simplest description of this process is due to Michaelis and Menten[§§§] who proposed a two-step process in which a substrate first binds *reversibly* with an enzyme and then reacts *irreversibly* to form a product. In this development we first consider the substrate A, the enzyme E, and the product D. To begin with, the enzyme E forms a complex with substrate A in a *reversible* manner as indicated by the combination of Eqs. 9.63a and 9.64a.

Elementary chemical kinetic schema I: $\qquad E + A \xrightarrow{\ k_1\ } EA$ $\qquad$ (9.63a)

Elementary stoichiometry I: $\qquad R_E^{\mathrm{I}} = -R_{EA}^{\mathrm{I}}, \quad R_E^{\mathrm{I}} = R_A^{\mathrm{I}}$ $\qquad$ (9.63b)

[‡‡‡] Segel, I. 1993, *Enzyme Kinetics: Behavior and Analysis of Rapid Equilibrium and Steady-State Enzyme Systems*, Wiley-Interscience, New York.
[§§§] Michaelis, L. and Menten, M.L. 1913, Die Kinetik der Invertinwirkung, Biochem. Z. **49**, 333–369.

Elementary chemical reaction rate equation I: $R_E^I = -k_I c_E c_A$ (9.63c)

Elementary chemical kinetic schema II: $EA \xrightarrow{k_{II}} E + A$ (9.64a)

Elementary stoichiometry II: $R_{EA}^{II} = -R_E^{II}, \quad R_{EA}^{II} = -R_A^{II}$ (9.64b)

Elementary chemical kinetic rate equation II: $R_{EA}^{II} = -k_{II} c_{EA}$ (9.64c)

In the final step, the complex EA reacts *irreversibly* to form the enzyme E and the product D according to

Elementary chemical kinetic schema III: $EA \xrightarrow{k_{III}} E + D$ (9.65a)

Elementary stoichiometry III: $R_{EA}^{III} = -R_E^{III}, \quad R_{EA}^{III} = -R_D^{III}$ (9.65b)

Elementary chemical reaction rate equation III: $R_{EA}^{III} = -k_{III} c_{EA}$ (9.65c)

In all these elementary steps, we assume that the stoichiometric schemas are identical in form to the chemical kinetic schemas. In the shorthand nomenclature of biochemical engineering, Michaelis-Menten kinetics are represented by

$$E + A \underset{k_{II}}{\overset{k_I}{\rightleftarrows}} EA \xrightarrow{k_{III}} E + D \qquad (9.66)$$

Our objective at this point is to develop an expression for the net rate of production of species D in terms of the concentration of species A. The net rate of production for species D takes the form

$$R_D = R_D^I + R_D^{II} + R_D^{III} = k_{III} c_{EA} \qquad (9.67)$$

and the net rates of production of the other species are given by

$$R_A = R_A^I + R_A^{II} + R_A^{III} = -k_I c_E c_A + k_{II} c_{EA} \qquad (9.68a)$$

$$R_E = R_E^I + R_E^{II} + R_E^{III} = -k_I c_E c_A + k_{II} c_{EA} + k_{III} c_{EA} \qquad (9.68b)$$

$$R_{EA} = R_{EA}^I + R_{EA}^{II} + R_{EA}^{III} = k_I c_E c_A - k_{II} c_{EA} - k_{III} c_{EA} \qquad (9.68c)$$

Since a catalyst only facilitates a reaction and is neither consumed nor produced by the reaction, we can assume that the total concentration of the enzyme catalyst is constant. We express this idea as

$$c_E + c_{EA} = c_E^o \qquad (9.69)$$

in which c_E^o is the initial concentration of the enzyme in the reactor. In addition, the net rate of production of the enzyme catalyst should be zero and we express this idea as

$$R_E = 0 \qquad (9.70)$$

Use of Eq. 9.70 with Eq. 9.68b leads to a constraint on the rates of reaction given by

$$0 = -k_I c_E c_A + k_{II} c_{EA} + k_{III} c_{EA} \qquad (9.71)$$

This can be arranged in the form

$$\frac{k_I c_E c_A}{k_{II} + k_{III}} = c_{EA} \qquad (9.72)$$

and use of the constraint on the total concentration of enzyme given by Eq. 9.70 provides

$$\frac{k_I (c_E^o - c_{EA}) c_A}{k_{II} + k_{III}} = c_{EA} \qquad (9.73)$$

Solving for the concentration of the enzyme complex gives

$$c_{EA} = \frac{c_E^o c_A}{\left[(k_{II} + k_{III})/k_I \right] + c_A} \qquad (9.74)$$

This result can be used in Eq. 9.67 to represent the net rate of production of the desired product as

$$R_D = \frac{(k_{III} c_E^o) c_A}{K_A + c_A} \qquad (9.75)$$

Here K_A is defined by

$$K_A = (k_{II} + k_{III})/k_I \qquad (9.76)$$

The maximum net rate of production of species D occurs when $c_A \gg K_A$ and this indicates that Eq. 9.75 can be expressed as

Michaelis-Menten kinetics: $\qquad R_D = \dfrac{\mu_{max} c_A}{K_A + c_A} \qquad (9.77)$

This *microscopic* result is identical *in form* to the *macroscopic* Monod equation for cell mass production (see Eq. 8.58); however, the production of cells shown earlier in Figure 8.8 is not the same as the production of species D illustrated in Figure 9.9. Certainly, there is a connection between the production of cells and the production of intercellular material, and this connection has been explored by Ramkrishna and Song[****].

It is of some interest to note that when the rate of production of species D is completely controlled by the reaction illustrated by Eq. 9.65a, we have a situation in which

$$k_{III} \ll k_{II} \qquad (9.78)$$

[****] Ramkrishna, D. and Song, H-S, 2008, A Rationale for Monod's Biochemical Growth Kinetics, Ind. Eng. Chem. Res. 47, 9090–9098.

and the parameter K_A in Eq. 9.76 simplifies to

$$K_A \rightarrow k_{II}/k_1 = K_{eq}^{-1}, \quad k_{III} \ll k_{II} \tag{9.79}$$

In this case, K_A becomes the inverse of a true *equilibrium coefficient*. Since the imposition of Eq. 9.78 has no effect on the *form* of Eq. 9.77, there is often confusion concerning the precise nature of K_A.

9.3   MECHANISTIC MATRIX

In this section, we explore in more detail the reaction rates associated with chemical kinetic schema of the type studied in the previous two sections. The *mechanistic matrix*[††††] will be introduced as a convenient method of organizing information about reaction rates. This matrix is different than the *pivot matrix* discussed in Chapter 6, and we need to be very clear about the similarities and differences between these two matrices, both of which contain coefficients that are often referred to as *stoichiometric coefficients*. In some cases, the mechanistic matrix is identical to the *stoichiometric matrix* and in some cases, it consists of both a *stoichiometric matrix* and a *Bodenstein matrix*.

We begin by considering a system in which there are five species and three chemical kinetic schemas described by

Elementary chemical kinetic schema I:

$$A + B \xrightarrow{\ k_I\ } C + D, \quad D \text{ is a by-product} \tag{9.80a}$$

Elementary chemical kinetic schema II:

$$C + B \xrightarrow{\ k_{II}\ } E, \quad E \text{ is the product} \tag{9.80b}$$

Elementary chemical kinetic schema III:

$$C + D \xrightarrow{\ k_{III}\ } A + B, \quad \text{reverse of schema I} \tag{9.80c}$$

In this example, we *assume* that the *stoichiometric schemas* are identical in form to the *chemical kinetic schema*, and we carefully follow the structure outlined in Sec. 9.1.1 in order to avoid algebraic errors. We begin with the first elementary step indicated by Eq. 9.80a, and our analysis of this step leads to

Elementary chemical kinetic schema I: $A + B \xrightarrow{\ k_I\ } C + D$ \hfill (9.81a)

Elementary stoichiometry I: $R_A^I = R_B^I, \quad R_A^I = -R_C^I, \quad R_A^I = -R_D^I$ \hfill (9.81b)

[††††] Bjornbom, P.H. 1977, The relation between the reaction mechanism and the stoichiometric behavior of chemical reactions, AIChE J. **23**, 285–288.

Elementary chemical reaction rate equation I: $R_A^I = -k_I c_A c_B$ (9.81c)

Elementary reference chemical reaction rate I: $r_I \equiv k_I c_A c_B$ (9.81d)

In Eq. 9.81d, we have defined an *elementary reference chemical reaction rate* that is designated by r_I, and we will choose a similar *reference quantity* for each chemical kinetic schema. The units of these reference quantities are *moles/(volume × time)* and they will be designated by r_I, r_{II}, and r_{III}. These reference chemical reaction rates will always be *positive*, and they must be distinguished from r_A, r_B, r_C, etc. that were used in Chapter 4 (see Eq. 4.6) to represent the net *mass* rate of production of species A, B, C, etc.

Moving on to the *second elementary step* indicated by Eq. 9.80b, we create a set of results analogous to Eqs. 9.81 that are given by:

Elementary chemical kinetic schema II: $C + B \xrightarrow{\;k_{II}\;} E$ (9.82a)

Elementary stoichiometry II: $R_C^{II} = R_B^{II}, \quad R_C^{II} = -R_E^{II}$ (9.82b)

Elementary chemical reaction rate equation II: $R_C^{II} = -k_{II} c_B c_C$ (9.82c)

Elementary reference chemical reaction rate II: $r_{II} \equiv k_{II} c_B c_C$ (9.82d)

Finally, we examine the *third elementary step* indicated by Eq. 9.80c in order to obtain the following relations

Elementary chemical kinetic schema III: $C + D \xrightarrow{\;k_{III}\;} A + B$ (9.83a)

Elementary stoichiometry III: $R_C^{III} = R_D^{III}, \quad R_C^{III} = -R_A^{III}, \quad R_C^{III} = -R_B^{III}$ (9.83b)

Elementary chemical reaction rate equation III: $R_C^{III} = -k_{III} c_C c_D$ (9.83c)

Elementary reference chemical reaction rate III: $r_{III} \equiv k_{III} c_C c_D$ (9.83d)

The net rate of production for each molecular species is given in terms of the elementary rates of production according to

Species A: $R_A = R_A^I + R_A^{II} + R_A^{III}$ (9.84a)

Species B: $R_B = R_B^I + R_B^{II} + R_B^{III}$ (9.84b)

Species C: $R_C = R_C^I + R_C^{II} + R_C^{III}$ (9.84c)

Species D: $R_D = R_D^I + R_D^{II} + R_D^{III}$ (9.84d)

Species E: $R_E = R_E^I + R_E^{II} + R_E^{III}$ (9.84e)

At this point, we can use the *elementary chemical reaction rates* to express the net rates of production according to

Species A: $$R_A = -r_{\text{I}} + 0 + r_{\text{III}} \tag{9.85a}$$

Species B: $$R_B = -r_{\text{I}} - r_{\text{II}} + r_{\text{III}} \tag{9.85b}$$

Species C: $$R_C = r_{\text{I}} - r_{\text{II}} - r_{\text{III}} \tag{9.85c}$$

Species D: $$R_D = r_{\text{I}} + 0 - r_{\text{III}} \tag{9.85d}$$

Species E: $$R_E = 0 + r_{\text{II}} + 0 \tag{9.85e}$$

In matrix form, these representations for the net rates of production are given by

$$
\begin{bmatrix} R_A \\ R_B \\ R_C \\ R_D \\ R_E \end{bmatrix}
=
\begin{bmatrix}
-1 & 0 & 1 \\
-1 & -1 & 1 \\
1 & -1 & -1 \\
1 & 0 & -1 \\
0 & 1 & 0
\end{bmatrix}
\begin{bmatrix} r_{\text{I}} \\ r_{\text{II}} \\ r_{\text{III}} \end{bmatrix}
\tag{9.86}
$$

Often it is convenient to express this result in the following compact form

$$R_{\text{M}} = M r \tag{9.87}$$

Here R_{M} is the column matrix of all the net rates of production, M is the mechanistic matrix and r is the column matrix of elementary chemical reactions. These quantities are defined explicitly by

$$
R_{\text{M}} = \begin{bmatrix} R_A \\ R_B \\ R_C \\ R_D \\ R_E \end{bmatrix}, \quad
M = \underbrace{\begin{bmatrix}
-1 & 0 & 1 \\
-1 & -1 & 1 \\
1 & -1 & -1 \\
1 & 0 & -1 \\
0 & 1 & 0
\end{bmatrix}}_{\text{mechanistic matrix}}, \quad
r = \begin{bmatrix} r_{\text{I}} \\ r_{\text{II}} \\ r_{\text{III}} \end{bmatrix}
\tag{9.88}
$$

When reactive intermediates, or Bodenstein products, are present, the mechanistic matrix is decomposed into a *stoichiometric matrix*, and a *Bodenstein matrix* and we give an example of this situation in the following paragraphs. Here, it is crucial to understand that the column matrix on the left-hand side of Eq. 9.86 consists of the *net molar rates of production* of all species including the reactive intermediates or Bodenstein products. It is equally important to understand that the column matrix on the right-hand side of Eq. 9.86 consists of *chemical reaction rates* that are *not* net molar rates of production. Instead, they are chemical reaction rates defined by Eqs. 9.81d, 9.82d, and 9.83d. The definitions of these chemical reaction rates can be expressed explicitly as

$$r = \begin{bmatrix} r_{\mathrm{I}} \\ r_{\mathrm{II}} \\ r_{\mathrm{III}} \end{bmatrix} \equiv \begin{bmatrix} k_{\mathrm{I}}\, c_A\, c_B \\ k_{\mathrm{II}}\, c_B\, c_C \\ k_{\mathrm{III}}\, c_C\, c_D \end{bmatrix} \tag{9.89}$$

The matrix representations given by Eq. 9.86 and Eq. 9.89 can be used to extract the individual expressions for R_A, R_B, R_C, R_D, and R_E that are given by

Species A: $$R_A = -k_{\mathrm{I}}\, c_A\, c_B + k_{\mathrm{III}}\, c_C\, c_D \tag{9.90a}$$

Species B: $$R_B = -k_{\mathrm{I}}\, c_A\, c_B - k_{\mathrm{II}}\, c_B\, c_C + k_{\mathrm{III}}\, c_C\, c_D \tag{9.90b}$$

Species C: $$R_C = k_{\mathrm{I}}\, c_A\, c_B - k_{\mathrm{II}}\, c_B\, c_C - k_{\mathrm{III}}\, c_C\, c_D \tag{9.90c}$$

Species D: $$R_D = k_{\mathrm{I}}\, c_A\, c_B - k_{\mathrm{III}}\, c_C\, c_D \tag{9.90d}$$

Species E: $$R_E = k_{\mathrm{II}}\, c_B\, c_C \tag{9.90e}$$

In addition to extracting these results directly from Eq. 9.86 and Eq. 9.89, we can also obtain them from the schemas illustrated by Eq. 9.80 in the same manner that was used in Sec. 9.1.1. In Chapter 6, we made use of the *pivot matrix* that maps the net rates of product of the *pivot species* onto the net rates of production of the *non-pivot species*. In this development, we see that the *mechanistic matrix* maps the elementary chemical reaction rates onto *all* the net rates of production.

At this point, we note that the row reduced echelon form of the mechanistic matrix illustrated in Eq. 9.88 is given by

$$M^* = \begin{bmatrix} 1 & 0 & -1 \\ 0 & 1 & 0 \\ 0 & 0 & 1 \\ 0 & 0 & 0 \\ 0 & 0 & 0 \end{bmatrix} \tag{9.91}$$

This indicates that two of the net rates of production are linearly dependent on the other three. From Eq. 9.90, we obtain

$$R_D = -R_A \tag{9.92a}$$

$$R_C = -R_A - R_E \tag{9.92b}$$

while the net rates of production for species A, B, and E are repeated here as

$$R_A = -k_{\text{I}}\, c_A\, c_B + k_{\text{III}}\, c_C\, c_D \tag{9.93a}$$

$$R_B = -k_{\text{I}}\, c_A\, c_B - k_{\text{II}}\, c_B\, c_C + k_{\text{III}}\, c_C\, c_D \tag{9.93b}$$

$$R_E = k_{\text{II}}\, c_B\, c_C \tag{9.93c}$$

These net rates of production can be used with Axiom I to analyze chemical reactors such as the batch reactors studied in Sec. 8.2.

9.3.1 HYDROGEN BROMIDE REACTION

At this point, we return to the hydrogen bromide reaction described briefly in Sec. 9.1 where Axiom II provided the result given by

Axiom II:
$$\begin{bmatrix} R_{H_2} \\ R_{Br_2} \end{bmatrix} = \underbrace{\begin{bmatrix} -1/2 \\ -1/2 \end{bmatrix}}_{pivot\ matrix} \begin{bmatrix} R_{HBr} \end{bmatrix} \tag{9.94}$$

Use of local stoichiometry along with the chemical kinetic schema given by Eq. 9.15 did not lead to a chemical reaction rate equation that was in agreement with the experimental result indicated by Eq. 9.19. Clearly, the molecular process suggested by Figure 9.3 is not an acceptable representation of the reaction kinetics and we need to explore the impact of *reactive intermediates* on the hydrogen bromide reaction. To do so, we propose the following chemical kinetic schemas:

Elementary chemical kinetic schema I: $\quad Br_2 \xrightarrow{\ k_{\text{I}}\ } 2Br\bullet \tag{9.95a}$

Elementary chemical kinetic schema II: $\quad Br\bullet + H_2 \xrightarrow{\ k_{\text{II}}\ } HBr + H\bullet \tag{9.95b}$

Elementary chemical kinetic schema III: $\quad H\bullet + Br_2 \xrightarrow{\ k_{\text{III}}\ } HBr + Br\bullet \tag{9.95c}$

Elementary chemical kinetic schema IV: $\quad H\bullet + HBr \xrightarrow{\ k_{\text{IV}}\ } H_2 + Br\bullet \tag{9.95d}$

Elementary chemical kinetic schema V: $\quad 2Br\bullet \xrightarrow{\ k_{\text{V}}\ } Br_2 \tag{9.95e}$

Here we note that Eq. 9.95 are simply a more complex example of Eqs. 9.81, thus we can follow the procedure outlined in the previous paragraphs assuming that the *stoichiometric schemas* are identical to the *chemical kinetic schemas*. We begin with the first elementary schema indicated by Eq. 9.95a and our analysis of this schema leads to

9.3.1.1   Schema I

Elementary chemical kinetic schema I: $\qquad\qquad Br_2 \xrightarrow{\ k_I\ } 2Br\bullet$ (9.96a)

Elementary stoichiometry I: $\qquad\qquad\qquad\qquad R^I_{Br_2} = -\dfrac{R^I_{Br\bullet}}{2}$ (9.96b)

Elementary chemical reaction rate equation I: $\qquad R^I_{Br_2} = -k_I\, c_{Br_2}$ (9.96c)

Elementary reference chemical reaction rate I: $\qquad r_I \equiv k_I\, c_{Br_2}$ (9.96d)

The remaining schemas lead to an analogous set of equations given by

9.3.1.2   Schema II

Elementary chemical kinetic schema II: $\qquad Br\bullet + H_2 \xrightarrow{\ k_{II}\ } HBr + H\bullet$ (9.97a)

Elementary stoichiometry II:
$$R^{II}_{Br\bullet} = R^{II}_{H_2}, \quad R^{II}_{Br\bullet} = -R^{II}_{HBr}$$
$$R^{II}_{Br\bullet} = -R^{II}_{H\bullet}$$
(9.97b)

Elementary chemical reaction rate equation II: $\qquad R^{II}_{Br\bullet} = -k_{II}\, c_{Br\bullet}\, c_{H_2}$ (9.97c)

Elementary reference chemical reaction rate II: $\qquad r_{II} \equiv k_{II}\, c_{Br\bullet}\, c_{H_2}$ (9.97d)

9.3.1.3   Schema III

Elementary chemical kinetic schema III: $\qquad H\bullet + Br_2 \xrightarrow{\ k_{III}\ } HBr + Br\bullet$ (9.98a)

Elementary stoichiometry III:
$$R^{III}_{H\bullet} = R^{III}_{Br_2}, \quad R^{III}_{H\bullet} = -R^{III}_{HBr}$$
$$R^{III}_{H\bullet} = -R^{III}_{Br\bullet}$$
(9.98b)

Elementary chemical reaction rate equation III: $\qquad R^{III}_{H\bullet} = -k_{III}\, c_{H\bullet}\, c_{Br_2}$ (9.98c)

Elementary reference chemical reaction rate III: $\qquad r_{III} \equiv k_{III}\, c_{H\bullet}\, c_{Br_2}$ (9.98d)

9.3.1.4   Schema IV

Elementary chemical kinetic schema IV: $\qquad H\bullet + HBr \xrightarrow{\ k_{IV}\ } H_2 + Br\bullet$ (9.99a)

Elementary stoichiometry IV:
$$R^{IV}_{H\bullet} = R^{IV}_{HBr}, \quad R^{IV}_{H\bullet} = -R^{IV}_{H_2}$$
$$R^{IV}_{H\bullet} = -R^{IV}_{Br\bullet}$$
(9.99b)

Elementary chemical reaction rate equation IV: $\qquad R^{IV}_{H\bullet} = -k_{IV}\, c_{H\bullet}\, c_{HBr}$ (9.99c)

Elementary reference chemical reaction rate IV: $\qquad r_{IV} \equiv k_{IV}\, c_{H\bullet}\, c_{HBr}$ (9.99d)

9.3.1.5 Schema V

Elementary chemical kinetic schema V: $\quad\quad 2\,Br\bullet \xrightarrow{\; k_V \;} Br_2 \quad\quad$ (9.100a)

Elementary stoichiometry V: $\quad\quad\quad\quad \dfrac{R^V_{Br\bullet}}{2} = -R^V_{Br_2} \quad\quad$ (9.100b)

Elementary chemical reaction rate equation V: $\quad R^V_{Br\bullet} = -k_V\, c^2_{Br} \quad\quad$ (9.100c)

Elementary reference chemical reaction rate V: $\quad r_V \equiv k_V\, c^2_{Br\bullet} \quad\quad$ (9.100d)

We begin our analysis of Eq. 9.95 by listing the net molar rates of production of all five species in terms of the elementary rates of reaction according to

$$R_{Br_2} = R^I_{Br_2} + R^{II}_{Br_2} + R^{III}_{Br_2} + R^{IV}_{Br_2} + R^V_{Br_2} \quad\quad (9.101a)$$

$$R_{H_2} = R^I_{H_2} + R^{II}_{H_2} + R^{III}_{H_2} + R^{IV}_{H_2} + R^V_{H_2} \quad\quad (9.101b)$$

$$R_{HBr} = R^I_{HBr} + R^{II}_{HBr} + R^{III}_{HBr} + R^{IV}_{HBr} + R^V_{HBr} \quad\quad (9.101c)$$

$$R_{H\bullet} = R^I_{H\bullet} + R^{II}_{H\bullet} + R^{III}_{H\bullet} + R^{IV}_{H\bullet} + R^V_{H\bullet} \quad\quad (9.101d)$$

$$R_{Br\bullet} = R^I_{Br\bullet} + R^{II}_{Br\bullet} + R^{III}_{Br\bullet} + R^{IV}_{Br\bullet} + R^V_{Br\bullet} \quad\quad (9.101e)$$

At this point, we can use the elementary chemical reaction rates to express the net rates of production according to

$$R_{Br_2} = -r_I + 0 - r_{III} + 0 + \frac{1}{2} r_V \quad\quad (9.102a)$$

$$R_{H_2} = 0 - r_{II} + 0 + r_{IV} + 0 \quad\quad (9.102b)$$

$$R_{HBr} = 0 + r_{II} + r_{III} - r_{IV} + 0 \quad\quad (9.102c)$$

$$R_{H\bullet} = 0 + r_{II} - r_{III} - r_{IV} + 0 \quad\quad (9.102d)$$

$$R_{Br\bullet} = 2\,r_I - r_{II} + r_{III} + r_{IV} - r_V \quad\quad (9.102e)$$

In matrix form these representations for the net rates of production are given by

$$
\begin{bmatrix} R_{Br_2} \\ R_{H_2} \\ R_{HBr} \\ R_{H\bullet} \\ R_{Br\bullet} \end{bmatrix}
=
\begin{bmatrix}
-1 & 0 & -1 & 0 & 1/2 \\
0 & -1 & 0 & 1 & 0 \\
0 & 1 & 1 & -1 & 0 \\
0 & 1 & -1 & -1 & 0 \\
2 & -1 & 1 & 1 & -1
\end{bmatrix}
\begin{bmatrix} r_I \\ r_{II} \\ r_{III} \\ r_{IV} \\ r_V \end{bmatrix}
\quad\quad (9.103)
$$

The compact form of this result can be expressed as

$$R_M = M\, r \tag{9.104}$$

Here R_M is the column matrix of all net rates of production, M is the *mechanistic matrix*, and r is the column matrix of elementary chemical reaction rates. These quantities are defined explicitly by

$$R_M = \begin{bmatrix} R_{Br_2} \\ R_{H_2} \\ R_{HBr} \\ R_{H\cdot} \\ R_{Br\cdot} \end{bmatrix}, \quad M = \begin{bmatrix} -1 & 0 & -1 & 0 & 1/2 \\ 0 & -1 & 0 & 1 & 0 \\ 0 & 1 & 1 & -1 & 0 \\ 0 & 1 & -1 & -1 & 0 \\ 2 & -1 & 1 & 1 & -1 \end{bmatrix} \tag{9.105}$$

$$r = \begin{bmatrix} r_I \\ r_{II} \\ r_{III} \\ r_{IV} \\ r_V \end{bmatrix} \equiv \begin{bmatrix} k_I\, c_{Br_2} \\ k_{II}\, c_{Br\cdot}\, c_{H_2} \\ k_{III}\, c_{H\cdot}\, c_{Br_2} \\ k_{IV}\, c_{H\cdot}\, c_{HBr} \\ k_V\, c_{Br\cdot}^2 \end{bmatrix}$$

Here we note that the row reduced echelon form of the mechanistic matrix is given by

$$M^* = \begin{bmatrix} 1 & 0 & 2 & 0 & -2 \\ 0 & 1 & 0 & -1 & 0 \\ 0 & 0 & 1 & 0 & 0 \\ 0 & 0 & 0 & 0 & 0 \\ 0 & 0 & 0 & 0 & 0 \end{bmatrix} \tag{9.106}$$

and this indicates that two of the net rates of production are linearly dependent on the other three. Some algebra associated with Eq. 9.102 indicates that this dependence can be expressed in the form

$$2R_{H_2} + R_{H\cdot} + R_{HBr} = 0 \tag{9.107a}$$

$$2R_{Br_2} + R_{Br\cdot} - 2R_{H_2} - R_{H\cdot} = 0 \tag{9.107b}$$

Useful representations for the three independent net rates of production can be extracted from Eq. 9.103; however, the analysis can be greatly simplified if we designate $H\cdot$ and $Br\cdot$ as *reactive intermediates* or *Bodenstein products* and then impose the condition of *local reaction equilibrium* expressed as

$$R_{H\cdot} = 0, \quad R_{Br\cdot} = 0 \tag{9.108}$$

In order to make use of this simplification, it is convenient to represent Eq. 9.104 in terms of the chemical reaction rate expressions and then apply a *row/row* partition (see Sec. 6.2.6, Problem 6.22 and Sec. C.1 of Appendix C) to obtain

$$
\begin{bmatrix} R_{Br_2} \\ R_{H_2} \\ R_{HBr} \\ \hline R_{H\cdot} \\ R_{Br\cdot} \end{bmatrix} = \left[\begin{array}{ccccc} -1 & 0 & -1 & 0 & 1/2 \\ 0 & -1 & 0 & 1 & 0 \\ 0 & 1 & 1 & -1 & 0 \\ \hline 0 & 1 & -1 & -1 & 0 \\ 2 & -1 & 1 & 1 & -1 \end{array} \right] \begin{bmatrix} k_I\, c_{Br_2} \\ k_{II}\, c_{Br}\, c_{H_2} \\ k_{III}\, c_H\, c_{Br_2} \\ k_{IV}\, c_H\, c_{HBr} \\ k_V\, c_{Br}^2 \end{bmatrix} \tag{9.109}
$$

Here the first partition takes the form

$$
\begin{bmatrix} R_{Br_2} \\ R_{H_2} \\ R_{HBr} \end{bmatrix} = \underbrace{\left[\begin{array}{ccccc} -1 & 0 & -1 & 0 & 1/2 \\ 0 & -1 & 0 & 1 & 0 \\ 0 & 1 & 1 & -1 & 0 \end{array} \right]}_{\text{stoichiometric matrix}} \begin{bmatrix} k_I\, c_{Br_2} \\ k_{II}\, c_{Br\cdot}\, c_{H_2} \\ k_{III}\, c_{H\cdot}\, c_{Br_2} \\ k_{IV}\, c_{H\cdot}\, c_{HBr} \\ k_V\, c_{Br}^2 \end{bmatrix} \tag{9.110}
$$

in which the matrix of coefficients is the *stoichiometric matrix*. The second partition is given by

$$
\begin{bmatrix} R_{H\cdot} \\ R_{Br\cdot} \end{bmatrix} = \underbrace{\left[\begin{array}{ccccc} 0 & 1 & -1 & -1 & 0 \\ 2 & -1 & 1 & 1 & -1 \end{array} \right]}_{\text{Bodenstein matrix}} \begin{bmatrix} k_I\, c_{Br_2} \\ k_{II}\, c_{Br\cdot}\, c_{H_2} \\ k_{III}\, c_{H\cdot}\, c_{Br_2} \\ k_{IV}\, c_{H\cdot}\, c_{HBr} \\ k_V\, c_{Br\cdot}^2 \end{bmatrix} \tag{9.111}
$$

in which this matrix of coefficients is the *Bodenstein matrix* that maps the rates of reaction onto the net rates of production of the *Bodenstein* products[####]. It is important to note that the stoichiometric matrix maps an array of chemical kinetic expressions onto the column matrix of the net rates of production of the three *stable molecular species*. This mapping process carried out by the *stoichiometric matrix* is quite different than the mapping process carried out by the *pivot matrix* that is illustrated by Eq. 9.94.

[####] Bodenstein, M. and Lind, S.C. 1907, Geschwindigkeit der Bildung des Bromwasserstoffes aus sienen Elementen, Z. physik. Chem. **57**, 168–192.

If we impose the condition of *local reaction equilibrium* indicated by Eq. 9.108, we obtain the following two constraints on the reaction rates:

$$R_{H\cdot} = 0: k_{II} c_{Br\cdot} c_{H_2} - k_{III} c_{H\cdot} c_{Br_2} - k_{IV} c_{H\cdot} c_{HBr} = 0 \qquad (9.112a)$$

$$R_{Br\cdot} = 0: \quad \begin{array}{l} 2 k_I c_{Br_2} - k_{II} c_{Br\cdot} c_{H_2} + k_{III} c_{H\cdot} c_{Br_2} \\ + k_{IV} c_{H\cdot} c_{HBr} - k_V c_{Br\cdot}^2 = 0 \end{array} \qquad (9.112b)$$

These two results can be used to determine the concentrations of the *Bodenstein products* that are given by

$$c_{H\cdot} = \frac{k_{II} c_{H_2} \sqrt{2k_I/k_V} \sqrt{c_{Br_2}}}{\left(k_{III} c_{Br_2} + k_{IV} c_{HBr}\right)}, \quad c_{Br\cdot} = \sqrt{2k_I/k_V} \sqrt{c_{Br_2}} \qquad (9.113)$$

On the basis of Eqs. 9.107 and Eqs. 9.108, we see that there is only a single independent equation associated with Eqs. 9.110 and we can use that equation to determine the net rate of production of hydrogen bromide as

$$R_{HBr} = \frac{\left(2k_{II} \sqrt{2k_I/k_V}\right) c_{H_2} \sqrt{c_{Br_2}}}{1 + (k_{IV}/k_{III})(c_{HBr}/c_{Br_2})} \qquad (9.114)$$

A little thought will indicate that this result has exactly the same form as the experimentally determined reaction rate expression given by Eq. 9.18.

In this section, we have illustrated the use of the *mechanistic matrix* to provide a compact representation of chemical reaction rate equations. When reactive intermediates (Bodenstein products) are involved in the reaction process, and local reaction equilibrium is assumed, it is convenient to represent the *mechanistic matrix* in terms of the *stoichiometric matrix* and the *Bodenstein matrix* as illustrated by Eqs. 9.109–9.111.

9.4 MATRICES

In this text, we have made use of matrix methods to solve problems and to clarify concepts. Here we summarize our knowledge of the matrices associated with the conservation of atomic species and the matrices associated with the analysis of chemical reaction rate phenomena.

9.4.1 ATOMIC MATRIX

The atomic matrix, A, was introduced in Sec. 6.2 in order to clearly identify the atoms and molecules involved in a particular process, and to provide a compact representation of Axiom II. The construction of the atomic matrix represents a key step in the analysis of chemical reactions since it identifies the molecular and atomic species that we *assume* are involved in the process under consideration. As an example, we consider the atomic matrix for the partial oxidation of ethane. The analysis begins with the following visual representation (see Example 6.4) of the molecules and atoms involved in this process.

$$\text{Molecular Species} \rightarrow \begin{array}{cccccc} C_2H_6 & O_2 & H_2O & CO & CO_2 & C_2H_4O \end{array}$$

$$\begin{array}{c} carbon \\ hydrogen \\ oxygen \end{array} \begin{bmatrix} 2 & 0 & 0 & 1 & 1 & 2 \\ 6 & 0 & 2 & 0 & 0 & 4 \\ 0 & 2 & 1 & 1 & 2 & 1 \end{bmatrix} \qquad (9.115)$$

In this case the atomic matrix is given by

$$\text{Atomic matrix:} \qquad A = \begin{bmatrix} 2 & 0 & 0 & 1 & 1 & 2 \\ 6 & 0 & 2 & 0 & 0 & 4 \\ 0 & 2 & 1 & 1 & 2 & 1 \end{bmatrix} \qquad (9.116)$$

and the column matrix of net molar rates of production takes the form

$$R = \begin{bmatrix} R_{C_2H_6} \\ R_{O_2} \\ R_{H_2O} \\ R_{CO} \\ R_{CO_2} \\ R_{C_2H_4O} \end{bmatrix} \qquad (9.117)$$

In terms of these two matrices Axiom II is given by

$$\text{Axiom II:} \qquad A\,R = 0 \qquad (9.118)$$

This represents a compact statement that atomic species are neither created nor destroyed by chemical reactions. The atomic matrix can always be expressed in *row reduced echelon form* (see Sec. 6.2.5) and this allows us to express Eq. 9.118 as

$$\text{Axiom II:} \qquad A^*R = 0 \qquad (9.119)$$

For the atomic matrix represented by Eq. 9.116, the *row reduced echelon form* is given by

$$A^* = \begin{bmatrix} 1 & 0 & 0 & 1/2 & 1/2 & 1 \\ 0 & 1 & 0 & 5/4 & 7/4 & 1 \\ 0 & 0 & 1 & -3/2 & -3/2 & -1 \end{bmatrix} \qquad (9.120)$$

The primary application of Axiom II takes the form of the *pivot theorem* that involves the *pivot matrix*.

9.4.2 PIVOT MATRIX

For the partial oxidation of ethane represented by Eq. 9.115, we can use Eqs. 9.117 and 9.120 to represent Eq. 9.119 in the form

$$
\begin{bmatrix}
1 & 0 & 0 & 1/2 & 1/2 & 1 \\
0 & 1 & 0 & 5/4 & 7/4 & 1 \\
0 & 0 & 1 & -3/2 & -3/2 & -1
\end{bmatrix}
\begin{bmatrix}
R_{C_2H_6} \\
R_{O_2} \\
R_{H_2O} \\
R_{CO} \\
R_{CO_2} \\
R_{C_2H_4O}
\end{bmatrix}
=
\begin{bmatrix}
0 \\
0 \\
0
\end{bmatrix}
\tag{9.121}
$$

Referring to the developments presented in Sec. 6.2.5, we note that a column/row partition of this result can be expressed as

$$
\begin{bmatrix}
1 & 0 & 0 & \vdots & 1/2 & 1/2 & 1 \\
0 & 1 & 0 & \vdots & 5/4 & 7/4 & 1 \\
0 & 0 & 1 & \vdots & -3/2 & -3/2 & -1
\end{bmatrix}
\begin{bmatrix}
R_{C_2H_6} \\
R_{O_2} \\
R_{H_2O} \\
\hline
R_{CO} \\
R_{CO_2} \\
R_{C_2H_4O}
\end{bmatrix}
=
\begin{bmatrix}
0 \\
0 \\
0
\end{bmatrix}
\tag{9.122}
$$

Carrying out the matrix multiplication illustrated by this column/row partition leads to a special case of the *pivot theorem* given by

$$
\begin{bmatrix}
R_{C_2H_6} \\
R_{O_2} \\
R_{CO}
\end{bmatrix}
=
\begin{bmatrix}
-1/2 & -1/2 & -1 \\
-5/4 & -7/4 & -1 \\
3/2 & 3/2 & 1
\end{bmatrix}
\begin{bmatrix}
R_{CO} \\
R_{CO_2} \\
R_{C_2H_4O}
\end{bmatrix}
\tag{9.123}
$$

The general representation of the pivot theorem takes the form

Pivot theorem: $R_{NP} = P\,R_P$ (9.124)

in which P is the pivot matrix. The pivot theorem is ubiquitous in the application of the concept that atomic species are neither created nor destroyed by chemical reactions. In the *analysis* of chemical reactors presented in Chapter 7, the global form of Eq. 9.124 was applied repeatedly, and we can express the global form as

Global pivot theorem: $$\mathcal{R}_{NP} = P\,\mathcal{R}_P \qquad (9.125)$$

Here $\mathcal{R}_{NP}$ represents the column matrix of non-pivot species *global* net rates of production while $\mathcal{R}_P$ represents the column matrix of pivot species *global* net rates of production.

9.4.3 MECHANISTIC MATRIX

In the *design* of chemical reactors, one needs to know how the *local* net rates of production are related to the concentration of the chemical species involved in the reaction. In the development of this relation, we encountered the *mechanistic matrix* that maps reference *chemical reaction rates* (see Eq. 9.87) onto *all net rates of production*. The general form is given by

$$R_M = M\,r \qquad (9.126)$$

in which R_M is the column matrix of all net rates of production, M is the *mechanistic matrix*, and r is the column matrix of elementary chemical reaction rates. For the hydrogen bromide reaction, Eq. 9.126 provides the detailed representation given by

$$
\underbrace{
\begin{bmatrix}
R_{Br_2} \\
R_{H_2} \\
R_{HBr} \\
R_{H\cdot} \\
R_{Br\cdot}
\end{bmatrix}
}_{\text{all species}}
=
\underbrace{
\begin{bmatrix}
-1 & 0 & -1 & 0 & 1/2 \\
0 & -1 & 0 & 1 & 0 \\
0 & 1 & 1 & -1 & 0 \\
0 & 1 & -1 & -1 & 0 \\
2 & -1 & 1 & 1 & -1
\end{bmatrix}
}_{\text{mechanistic matrix}}
\underbrace{
\begin{bmatrix}
k_I\,c_{Br_2} \\
k_{II}\,c_{Br\cdot}\,c_{H_2} \\
k_{III}\,c_{H\cdot}\,c_{Br_2} \\
k_{IV}\,c_{H\cdot}\,c_{HBr} \\
k_V\,c_{Br\cdot}^2
\end{bmatrix}
}_{\substack{\text{chemical} \\ \text{reaction rates}}}
\qquad (9.127)
$$

In many texts on chemical reactor design, the mechanistic matrix is referred to as the stoichiometric matrix. However, when Bodenstein products are present, and they usually are, it is appropriate to partition the mechanistic matrix into a *stoichiometric matrix* and a *Bodenstein matrix* as indicated by Eqs. 9.109–9.111. The general partitioning of Eq. 9.126 can be expressed as

$$
R_M =
\begin{bmatrix}
R \\
R_B
\end{bmatrix}
\qquad
Mr =
\begin{bmatrix}
S \\
B
\end{bmatrix}
r
\qquad (9.128)
$$

and this leads to forms analogous to Eqs. 9.110 and 9.111. We list the first of these results as

Stoichiometric matrix: $$R = Sr \qquad\qquad (9.129)$$

in which S is the stoichiometric matrix composed of *stoichiometric coefficients*. The second of Eqs. 9.128 is given by

Bodenstein matrix: $$R_B = Br \qquad\qquad (9.130)$$

in which B represents the Bodenstein matrix. In general, the Bodenstein products are subject to the approximation of local reaction equilibrium that is expressed as

Local reaction equilibrium: $$R_B = 0 \qquad\qquad (9.131)$$

This result allows one to extract additional constraints on the elementary chemical reaction rates. The stoichiometric matrix given by Eq. 9.129 represents a key aspect of reactor design that can be expressed in more detailed form by

$$R_A = \sum_{B=1}^{B=K} S_{AB}\, r_B, \quad A = 1, 2, \ldots, N \qquad\qquad (9.132)$$

Here S_{AB} represents the stoichiometric coefficients, r_B represents the elementary chemical reaction rates, and K represents the number of elementary reactions as indicated by Eq. 9.31.

9.5   PROBLEMS

SECTION 9.1

9.1. Apply a column/row partition to show how Eq. 9.6 is obtained from Eq. 9.5.

9.2. Illustrate how a row/row partition leads from Eq. 9.22 to Eq. 9.23.

9.3. Use Eqs. 9.34–9.42 to obtain a local chemical reaction rate equation for species D.

9.4. Develop the local chemical reaction rate equation for species C on the basis of Eqs.9.34–9.42.

9.5. Re-write Eqs. 9.39 and 9.41 using the stoichiometric coefficients, v_{AI}, v_{DII}, etc., in place of α, β, γ, etc. Show that this change in the nomenclature leads to the *form* encountered in Eq. 6.38.

9.6. Develop the representation for R_B given in Eqs. 9.42.

9.7. It is difficult to find a real system containing three species for which the reactions can be described by

$$A \xrightarrow{k_1} B \xrightarrow{k_2} C \tag{1}$$

however, this represents a useful model for the exploration of the condition of local reaction equilibrium. The stoichiometric constraint for this series of first order reactions is given by

$$R_A + R_B + R_C = 0 \tag{2}$$

since the three molecular species must all contain the same atomic species. For a constant volume batch reactor, and the initial conditions given by

I.C. $$\qquad c_A = c_A^o, \quad c_B = 0, \quad c_C = 0, \quad t = 0 \tag{3}$$

one can determine the concentrations and reaction rates of all three species as a function of time. If one thinks of species B as a *reactive intermediate*, the condition of local equilibrium takes the form

Local reaction equilibrium: $$\qquad R_B = 0 \tag{4}$$

In reality, the reaction rate for species B cannot be exactly zero; however, we can have a situation in which

$$R_B \ll |R_A| \tag{5}$$

Often this condition is associated with a *very large* value of k_2, and in this problem you are asked to develop and use the exact solution for this process to determine how large is *very large*. You can also use the exact solution to see why the condition of local reaction equilibrium might be referred to as the *steady-state assumption* for batch reactors.

Section 9.2

9.8. In our study of Michaelis-Menten kinetics, we simplified the analysis by ignoring the influence of the substrate B, or any other substrate, on the enzyme catalyzed reaction of species A. When multiple substrates are considered, the analysis becomes very complex[§§§§]; however, if we assume that the *reversible binding steps are at equilibrium*, the analysis of two substrates becomes quite tractable. The binding between enzyme E and substrate A is described by

$$E + A \underset{k_{\mathrm{II}}}{\overset{k_1}{\rightleftharpoons}} EA \tag{1}$$

[§§§§] Rodgers, A. and Gibon, Y. 2009, Enzyme Kinetics: Theory and Practice, Chapter 4 in *Plant Metabolic Networks*, edited by J. Schwender, Springer, New York.

and when species A is in *equilibrium* with species EA, we can replace Eq. 9.63 and Eq. 9.64 with the equilibrium relation given by

$$c_{EA} = \frac{k_I}{k_{II}} c_E c_A = K_{eq}^I c_E c_A \tag{2}$$

The reversible binding between enzyme E and substrate B is similarly described by

$$E + B \rightleftarrows EB \tag{3}$$

and we express the equilibrium relation as

$$c_{EB} = K_{eq}^{II} c_E c_B \tag{4}$$

In this model, we assume that once substrate B has reacted with enzyme E to form the complex EB, no additional reaction with substrate A can take place. In this step, substrate B acts as an *inhibitor* since it removes some enzyme E from the system. However, the complex EA can *further react* with substrate B to form the complex EAB as indicated by

$$EA + B \rightleftarrows EAB \tag{5}$$

The equilibrium relation associated with this process is given by

$$c_{EAB} = K_{eq}^{III} c_{EA} c_B \tag{6}$$

In this step, substrate B acts as a reactant since it produces the complex EAB that is the source of the product D.

Because of the assumed equilibrium relations indicated by Eqs. 2, 4, and 6, we have only a single rate equation based on an *irreversible* reaction. This irreversible reaction is described by

Elementary chemical kinetic schema III: $EAB \xrightarrow{\ k_{III}\ } E + D$ $\qquad$ (7a)

Elementary stoichiometry III: $R_{EAB}^{III} = -R_E^{III}, \quad R_D^{III} = R_E^{III}$ $\qquad$ (7b)

Elementary chemical kinetic rate equation III: $R_D^{III} = k_{III} c_{EAB}$ $\qquad$ (7c)

Since the product D is involved in only this single reaction, we express the rate of production as

$$R_D = k_{III} c_{EAB} \tag{8}$$

In the absence of significant cell death, one can assume that the enzyme E remains within the cells, thus the total concentration of enzyme is constant as indicated by

$$c_E + c_{EA} + c_{EB} + c_{EAB} = c_E^o \tag{9}$$

In this problem, you are asked to show that Eqs. 2–9 lead to an equation having the form of

$$R_D = \mu_{max} \left(\frac{c_A}{K_A + c_A} \right) \left(\frac{c_B}{K_B + c_B} \right) \tag{10}$$

provided that you impose the special condition given by

$$K_{eq}^{III} = K_{eq}^{II} \tag{11}$$

This restriction is based on the idea that B binds with EA in the same manner that B binds with E.

9.9. In Problem 9.8, we considered a case in which the substrate B acted both as a reactant in the production of species D and as an inhibitor in that process. In some cases, we can have an analog of the substrate that acts as a *pure inhibitor* and the Michaelis-Menten process takes the form

$$E + A \underset{k_I}{\overset{k_{II}}{\rightleftarrows}} EA \overset{k_{III}}{\longrightarrow} E + D \tag{1}$$

$$E + H \underset{k_{IV}}{\overset{k_V}{\rightleftarrows}} EH \tag{2}$$

in which we have used H to represent the inhibitor.

In this problem, we assume that $k_{III} \ll k_{II}$ in order to utilize (as an approximation) the equilibrium relation given by

$$c_{EA} = K_{eq}^I c_E c_A \tag{3}$$

In addition, we assume that the process illustrated by Eq. 2 can be approximated by the following equilibrium relation

$$c_{EH} = K_{eq}^{II} c_E c_H \tag{4}$$

In this problem, you are asked to make use of the condition given by

$$c_E + c_{EA} + c_{EH} = c_E^o \tag{5}$$

in order to develop an expression for R_D in which the concentration of the inhibitor, c_H, is an unknown. Consider the special cases that occur when $K_{eq}^{II} c_H / K_{eq}^I$ becomes both *large* and *small* relative to some appropriate parameter.

Section 9.3

9.10. Indicate how Eqs. 9.85 are obtained from Eqs. 9.84.

9.11. Beginning with the second of Eqs. 9.88 derive Eq. 9.91 using elementary row operations.

9.12. In the analysis of the hydrogen bromide reaction described by Eqs. 9.95, the concept of *local reaction equilibrium* was imposed on the reactive intermediates, H• and Br•, according to

$$\text{Local reaction equilibrium:} \quad R_{H\bullet} = 0, \quad R_{Br\bullet} = 0 \tag{1}$$

Use of these simplifications, along with the chemical kinetic schemata and the associated mass action kinetics given by Eqs. 9.96–9.100, leads to the net rate of production of hydrogen bromide given by Eq. 9.114. In reality, $R_{H\bullet}$ and $R_{Br\bullet}$ will not be zero but they may be *small enough* to recover Eq. 9.114. The concept that something is small enough so that it can be set equal to zero is explored by Eqs. 9.59–9.62. In this problem, you are asked to develop an analysis indicating that Eq. 1 listed above are acceptable approximations when the following inequalities are satisfied:

$$R_{Br\bullet} \ll 2 k_1 c_{Br_2} / k_V, R_{H\bullet} \ll 2 k_1 c_{Br_2} / k_V$$

$$R_{H\bullet} \ll k_{II} c_{H_2} \sqrt{2 k_1 c_{Br_2} / k_V} \tag{2}$$

One should think of Eq. 1 as being *assumptions* concerning the rates of production of the reactive intermediates, while Eq. 2 should be thought of as *restrictions* on the magnitude of these rates.

9.13. Use Eq. 9.101a to verify Eq. 9.102a.

9.14. The global *stoichiometric schema* associated with the decomposition of N_2O_5 to produce NO_2 and O_2 can be expressed as

$$N_2O_5 \rightarrow 2NO_2 + \frac{1}{2}O_2 \tag{1}$$

and experimental studies indicate that the reaction can be modeled as first order in N_2O_5. Show why the following elementary chemical kinetic schemata give rise to a first order decomposition of N_2O_5.

Elementary chemical kinetic schema I: $\quad N_2O_5 \xrightarrow{k_I} NO_2 + NO_3\bullet \tag{2}$

Elementary chemical kinetic schema II: $\quad NO_2 + NO_3 \xrightarrow{k_{II}} NO_2 + O_2 + NO\bullet \tag{3}$

Elementary chemical kinetic schema III: $\quad NO\bullet + NO_3\bullet \xrightarrow{k_{III}} 2NO_2 \tag{4}$

Elementary chemical kinetic schema IV: $\quad NO_2 + NO_3\bullet \xrightarrow{k_{IV}} N_2O_5 \tag{5}$

Since neither NO• nor NO$_3$• appear in the global stoichiometric schema given by Eq. 1, one can assume that these two compounds are *reactive intermediates* or *Bodenstein products* and their rates of production can be set equal to zero *as an approximation*

9.15. In Problem 9.14, we considered the decomposition of N$_2$O$_5$ to produce NO$_2$ and O$_2$ by means of the kinetic schemas illustrated by Eqs. 2–5 in Problem 9.14. The reactive intermediates were identified as NO• and NO$_3$•. The rate equations for these two reactive intermediates and for N$_2$O$_5$ are given by

$$R_{NO_3•} = k_1 c_{N_2O_5} - k_{II} c_{NO_3•} c_{NO_2} - k_{III} c_{NO•} c_{NO_3•} - k_{IV} c_{NO_2} c_{NO_3•}. \tag{1}$$

$$R_{NO•} = k_{II} c_{NO_3•} c_{NO_2} - k_{III} c_{NO•} c_{NO_3•}. \tag{2}$$

$$R_{N_2O_5} = -k_1 c_{N_2O_5} + k_{IV} c_{NO_2} c_{NO_3•}. \tag{3}$$

If the condition of local reaction equilibrium is imposed according to

Local reaction equilibrium: $R_{NO_3•} = 0, \quad R_{NO•} = 0$ (4)

one can obtain a simple representation for $R_{N_2O_5}$ in terms of only $c_{N_2O_5}$. In this problem, you are asked to replace the *assumptions* given by Eq. 4 with *restrictions* indicating that $R_{NO_3•}$ and $R_{NO•}$ are *small enough* so that Eq. 4 become acceptable approximations.

Section 9.4

9.16. If species C in Eq. 9.86 is the only *reactive intermediate* or *Bodenstein product*, identify the stoichiometric matrix and the Bodenstein matrix associated with Eq. 9.86. Impose the condition of local reaction equilibrium on the Bodenstein product in order to derive an expression for R_E in terms of c_A, c_B, and c_D.

Appendix A: Material Balances for Chemical Reacting Systems

A.1 ATOMIC MASS OF COMMON ELEMENTS REFERRED TO CARBON-12

Element	Symbol	Atomic Mass (g/mol)
Aluminum	Al	26.9815
Antimony	Sb	121.75
Argon	Ar	39.948
Arsenic	As	74.9216
Barium	Ba	137.34
Beryllium	Be	9.0122
Bismuth	Bi	208.980
Boron	B	10.811
Bromine	Br	79.904
Cadmium	Cd	112.40
Calcium	Ca	40.08
Carbon	C	12.01
Cerium	Ce	140.12
Cesium	Cs	132.905
Chlorine	Cl	35.453
Chromium	Cr	51.996
Cobalt	Co	58.9332
Copper	Cu	63.546
Fluorine	F	18.9984
Gallium	Ga	69.72
Germanium	Ge	72.59
Gold	Au	196.967
Hafnium	Hf	178.49
Helium	He	4.0026
Hydrogen	H	1.00797
Indium	In	114.82
Iodine	I	126.904
Iridium	Ir	192.2
Iron	Fe	55.847
Krypton	Kr	83.80
Lead	Pb	207.19
Lithium	Li	6.939
Magnesium	Mg	24.312
Manganese	Mn	54.938
Mercury	Hg	200.59

(Continued)

Element	Symbol	Atomic Mass (g/mol)
Molybdenum	Mo	95.94
Neon	Ne	20.183
Nickel	Ni	58.71
Niobium	Nb	92.906
Nitrogen	N	14.0067
Oxygen	O	15.9994
Palladium	Pd	106.4
Phosphorus	P	30.9738
Platinum	Pt	195.09
Plutonium	Pu	242
Potassium	K	39.102
Radium	Ra	226
Radon	Rn	222
Rhodium	Rh	102.905
Rubidium	Rb	85.47
Selenium	Se	78.96
Silicon	Si	28.086
Silver	Ag	107.868
Sodium	Na	22.9898
Strontium	Sr	87.62
Sulfur	S	32.064
Tantalum	Ta	180.948
Tellurium	Te	127.60
Thallium	Tl	204.37
Thorium	Th	232.038
Tin	Sn	118.69
Titanium	Ti	47.90
Tungsten	W	183.85
Uranium	U	238.03
Vanadium	V	50.942
Xenon	Xe	131.30
Yttrium	Y	88.905
Zinc	Zn	65.37
Zirconium	Zr	91.22

Source: Details are available at http://www.nist.gov/physlab/data/comp.cfm.

A.2 PHYSICAL PROPERTIES OF VARIOUS CHEMICAL COMPOUNDS

Name	Formula	Molecular Mass (g/mol)	Liquid @ T(K) Density (g/L)		$T_{melting}$ (K)	$T_{boiling}$ (K)
Argon	Ar	39.948			83.8	87.3
Acetaldehyde	C_2H_4O	44.054	778	293	150.2	293.6
Acetic acid	$C_2H_4O_2$	60.052	1,049	293	289.8	391.1
Acetone	C_3H_6O	58.08	790	293	178.2	329.4

(Continued)

Name	Formula	Molecular Mass (g/mol)	Liquid @ T(K) Density (g/L)		$T_{melting}$ (K)	$T_{boiling}$ (K)
Acetylene	C_2H_2	26.038			192.4	189.2
Acrylic acid	$C_3H_4O_2$	72.064	1,051	293	285	414
Ammonia	NH_3	17.031			195.4	239.7
Aniline	C_6H_7N	93.129	1,022	293	267	457.5
Benzaldehide	C_7H_6O	106.124	1,045	293	216	452
Benzene	C_6H_6	78.114	885	289	278.7	353.3
Benzoic acid	$C_7H_6O_2$	122.124	1,075	403	395.6	523
Bromine	Br_2	159.808	3,119	293	266	331.9
1,2-Butadiene	C_4H_6	54.092			137	284
1,3-Butadiene	C_4H_6	54.092			164.3	268.7
n-Butane	C_4H_{10}	58.124			134.8	272.7
i-Butane	C_4H_{10}	58.124			113.6	261.3
n-Butanol	$C_4H_{10}O$	74.123	810	293	183.9	390.9
1-Butene	C_4H_8	56.108			87.8	266.9
i-Butene	C_4H_8	56.108			132.8	266.3
Carbon	C	12.01				
tetrachloride	CCl_4	153.823	1,584	298	250	349.7
Carbon dioxide	CO_2	44.01			216.6	194.7
Carbon monoxide	CO	28.01			68.1	81.7
Chlorine	Cl_2	70.906			172.2	238.7
Chlorobenzene	C_6H_5Cl	112.559	1,106	293.	227.6	404.9
Chloroform	$CHCl_3$	119.378	1,498	293	209.6	334.3
Cyclobutane	C_4H_8	56.108			182.4	285.7
Cyclohexane	C_6H_{12}	84.162	779	293	279.7	353.9
Cyclohexanol	$C_6H_{12}O$	100.161	942	303	298	434.3
Cyclopentane	C_5H_{10}	70.135	745	293	179.3	322.4
Cyclopentene	C_5H_8	68.119	772	293	138.1	317.4
Ethane	C_2H_6	30.07			89.9	184.5
Ethanol	C_2H_6O	46.069	789	293	159.1	351.5
Ethyl amine	C_2H_7N	45.085	683	293	192	289.7
Ethyl acetate	$C_4H_8O_2$	88.107	901	293	89.6	350.3
Ethylbenzene	C_8H_{10}	106.168	867	293	178.2	409.3
Ethylendiamine	$C_2H_8N_2$	60.099	896	293	284	390.4
Ethyl ether	$C_4H_{10}O$	74.123	713	293	156.9	307.7
Ethyl propionate	$C_5H_{10}O_2$	102.134	895	293	199.3	372
Ethylene	C_2H_4	28.054			104	169.4
Ethylene Glycol	$C_2H_6O_2$	62.069	1,114	293	260.2	470.4
Ethylene oxide	C_2H_4O	44.054			161	283.5
Fluorine	F_2	37.997			53.5	85
Formaldehide	CH2O	30.026			156	254
Formic acid	CH_2O_2	46.025	1,226	288	281.5	373.8
Glycerol	$C_3H_8O_3$	92.095	1,261	293	291	563
n-Heptane	C_7H_{16}	100.205	684	293	182.6	371.6
1-Heptanol	$C_7H_{16}O$	116.204	822	293	239.2	449.5
1-Heptene	C_7H_{14}	98.189	697	293	154.3	366.8
n-Hexane	C_6H_{14}	86.178	659	293	177.8	341.9

(Continued)

Name	Formula	Molecular Mass (g/mol)	Liquid @ T(K) Density (g/L)		$T_{melting}$ (K)	$T_{boiling}$ (K)
1-Hexanol	$C_6H_{14}O$	102.177	819	293	229.2	430.2
Hydrogen	H_2	2.016			14.0	20.4
Hydrogen bromide	HBr	80.912			187.1	206.1
Hydrogen chloride	HCl	36.461			159.0	188.1
Hydrogen cyanide	CHN	27.026	688	293	259.9	298.9
Hydrogen sulfide	H_2S	34.08			187.6	212.8
Iodine	I_2	253.808	3,740	453	386.8	457.5
Isopropyl alcohol	C_3H_8O	60.096	786	293	184.7	355.4
Maleic anhydride	$C_4H_2O_3$	98.058	1,310	333	326	472.8
Methane	CH_4	16.043			90.7	111.7
Mercury	Hg	200.59	13,546	293	234.3	630.1
Methanol	CH_4O	32.042	791	293	175.5	337.8
Methyl acetate	$C_3H_6O_2$	74.08	934	293	175	330.1
Methyl acrylate	$C_4H_7O_2$	86.091	956	293	196.7	353.5
Methyl amine	CH_5N	31.058			179.7	266.8
Methyl benzoate	$C_8H_8O_2$	136.151	1,086	293	260.8	472.2
Methyl ethyl ketone	C_4H_8O	72.107	805	293	186.5	352.8
Naphtalene	$C_{10}H_8$	128.174	971	363	353.5	491.1
Nitric oxide	NO	30.006			109.5	121.4
Nitrogen	N_2	28.013			63.3	77.4
Nitrogen dioxide	NO	30.01			112.2	122.2
Nitrogen tetroxide	N_2O_4	46.006			261.9	294.3
Nitrous oxide	N_2O	44.013			182.3	184.7
Oxygen	O_2	31.999			54.4	90.2
n-Pentane	C_5H_{12}	72.151	626	293	143.4	309.2
1-Pentanol	$C_5H_{12}O$	88.15	815	293	195	411
1-Pentyne	C_5H_8	68.119	690	293	167.5	313.3
1-Pentene	C_5H_{10}	70.135	640	293	107.9	303.1
Phenol	C_6H_6O	94.113	1,059	314	313	455
Propane	C_3H_8	44.097			85.5	231.1
1-Propanol	C_3H_8O	60.096	804	293	146.9	370.4
Propionic acid	$C_3H_6O_2$	74.08	993	293	252.5	414
Propylene	C_3H_6	42.081			87.9	225.4
Propylene oxide	C_3H_6O	58.08	829	293	161	307.5
Styrene	C_8H_8	104.152	906	293	242.5	418.3
Succinic acid	$C_4H_6O_4$	118.09			456	508
Sulfur dioxide	SO_2	64.063			197.7	263
Sulfur trioxide	SO_3	80.058	1,780	318	290	318
Toluene	C_7H_8	92.141	867	293	178	383.8
Trimethyl amine	C_3H_9N	59.112			156	276.1
Vinyl chloride	C_2H_3Cl	62.499			119.4	259.8
Water	H_2O	18.015	998	293	273.2	373.2
o-Xylene	C_8H_{10}	106.168	880	293	248	417.6
m-Xylene	C_8H_{10}	106.168	864	293	225.3	412.3
p-Xylene	C_8H_{10}	106.168	861	293	286.4	411.5

A.3   CONSTANTS FOR ANTOINE'S EQUATION

$\log p_{vap} = A - B/(\theta + T)$, p_{vap} is in mm Hg and T is in $^\circ C$

Compound	Formula	A	B	θ
Acetaldehyde	CH_3CHO	7.0565	1,070.6	236.0
Acetic acid	CH_3COOH	7.29963	1,479.02	216.81
Acetone	CH_3COCH_3	7.23157	1,277.03	237.23
Acetylene	C_2H_2	7.0949	709.1	253.2
Acrylic acid	C_2H_3COOH	7.1927	1,441.5	192.66
Ammonia	NH_3	7.36050	926.132	240.17
Aniline	C_6H_7N	7.2418	1,675.3	200.01
Benzaldehide	C_7H_6O	7.1007	1,628.00	207.04
Benzene	C_6H_6	6.90565	1,211.03	220.790
Benzoic acid	$C_7H_6O_2$	7.45397	1,820	147.96
1,2-Butadiene	C_4H_6	7.1619	1,121.0	251.00
1,3-Butadiene	C_4H_6	6.85941	935.531	239.554
n-Butane	C_4H_{10}	6.83029	945.90	240.00
n-Butanol	$C_3H_7CH_2OH$	7.4768	1,632.39	178.83
i-Butane	C_4H_{10}	6.74808	882.80	240.00
n-Butene	C_4H_8	6.84290	926.10	240.00
i-Butene	C_4H_8	6.84134	923.200	240.00
Carbon	C			
tetrachloride	CCl_4	6.8941	1,219.58	227.17
Chlorobenzene	C_6H_5Cl	6.9781	1,431.05	217.56
Chloroform	$CHCl_3$	6.93707	1,171.2	236.01
Cyclobutane	C_4H_8	6.92804	1,024.54	241.38
Cyclohexane	C_6H_{12}	6.84498	1,203.526	222.863
Cyclopentane	C_5H_{10}	6.88678	1,124.16	231.37
Cyclopentene	C_5H_8	6.92704	1,121.81	233.46
Ethane	C_2H_6	6.80266	656.40	256.00
Ethanol	CH_3CH_2OH	8.16290	1,623.22	228.98
Ethyl amine	C_2H_7N	7.38618	1,137.3	235.86
Ethyl acetate	$C_4H_8O_2$	7.01455	1,211.9	216.01
Ethene (Ethylene)	C_2H_4	6.74756	585.00	255.00
Ethylbenzene	$C_6H_5C_2H_5$	6.95719	1,424.55	213.206
Ethylenediamine	$C_2H_8N_2$	7.12599	1,350.0	201.03
Ethyl ether	$C_4H_{10}O$	6.98467	1,090.64	231.21
Ethyl propionate	$C_5H_{10}O_2$	7.01907	1,274.7	209.0
Ethylene glycol	$C_2H_6O_2$	8.7945	2,615.4	244.91
Formaldehyde	$HCHO$	7.1561	957.24	243.0
Formic acid	$HCOOH$	7.37790	1,563.28	247.06
Glycerol	$C_3H_8O_3$	7.48689	1,948.7	132.96
n-Heptane	C_7H_{16}	6.90240	1,268.115	216.900
1-Heptanol	$C_7H_{16}O$	6.64766	1,140.64	126.56
1-Heptene	C_7H_{14}	6.90068	1,257.5	219.19

(Continued)

Compound	Formula	*A*	*B*	θ
n-Hexane	C_6H_{14}	6.87776	1,171.530	224.366
Hydrogen bromide	HBr	6.28370	539.62	225.30
Hydrogen chloride	HCl	7.167160	744.49	258.704
Hydrogen cyanide	HCN	7.17185	1,123.0	236.01
Hydrogen sulfide	H_2S	6.99392	768.13	247.093
Iodine (c)	I_2	9.8109	2,901.0	256.00
Isopropyl alcohol	C_3H_8O	8.11822	1,580.92	219.62
Maleic anhydride	$C_4H_2O_3$	7.06801	1,635.4	191.01
Methane	CH_4	6.61184	389.93	266.00
Methanol	CH_3OH	8.07246	1,574.99	238.86
Methyl acetate	$C_3H_6O_2$	7.00495	1,130.0	217.01
Methyl acrylate	$C_4H_7O_2$	6.99596	1,211.0	214.01
Methyl amine	CH_5N	7.49688	1,079.15	240.24
Methyl benzoate	$C_8H_8O_2$	7.04738	1,629.4	192.01
Methyl ethyl ketone	C_4H_8O	7.20868	1,368.21	236.51
Napthalene	$C_{10}H_8$	6.84577	1,606.5	187.227
Nitric oxide	NO	8.74300	682.94	268.27
Nitrogen tetroxide	N_2O_4	7.38499	1,185.72	234.18
Nitrous oxide	N_2O	7.00394	654.26	247.16
n-Pentane	C_5H_{12}	6.85221	1,064.63	232.000
1-Pentanol	$C_5H_{12}O$	7.17758	1,314.56	168.16
1-Pentyne	C_5H_8	6.96734	1,092.52	227.19
1-Pentene	C_5H_{10}	6.84648	1,044.9	233.53
Phenol	C_6H_5OH	7.13457	1,516.07	174.569
Phosphorus trichloride	PCl_3	6.8267	1,196	227.0
Phosphine	PH_3	6.71559	645.512	256.066
Propane	C_3H_8	6.82973	813.20	248.00
1-Propanol	$CH_3CH_2CH_2OH$	6.79498	969.27	150.42
Propionic acid	$C_3H_6O_2$	7.57456	1,617.06	205.68
Propene (Propylene)	C_3H_6	6.81960	785.00	247.00
Propylene oxide	C_3H_6O	6.65456	915.31	208.29
n-Propionic Acid	CH_3CH_2COOH	7.54760	1,617.06	205.67
Styrene	C_8H_8	6.95709	1,445.58	209.44
Sulfur dioxide	SO_2	7.28228	999.900	237.190
Sulfur trioxide	SO_3	9.05085	1,735.31	236.50
Toluene	$C_6H_5CH_3$	6.95464	1,344.800	219.482
Trimethyl amine	C_3H_9N	6.97038	968.7	234.01
Vinyl chloride	C_2H_3Cl	6.48709	783.4	230.01
Water	H_2O	7.94915	1,657.46	227.03
o-Xylene	$C_6H_5(CH_3)_2$	6.99891	1,474.679	213.686
p-Xylene	$C_6H_5(CH_3)_2$	6.99052	1,453.430	215.307

Appendix B: Iteration Methods

B.1 BISECTION METHOD

Given some function of x such as $H(x)$, we are interested in the solution of the equation

$$H(x) = 0, \quad x = x^*$$

(B.1)

Here, we have used x^* to represent the solution. For simple functions such as $H(x) = x - b$, we obtain a single solution given by $x^* = b$, while for a more complex function such as $H(x) = x^2 - b$, we obtain more than one solution as indicated by $x^* = \pm\sqrt{b}$. In many cases, there is no *explicit* solution for Eq. B.1. For example, if $H(x)$ is given by

$$H(x) = a\sin(x\pi/2) + b\cos(2\pi x)$$

(B.2)

we need to use iterative methods to determine the solution $x = x^*$.

The simplest iterative method is the bisection method (Corliss, 1977) that is illustrated in Figure B.1. This method begins by locating x_0 and x_1 such that $H(x_0)$ and $H(x_1)$ have different signs. In Figure B.1, we see that x_0 and x_1 have been chosen so that there is a change of sign for $H(x)$, i.e.,

$$H(x_0) > 0, \quad H(x_1) < 0$$

(B.3)

Thus, if $H(x)$ is a continuous function we know that a solution $H(x^*) = 0$ exists somewhere between x_0 and x_1. We attempt to locate that solution by means of a guess (i.e., the bisection) indicated by

$$x_2 = \frac{x_0 + x_1}{2}$$

(B.4)

As illustrated in Figure B.1, this guess is closer to the solution, $x = x^*$, than either x_0 or x_1, and if we repeat this procedure, we will eventually find a value of x that produces a value of $H(x)$ that is arbitrarily close to zero.

In terms of the particular graph illustrated in Figure B.1, it is clear that x_3 will be located between x_1 and x_2; however, this need not be the case. For example, in Figure B.2, we have represented a slightly different function for which x_3 will be located between x_0 and x_2. The location of the next guess is based on the idea that the function $H(x)$ must change sign. In order to determine the location of the next guess, we examine the products $H(x_n) \times H(x_{n-1})$ and $H(x_n) \times H(x_{n-2})$ in order to make the following decisions:

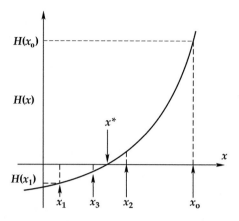

FIGURE B.1   Illustration of the bisection method.

$$\text{if}\quad H(x_n)\times H(x_{n-1})<0,\quad\text{then }x_{n+1}=\frac{x_n+x_{n-1}}{2}$$

$$\text{if}\quad H(x_n)\times H(x_{n-2})<0,\quad\text{then }x_{n+1}=\frac{x_n+x_{n-2}}{2}$$

(B.5)

Since these two choices are mutually exclusive, there is no confusion about the next choice of the dependent variable. The use of Eq. B.5 is crucial when the details of $H(x)$ are not clear, and a program is written to solve the implicit equation.

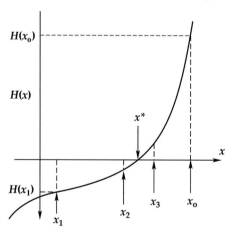

FIGURE B.2   Alternate choice for the second bisection.

B.2   FALSE POSITION METHOD

The false position method is also known as the *method of interpolation*[*] and it represents a minor variation of the bisection method. Instead of bisecting the distance between x_0 and x_1 in Figure B.1 in order to locate the point x_2, we use the straight line

---

[*] Wylie, C.R., Jr. 1951, *Advanced Engineering Mathematics*, McGraw-Hill Book Co., Inc., New York.

indicated in Figure B.3. Sometimes this line is called the secant line. The definition of the tangent of the angle φ provides

$$\tan \varphi = \frac{0 - H(x_1)}{x_2 - x_1} = \frac{H(x_o) - H(x_1)}{x_o - x_1} \tag{B.6}$$

and we can solve for x_2 to obtain

$$x_2 = x_1 - \frac{(x_o - x_1) H(x_1)}{H(x_o) - H(x_1)} \tag{B.7}$$

This replaces Eq. B.4 in the bisection method and it can be generalized to obtain

$$x_{n+2} = x_{n+1} - \frac{(x_n - x_{n+1}) H(x_{n+1})}{H(x_n) - H(x_{n+1})} \tag{B.8}$$

Application of successive iterations will lead to a value of x that approaches x^* shown in Figure B.3.

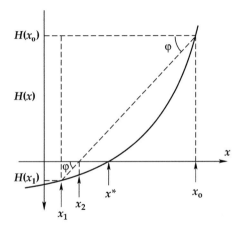

FIGURE B.3   False position construction.

B.3   NEWTON'S METHOD

Newton's method[†], which is also known as the Newton-Raphson method, is named for Sir Isaac Newton and is perhaps the best-known method for finding roots of real valued functions. The method is similar to the false position method in that a straight line is used to locate the next estimate of the root of an equation; however, in this case, it is a tangent line and not a secant line. This is illustrated in Figure B.4 where we have chosen x_0 as our first estimate of the solution to Eq. B.1 and we have constructed a tangent line to $H(x)$ at x_0. The slope of this tangent line is given by

$$\left. \frac{dH}{dx} \right|_{x=x_o} = \frac{H(x_o) - 0}{x_o - x_1} \tag{B.9}$$

[†] Ypma, T.J. 1995, Historical development of the Newton-Raphson method, SIAM Review **37**, 531–551.

and we can solve this equation to produce our next estimate of the root. This new estimate is given by

$$x_1 = x_o - \frac{H(x_o)}{(dH/dx)|_{x=x_o}} \tag{B.10}$$

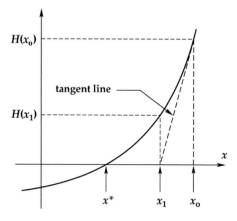

FIGURE B.4 First estimate using Newton's method.

and we use this result to determine $H(x_1)$ as indicated in Figure B.4.

Given $H(x_1)$ and x_1 we can construct a second estimate as indicated in Figure B.5, and this process can be continued to find the solution given by x^*. The general iterative procedure is indicated by

$$x_{n+1} = x_n - \frac{H(x_n)}{(dH/dx)|_{x=x_n}}, \quad n = 0,1,2,..., \tag{B.11}$$

Newton's method is certainly an attractive technique for finding solutions to implicit equations; however, it does require that one know both the function and its derivative.

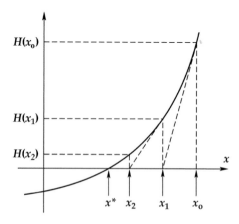

FIGURE B.5 Second estimate using Newton's method.

For complex functions, calculating the derivative at each step in the iteration may require more effort than that associated with the bisection method or the false position method. In addition, if the derivative of the function is zero in the region of interest, Newton's method will fail.

B.4   PICARD'S METHOD

Picard's method for solving Eq. B.1 begins by defining a new function according to

Definition:$\qquad\qquad\qquad f(x) = x + H(x)$$\qquad\qquad\qquad$(B.12)

Given any value of the dependent variable, x_n, we *define* a new value, x_{n+1}, by

Definition:$\qquad\qquad\qquad x_{n+1} = f(x_n), \quad n = 0, 1, 2, 3, \ldots$$\qquad\qquad$(B.13)

This represents Picard's method or the *method of direct substitution* or the *method of successive substitution*. If this procedure converges, we have

$$f(x^*) = x^* + H(x^*) = x^*$$(B.14)

In Eq. B.13, we note that the function $f(x_n)$, maps the point x_n to the new point x_{n+1}. If the function $f(x)$ maps the point x to itself, i.e., $f(x) = x$, then x is called the *fixed point* of $f(x)$. In Figure B.6, we again consider the function represented in Figures B.1, B.3, and B.4, and we illustrate the functions $f(x)$, $y(x)$, and $H(x)$. The graphical representation of the fixed point, x^*, is the intersection of the function of $f(x)$ with the line $y = x$.

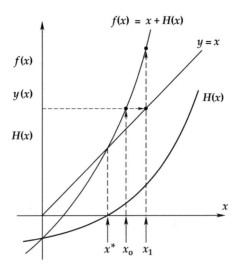

FIGURE B.6   Picard's method.

Note that not all functions have fixed points. For example, if $f(x)$ is parallel to the line $y = x$, there can be no intersection and no fixed point. Given our first estimate, x_o, we use Eq. B.13 to compute x_1 according to

$$x_1 = f(x_o) \tag{B.15}$$

Clearly, x_1 is further from the solution, x^*, than x_o and we can see from the graphical representation in Figure B.6 that Picard's method *diverges* for this case. If x_o were chosen to be less than the solution, x^*, we would also find that the iterative procedure diverges. If the slope of $f(x)$ were less than the slope of $y(x)$, we would find that Picard's method converges. This suggests that the method is useful for "weak" functions of x, i.e., $|df/dx| < 1$ and this is confirmed in Sec. B.6.

B.5   WEGSTEIN'S METHOD

In Figure B.7, we have illustrated the same function, $f(x)$, that appears in Figure B.6. For some point, x_1 in the neighborhood of x_o we can approximate the derivative of $f(x)$ according to

$$\frac{df}{dx} \approx \frac{f(x_1) - f(x_o)}{x_1 - x_o} \approx \text{slope} = S \tag{B.16}$$

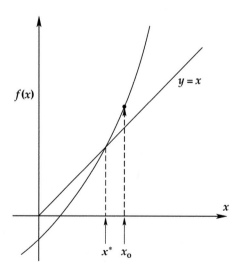

FIGURE B.7   Wegstein's method.

and we can use this result to obtain an approximation for the function $f(x_1)$.

$$f(x_1) \approx f(x_o) + S(x_1 - x_o) \tag{B.17}$$

At this point, we recall Eq. B.14 in the form

$$f(x^*) = x^* \tag{B.18}$$

and note that if x_1 is in the neighborhood of x^* we obtain the approximation

$$f(x_1) \approx x_1 \tag{B.19}$$

We use this result in Eq. B.17 to produce an equation

$$x_1 = f(x_o) + S(x_1 - x_o) \tag{B.20}$$

in which S is an *adjustable parameter* that is used to determine the next step in the iterative procedure. It is traditional, but not necessary, to define a new *adjustable parameter* according to

$$q = \frac{S}{S-1} \tag{B.21}$$

Use of this representation in Eq. B.20 leads to

$$x_1 = (1-q) f(x_o) + q x_o \tag{B.22}$$

and we can generalize this result to Wegstein's method given by

$$x_{n+1} = (1-q) f(x_n) + q x_n, \quad n = 0, 1, 2, 3, \ldots \text{ etc} \tag{B.23}$$

When the adjustable parameter is equal to zero, $q = 0$, we obtain Picard's method described in Sec. B.4. When the adjustable parameter greater than zero and less than one, $0 < q < 1$, we obtain a *damped* successive substitution process that improves stability for nonlinear systems. When the adjustable parameter is negative, $q < 0$, we obtain an *accelerated* successive substitution that may lead to an unstable procedure.

B.6 STABILITY OF ITERATION METHODS

In this section, we consider the *linear* stability characteristics of Newton's method, Picard's method, and Wegstein's method that have been used to solve the implicit equation given by

$$H(x) = 0, \quad x = x^* \tag{B.24}$$

The constraint associated with the linear analysis will be listed below and it must be kept in mind when interpreting results such as those presented in Chapter 7. We begin by recalling the three iterative methods as

Newton's method: $\quad x_{n+1} = x_n - \dfrac{H(x_n)}{(dH/dx)\vert_{x=x_n}}, \quad n = 0, 1, 2, \ldots,$ $\tag{B.25}$

Picard's method: $\quad x_{n+1} = f(x_n), \quad n = 0, 1, 2, \ldots$ $\tag{B.26}$

Wegstein's method: $\quad x_{n+1} = (1-q)f(x_n) + q x_n, \quad n = 0,1,2,\ldots$ $\qquad$ (B.27)

in which the auxiliary function, $f(x)$, is defined by

Definition: $\qquad\qquad\qquad\qquad f(x) = x + H(x)$ $\qquad\qquad\qquad\qquad$ (B.28)

The general form of these three iterative methods is given by

$$x_{n+1} = G(x_n), \quad n = 0,1,2,\ldots \qquad\qquad\qquad (B.29)$$

and for each of the three methods on seeks to find the fixed point x^* of $G(x)$ such that

$$x^* = G(x^*) \qquad\qquad\qquad (B.30)$$

Our stability analysis of Eqs. B.25–B.27 is based on linearizing $G(x)$ about the fixed point x^*. We let Δx_n and Δx_{n+1} be small perturbations from the fixed point as indicated by

$$x_n = x^* + \Delta x_n, \quad x_{n+1} = x^* + \Delta x_{n+1} \qquad\qquad\qquad (B.31)$$

This allows us to express Eq. B.29 as

$$x^* + \Delta x_{n+1} = G(x^* + \Delta x_n), \quad n = 0,1,2,\ldots \qquad\qquad\qquad (B.32)$$

and a Taylor series expansion (See Problems 5.30 and 5.31 in Chapter 5) leads to

$$G(x^* + \Delta x_n) = G(x^*) + \Delta x_n \left(\frac{dG}{dx} \right)_{x^*} + \frac{1}{2} \Delta x_n^2 \left(\frac{d^2 G}{dx^2} \right)_{x^*} + \cdots \qquad (B.33)$$

On the basis of Eq. B.30, this infinite series simplifies to

$$G(x^* + \Delta x_n) = x^* + \Delta x_n \left(\frac{dG}{dx} \right)_{x^*} + \frac{1}{2} \Delta x_n^2 \left(\frac{d^2 G}{dx^2} \right)_{x^*} + \cdots \qquad (B.34)$$

and we can use Eq. B.32 to represent the left hand side in a simpler form to obtain

$$x^* + \Delta x_{n+1} = x^* + \Delta x_n \left(\frac{dG}{dx} \right)_{x^*} + \frac{1}{2} \Delta x_n^2 \left(\frac{d^2 G}{dx^2} \right)_{x^*} + \cdots \qquad (B.35)$$

At this point, we impose a constraint on the higher order terms expressed as

Constraint: $\qquad\qquad \Delta x_n \left(\frac{dG}{dx} \right)_{x^*} \gg \frac{1}{2} \Delta x_n^2 \left(\frac{d^2 G}{dx^2} \right)_{x^*} + \cdots \qquad$ (B.36)

so that Eq. B.35 takes the form

$$\Delta x_{n+1} = \Delta x_n \left(\frac{dG}{dx} \right)_{x^*}, \quad n = 0,1,2,3,4,\ldots \qquad (B.37)$$

If we write a few of these equations explicitly as

$$\Delta x_1 = \Delta x_o \left(\frac{dG}{dx} \right)_{x^*} \tag{B.38a}$$

$$\Delta x_2 = \Delta x_1 \left(\frac{dG}{dx} \right)_{x^*} \tag{B.38b}$$

$$\Delta x_3 = \Delta x_2 \left(\frac{dG}{dx} \right)_{x^*} \tag{B.38c}$$

$$\Delta x_n = \Delta x_{n-1} \left(\frac{dG}{dx} \right)_{x^*} \tag{B.38d}$$

it becomes clear that they can be used to provide a general representation given by

$$\Delta x_n = \Delta x_o \left[(dG/dx)_{x^*} \right]^n, \quad n = 0,1,2,\ldots \tag{B.39}$$

At this point, we see that $\Delta x_n \to 0$ when $n \to \infty$ provided that

$$\left| (dG/dx)_{x^*} \right| < 1 \tag{B.40}$$

When $\Delta x_n \to 0$ as $n \to \infty$ the system *converges* and one says that the fixed point x^* is *attracting*. The three special cases represented by Eq. B.39 can be expressed as

I. $\left| (dG/dx)_{x^*} \right| < 1$, the fixed point x^* is *attracting* and the iteration *converges* (B.41)

II. $\left| (dG/dx)_{x^*} \right| > 1$, the fixed point x^* is *repelling* and the iteration *diverges* (B.42)

III. $\left| (dG/dx)_{x^*} \right| = 1$, the fixed point x^* is *neither* attracting nor *repelling* (B.43)

It is extremely important to note that the stability analysis leading to these three results is based on the *linear approximation* associated with Eq. B.36. In this development, the word *attracting* is used for a system that converges since x_n *moves toward* x^* as n increases, while the word *repelling* is used for a system that diverges since x_n *moves away from* x^* as n increases. The case in which the fixed point is neither attracting nor repelling can lead to chaos[‡,§].

At this point, we are ready to return to Eqs. B.25–B.27 in order to determine the linear stability characteristics of Newton's method, Picard's method, and Wegstein's method.

[‡] Gleick, J. 1988 *Chaos: Making a New Science*, Penguin Books, New York.
[§] Peitgen, H.-O., Jürgens, H., and Saupe, D., 1992, *Chaos and Fractals. New Frontiers of Science*, Springer-Verlag, New York.

B.6.1 Newton's Method

In this case, we have

$$G(x) = x - \frac{H(x)}{(dH/dx)} \tag{B.44}$$

and the derivative that is required to determine the stability is given by

$$(dG/dx) = \frac{H(x)}{(dH/dx)^2} \frac{d^2H}{dx^2} \tag{B.45}$$

Evaluation of this derivative at the fixed point where $H(x^*) = 0$ leads to

$$(dG/dx)_{x^*} = 0 \tag{B.46}$$

This indicates that Newton's method will converge provided that $(dH/dx) \neq 0$ and provided that the initial guess, x_o, is close enough to x^* so that Eq. B.36 is satisfied. If Eq. B.36 is not satisfied, the linear stability analysis leading to Eqs. B.41–B.43 is not valid.

B.6.2 Picard's Method

In this case, Eqs. B.26 and B.29 provide $G(x) = f(x)$ and

$$(dG/dx)_{x^*} = (df/dx)_{x^*} \tag{B.47}$$

and from Eq. B.41, we conclude that Picard's method is stable when

$$\left|(df/dx)_{x^*}\right| < 1 \tag{B.48}$$

In Example 7.7 of Chapter 7 we used the fixed point iteration (see Eq. 17) that can be expressed as

$$x_{n+1} = f(x_n) = (1 - C)(1 + x_n), \quad C = 0.30, \quad n = 0, 1, 2, 3, \ldots \tag{B.49}$$

This leads to the condition

$$\left|(df/dx)\right| = \left|(1 - C)\right| < 1 \tag{B.50}$$

that produces the *stable* iteration illustrated in Table 7.7a. In Example 7.8 of Chapter 7, we find another example of Picard's method (see Eq. 40) that we repeat here as

$$x_{n+1} = f(x_n) = x_n + \left[37.6190 - \frac{15.1678 \, x_n}{0.8571 - x_n}\right], \quad n = 0, 1, 2, 3, \ldots \tag{B.51}$$

The solution is given by $x^* = 0.6108$ and this leads to

$$(df/dx)_{x^*} = -213.1733 \tag{B.52}$$

indicating that Picard's method is unstable for this particular fixed point iteration. This result is consistent with the entries in Table 7.8a.

B.6.3 WEGSTEIN'S METHOD

In this case, Eqs. B.27 and B.29 provide

$$G(x) = (1-q) f(x) + qx \tag{B.53}$$

which leads to

$$G(x) = (1-q) f(x) + qx \tag{B.54}$$

From this we have

$$\frac{dG}{dx} = (1-q)\frac{df}{dx} + q \tag{B.55}$$

and the stability condition given by Eq. B.41 indicates that Wegstein's method will converge provided that

$$(1-q)\left|\frac{df}{dx}\right| + q < 1 \tag{B.56}$$

Here, one can see that the adjustable parameter q can often be chosen so that this inequality is satisfied and Wegstein's method will converge as illustrated in Examples 7.7 and 7.8 of Chapter 7.

Appendix C: Matrices

C.1   MATRIX METHODS AND PARTITIONING

In order to support the results obtained for the *atomic matrix* studied in Chapter 6 and for the *mechanistic matrix* studied in Chapter 9, we need to consider that matter of partitioning matrices. All the information necessary for our studies of stoichiometry is contained in Eq. 6.22; however, that information can be presented in *different forms* depending on how the atomic matrix and the column matrix of net rates of production are partitioned. In our analysis of reaction kinetics, all the information we need is contained in the mechanistic matrix. However, that information can be presented in *different forms* depending on presence or absence of Bodenstein products. In this appendix, we review the methods required to develop the desired *different forms*.

C.1.1   MATRIX ADDITION

We begin our study of partitioning with the process of *addition* (or subtraction) as illustrated by the following matrix equation

$$
\begin{bmatrix}
a_{11} & a_{12} & a_{13} & a_{14} \\
a_{21} & a_{22} & a_{23} & a_{24} \\
a_{31} & a_{32} & a_{33} & a_{34} \\
a_{41} & a_{42} & a_{43} & a_{44}
\end{bmatrix}
+
\begin{bmatrix}
b_{11} & b_{12} & b_{13} & b_{14} \\
b_{21} & b_{22} & b_{23} & b_{24} \\
b_{31} & b_{32} & b_{33} & b_{34} \\
b_{41} & b_{42} & b_{43} & b_{44}
\end{bmatrix}
=
\begin{bmatrix}
c_{11} & c_{12} & c_{13} & c_{14} \\
c_{21} & c_{22} & c_{23} & c_{24} \\
c_{31} & c_{32} & c_{33} & c_{34} \\
c_{41} & c_{42} & c_{43} & c_{44}
\end{bmatrix}
\tag{C.1}
$$

This can be expressed in more compact nomenclature according to

$$
A + B = C
\tag{C.2}
$$

The fundamental meaning of Eqs. C.1 and C.2 is given by the following sixteen (16) equations:

$$
\begin{aligned}
a_{11} + b_{11} &= c_{11} & a_{21} + b_{21} &= c_{21} \\
a_{12} + b_{12} &= c_{12} & a_{22} + b_{22} &= c_{22} \\
a_{13} + b_{13} &= c_{13} & a_{23} + b_{23} &= c_{23} \\
a_{14} + b_{14} &= c_{14} & a_{24} + b_{24} &= c_{24} \\
\\
a_{31} + b_{31} &= c_{31} & a_{41} + b_{41} &= c_{41} \\
a_{32} + b_{32} &= c_{32} & a_{42} + b_{42} &= c_{42} \\
a_{33} + b_{33} &= c_{33} & a_{43} + b_{43} &= c_{43} \\
a_{34} + b_{34} &= c_{34} & a_{44} + b_{44} &= c_{44}
\end{aligned}
\tag{C.3}
$$

These equations represent a *complete partitioning* of the matrix equation given by Eq. C.1, and we can also represent this complete partitioning in the form

$$
\begin{bmatrix}
a_{11} & a_{12} & a_{13} & a_{14} \\
a_{21} & a_{22} & a_{23} & a_{24} \\
a_{31} & a_{32} & a_{33} & a_{34} \\
a_{41} & a_{42} & a_{43} & a_{44}
\end{bmatrix}
+
\begin{bmatrix}
b_{11} & b_{12} & b_{13} & b_{14} \\
b_{21} & b_{22} & b_{23} & b_{24} \\
b_{31} & b_{32} & b_{33} & b_{34} \\
b_{41} & b_{42} & b_{43} & b_{44}
\end{bmatrix}
=
\begin{bmatrix}
c_{11} & c_{12} & c_{13} & c_{14} \\
c_{21} & c_{22} & c_{23} & c_{24} \\
c_{31} & c_{32} & c_{33} & c_{34} \\
c_{41} & c_{42} & c_{43} & c_{44}
\end{bmatrix}
\tag{C.4}
$$

Here, we have shaded the particular partition that represents the first of Eq. C.3. The complete partitioning illustrated by Eq. C.4 is not particularly useful; however, there are other possibilities that we will find to be very useful and one example is the *row/column partition* given by

$$
\begin{bmatrix}
a_{11} & a_{12} & a_{13} & a_{14} \\
a_{21} & a_{22} & a_{23} & a_{24} \\
a_{31} & a_{32} & a_{33} & a_{34} \\
a_{41} & a_{42} & a_{43} & a_{44}
\end{bmatrix}
+
\begin{bmatrix}
b_{11} & b_{12} & b_{13} & b_{14} \\
b_{21} & b_{22} & b_{23} & b_{24} \\
b_{31} & b_{32} & b_{33} & b_{34} \\
b_{41} & b_{42} & b_{43} & b_{44}
\end{bmatrix}
=
\begin{bmatrix}
c_{11} & c_{12} & c_{13} & c_{14} \\
c_{21} & c_{22} & c_{23} & c_{24} \\
c_{31} & c_{32} & c_{33} & c_{34} \\
c_{41} & c_{42} & c_{43} & c_{44}
\end{bmatrix}
\tag{C.5}
$$

Each partitioned matrix can be expressed in the form

$$
A =
\begin{bmatrix}
A_{11} & A_{12} \\
A_{21} & A_{22}
\end{bmatrix}
=
\begin{bmatrix}
a_{11} & a_{12} & a_{13} & a_{14} \\
a_{21} & a_{22} & a_{23} & a_{24} \\
a_{31} & a_{32} & a_{33} & a_{34} \\
a_{41} & a_{42} & a_{43} & a_{44}
\end{bmatrix}
\tag{C.6}
$$

and the partitioned matrix equation is given by

$$
\begin{bmatrix}
A_{11} & A_{12} \\
A_{21} & A_{22}
\end{bmatrix}
+
\begin{bmatrix}
B_{11} & B_{12} \\
B_{21} & B_{22}
\end{bmatrix}
=
\begin{bmatrix}
C_{11} & C_{12} \\
C_{21} & C_{22}
\end{bmatrix}
\tag{C.7}
$$

We usually think of the elements of a matrix as numbers such as a_{11}, a_{12}, etc.; however, the elements of a matrix can also be matrices as indicated in Eq. C.7. The usual rules for matrix addition leads to

$$
A_{11} + B_{11} = C_{11} \tag{C.8a}
$$

$$
A_{12} + B_{12} = C_{12} \tag{C.8b}
$$

$$
A_{21} + B_{21} = C_{21} \tag{C.8c}
$$

$$
A_{22} + B_{22} = C_{22} \tag{C.8d}
$$

and the details associated with Eq. C.8a are given by

$$
\begin{bmatrix}
a_{11} & a_{12} \\
a_{21} & a_{22}
\end{bmatrix}
+
\begin{bmatrix}
b_{11} & b_{12} \\
b_{21} & b_{22}
\end{bmatrix}
=
\begin{bmatrix}
c_{11} & c_{12} \\
c_{21} & c_{22}
\end{bmatrix}
\tag{C.9}
$$

A little thought will indicate that this matrix equation represents the first four equations given in Eq. C.3. Other partitions of Eq. C.1 are obviously available and will be encountered in the following paragraphs.

C.1.2 MATRIX MULTIPLICATION

Multiplication of matrices can also be represented in terms of *submatrices*, provided that one is careful to follow the rules of matrix multiplication. As an example, we consider the following matrix equation

$$
\begin{bmatrix}
a_{11} & a_{12} & a_{13} & a_{14} \\
a_{21} & a_{22} & a_{23} & a_{24} \\
a_{31} & a_{32} & a_{33} & a_{34} \\
a_{41} & a_{42} & a_{43} & a_{44}
\end{bmatrix}
\begin{bmatrix}
b_{11} & b_{12} \\
b_{21} & b_{22} \\
b_{31} & b_{32} \\
b_{41} & b_{42}
\end{bmatrix}
=
\begin{bmatrix}
c_{11} & c_{12} \\
c_{21} & c_{22} \\
c_{31} & c_{32} \\
c_{41} & c_{42}
\end{bmatrix}
\tag{C.10}
$$

which conforms to the rule that the number of columns in the first matrix is equal to the number of rows in the second matrix. Equation C.10 represents the eight (8) individual equations given by

$$a_{11}b_{11} + a_{12}b_{21} + a_{13}b_{31} + a_{14}b_{41} = c_{11} \tag{C.11a}$$

$$a_{11}b_{12} + a_{12}b_{22} + a_{13}b_{32} + a_{14}b_{42} = c_{12} \tag{C.11b}$$

$$a_{21}b_{11} + a_{22}b_{21} + a_{23}b_{31} + a_{24}b_4 = c_{21} \tag{C.11c}$$

$$a_{21}b_{12} + a_{22}b_{22} + a_{23}b_{32} + a_{24}b_{42} = c_{22} \tag{C.11d}$$

$$a_{31}b_{11} + a_{32}b_{21} + a_{33}b_{31} + a_{34}b_{41} = c_{31} \tag{C.11e}$$

$$a_{31}b_{12} + a_{32}b_{22} + a_{33}b_{32} + a_{34}b_{42} = c_{32} \tag{C.11f}$$

$$a_{41}b_{11} + a_{42}b_{21} + a_{43}b_{31} + a_{44}b_{41} = c_{41} \tag{C.11g}$$

$$a_{41}b_{12} + a_{42}b_{22} + a_{43}b_{32} + a_{44}b_{42} = c_{42} \tag{C.11h}$$

which can also be expressed in compact form according to

$$AB = C \tag{C.12}$$

Here, the matrices A, B, and C are defined explicitly by

$$
A =
\begin{bmatrix}
a_{11} & a_{12} & a_{13} & a_{14} \\
a_{21} & a_{22} & a_{23} & a_{24} \\
a_{31} & a_{32} & a_{33} & a_{34} \\
a_{41} & a_{42} & a_{43} & a_{44}
\end{bmatrix}
\quad
B =
\begin{bmatrix}
b_{11} & b_{12} \\
b_{21} & b_{22} \\
b_{31} & b_{32} \\
b_{41} & b_{42}
\end{bmatrix}
\quad
C =
\begin{bmatrix}
c_{11} & c_{12} \\
c_{21} & c_{22} \\
c_{31} & c_{32} \\
c_{41} & c_{42}
\end{bmatrix}
\tag{C.13}
$$

In Eqs. C.1–C.9, we have illustrated that the process of addition and subtraction can be carried out in terms of partitioned matrices. Matrix multiplication can also be carried out in terms of partitioned matrices; however, in order to conform to the rules of matrix multiplication, we must partition the matrices properly. For example, a proper *row partition* of Eq. C.10 can be expressed as

$$
\left[
\begin{array}{cccc}
a_{11} & a_{12} & a_{13} & a_{14} \\
a_{21} & a_{22} & a_{23} & a_{24} \\
\hline
a_{31} & a_{32} & a_{33} & a_{34} \\
a_{41} & a_{42} & a_{43} & a_{44}
\end{array}
\right]
\left[
\begin{array}{cc}
b_{11} & b_{12} \\
b_{21} & b_{22} \\
b_{31} & b_{32} \\
b_{41} & b_{42}
\end{array}
\right]
=
\left[
\begin{array}{cc}
c_{11} & c_{12} \\
c_{21} & c_{22} \\
\hline
c_{31} & c_{32} \\
c_{41} & c_{42}
\end{array}
\right]
\tag{C.14}
$$

In terms of the submatrices defined by

$$
A_{11} =
\left[
\begin{array}{cccc}
a_{11} & a_{12} & a_{13} & a_{14} \\
a_{21} & a_{22} & a_{23} & a_{24}
\end{array}
\right], \quad
A_{21} =
\left[
\begin{array}{cccc}
a_{31} & a_{32} & a_{33} & a_{34} \\
a_{41} & a_{42} & a_{43} & a_{44}
\end{array}
\right]
\tag{C.15}
$$

$$
C_{11} =
\left[
\begin{array}{cc}
c_{11} & c_{12} \\
c_{21} & c_{22}
\end{array}
\right] \quad
C_{21} =
\left[
\begin{array}{cc}
c_{31} & c_{32} \\
c_{41} & c_{42}
\end{array}
\right]
$$

we can represent Eq. C.14 in the form

$$
\left[
\begin{array}{c}
A_{11} \\
A_{21}
\end{array}
\right] B =
\left[
\begin{array}{c}
A_{11}B \\
A_{21}B
\end{array}
\right] =
\left[
\begin{array}{c}
C_{11} \\
C_{21}
\end{array}
\right]
\tag{C.16}
$$

Often, it is useful to work with the separate matrix equations that we have created by the partition, and these are given by

$$
A_{11}B = C_{11}
\tag{C.17}
$$

$$
A_{21}B = C_{21}
\tag{C.18}
$$

The details of the first of these can be expressed as

$$
A_{11}B =
\left[
\begin{array}{cccc}
a_{11} & a_{12} & a_{13} & a_{14} \\
a_{21} & a_{22} & a_{23} & a_{24}
\end{array}
\right]
\left[
\begin{array}{cc}
b_{11} & b_{12} \\
b_{21} & b_{22} \\
b_{31} & b_{32} \\
b_{41} & b_{42}
\end{array}
\right], \quad
C_{11} =
\left[
\begin{array}{cc}
c_{11} & c_{12} \\
c_{21} & c_{22}
\end{array}
\right]
\tag{C.19a}
$$

Multiplication can be carried out to obtain

$$
\left[
\begin{array}{cc}
a_{11}b_{11} + a_{12}b_{21} + a_{13}b_{31} + a_{14}b_{41} & a_{11}b_{12} + a_{12}b_{22} + a_{13}b_{32} + a_{14}b_{42} \\
a_{21}b_{11} + a_{22}b_{21} + a_{23}b_{31} + a_{24}b_{41} & a_{21}b_{12} + a_{22}b_{22} + a_{23}b_{32} + a_{24}b_{42}
\end{array}
\right]
=
\left[
\begin{array}{cc}
c_{11} & c_{12} \\
c_{21} & c_{22}
\end{array}
\right]
\tag{C.19b}
$$

and equating the four elements of each matrix leads to

$$
\begin{aligned}
a_{11}b_{11} + a_{12}b_{21} + a_{13}b_{31} + a_{14}b_{41} &= c_{11} \\
a_{11}b_{12} + a_{12}b_{22} + a_{13}b_{32} + a_{14}b_{42} &= c_{12} \\
a_{21}b_{11} + a_{22}b_{21} + a_{23}b_{31} + a_{24}b_{41} &= c_{21} \\
a_{21}b_{12} + a_{22}b_{22} + a_{23}b_{32} + a_{24}b_{42} &= c_{22}
\end{aligned}
\tag{C.19c}
$$

Here, we see that these four individual equations (associated with the partitioned matrix equation) are those given originally by Eqs. C.11a–C.11d. A little thought will indicate that the matrix equation represented by Eq. C.18 contains the four individual equations represented by Eqs. C.11e–C.11h. All of the information available in Eq. C.10 is given explicitly in Eq. C.11 and partitioning of the original matrix equation does nothing more than arrange the information in a different form.

If we wish to obtain a *column partition* of the matrix A in Eq. C.10, we must also create a row partition of matrix B in order to conform to the rules of matrix multiplication. This *column/row partition* takes the form

$$
\begin{bmatrix}
a_{11} & a_{12} & \vdots & a_{13} & a_{14} \\
a_{21} & a_{22} & \vdots & a_{23} & a_{24} \\
a_{31} & a_{32} & \vdots & a_{33} & a_{34} \\
a_{41} & a_{42} & \vdots & a_{43} & a_{44}
\end{bmatrix}
\begin{bmatrix}
b_{11} & b_{12} \\
b_{21} & b_{22} \\
\hline
b_{31} & b_{32} \\
b_{41} & b_{42}
\end{bmatrix}
=
\begin{bmatrix}
c_{11} & c_{12} \\
c_{21} & c_{22} \\
c_{31} & c_{32} \\
c_{41} & c_{42}
\end{bmatrix}
\tag{C.20}
$$

and the submatrices are identified explicitly according to

$$
A_{11} =
\begin{bmatrix}
a_{11} & a_{12} \\
a_{21} & a_{22} \\
a_{31} & a_{32} \\
a_{41} & a_{42}
\end{bmatrix}
\qquad
A_{12} =
\begin{bmatrix}
a_{13} & a_{14} \\
a_{23} & a_{24} \\
a_{33} & a_{34} \\
a_{43} & a_{44}
\end{bmatrix}
\tag{C.21}
$$

$$
B_{11} =
\begin{bmatrix}
b_{11} & b_{12} \\
b_{21} & b_{22}
\end{bmatrix}
\qquad
B_{21} =
\begin{bmatrix}
b_{31} & b_{32} \\
b_{41} & b_{42}
\end{bmatrix}
$$

Use of these representations in Eq. C.20 leads to

$$
[A_{11} \ A_{12}]
\begin{bmatrix}
B_{11} \\
B_{21}
\end{bmatrix}
= C
\tag{C.22}
$$

and matrix multiplication in terms of the submatrices provides

$$
A_{11}B_{11} + A_{12}B_{21} = C
\tag{C.23}
$$

In some cases, we will make use of a *complete column partition* of the matrix A which requires a *complete row partition* of the matrix B. This partition is illustrated by

$$\begin{bmatrix} a_{11} & a_{12} & a_{13} & a_{14} \\ a_{21} & a_{22} & a_{23} & a_{24} \\ a_{31} & a_{32} & a_{33} & a_{34} \\ a_{41} & a_{42} & a_{43} & a_{44} \end{bmatrix} \begin{bmatrix} b_{11} & b_{12} \\ b_{21} & b_{22} \\ b_{31} & b_{32} \\ b_{41} & b_{42} \end{bmatrix} = \begin{bmatrix} c_{11} & c_{12} \\ c_{21} & c_{22} \\ c_{31} & c_{32} \\ c_{41} & c_{42} \end{bmatrix} \qquad \text{(C.24)}$$

In terms of the submatrices, it can be expressed as

$$\begin{bmatrix} A_{11} & A_{12} & A_{13} & A_{14} \end{bmatrix} \begin{bmatrix} B_{11} \\ B_{21} \\ B_{31} \\ B_{41} \end{bmatrix} = C \qquad \text{(C.25)}$$

and matrix multiplication leads to

$$A_{11}B_{11} + A_{12}B_{21} + A_{13}B_{31} + A_{14}B_{41} = C \qquad \text{(C.26)}$$

C.2   PROBLEMS

C.1. Given a matrix equation of the form $c = Ab$ having an explicit representation of the form,

$$\begin{bmatrix} c_1 \\ c_2 \\ c_3 \\ c_4 \\ c_5 \end{bmatrix} = \begin{bmatrix} a_{11} & a_{12} & a_{13} & a_{14} \\ a_{21} & a_{22} & a_{23} & a_{24} \\ a_{31} & a_{32} & a_{33} & a_{34} \\ a_{41} & a_{42} & a_{43} & a_{44} \\ a_{51} & a_{52} & a_{53} & a_{54} \end{bmatrix} \begin{bmatrix} b_1 \\ b_2 \\ b_3 \\ b_4 \end{bmatrix} \qquad \text{(1)}$$

develop a partition that will lead to an equation for the column vector represented by

$$\begin{bmatrix} c_1 \\ c_2 \\ c_3 \end{bmatrix} = ? \qquad \text{(2)}$$

If the elements of c are related to the elements of b according to

$$\begin{bmatrix} c_4 \\ c_5 \end{bmatrix} = \begin{bmatrix} b_3 \\ b_4 \end{bmatrix} \qquad \text{(3)}$$

what are the elements of the matrix normally identified as A_{21}?

C.2. Construct the complete column/row partition of the matrix equation given by

$$
\begin{bmatrix}
a_{11} & a_{12} & a_{13} & a_{14} \\
a_{21} & a_{22} & a_{23} & a_{24} \\
a_{31} & a_{32} & a_{33} & a_{34} \\
a_{41} & a_{42} & a_{43} & a_{44}
\end{bmatrix}
\begin{bmatrix}
b_1 \\ b_2 \\ b_3 \\ b_4
\end{bmatrix}
=
\begin{bmatrix}
c_1 \\ c_2 \\ c_3 \\ c_4
\end{bmatrix}
\tag{1}
$$

and show how it can be represented in the form

$$
\begin{bmatrix} \ \end{bmatrix} b_1 + \begin{bmatrix} \ \end{bmatrix} b_2 + \begin{bmatrix} \ \end{bmatrix} b_3 + \begin{bmatrix} \ \end{bmatrix} b_4 = \begin{bmatrix} \ \end{bmatrix}
\tag{2}
$$

Appendix D: Atomic Species Balances

Throughout this text, we have made use of macroscopic mass and mole balances to solve a variety of problems with and without chemical reactions. Our solutions have been based on the application of Axiom I and Axiom II, and our attention has been focused on mass flow rates, molar rates of production owing to chemical reaction, accumulation of mass or moles, etc. In this appendix, we show how problems can be solved using macroscopic atomic species balances. This approach has some advantages when carrying out calculations by hand since the number of atomic species balance equations is almost always less than the number of molecular species balance equations.

We begin our development of atomic species balance equations with Axioms I and II given by

Axiom I: $\qquad \dfrac{d}{dt}\displaystyle\int_V c_A dV + \int_A c_A \mathbf{v}_A \cdot \mathbf{n}dA = \int_V R_A dV, \quad A = 1,2,\ldots,N$ (D.1)

Axiom II: $\qquad \displaystyle\sum_{A=1}^{A=N} N_{JA} R_A = 0, \quad J = 1,2,\ldots,T$ (D.2)

Here, we should remember that the individual components of the atomic matrix, $[N_{JA}]$, are described by (see Sec. 6.2)

$$N_{JA} = \left\{ \begin{array}{c} \text{number of moles of} \\ J\text{-type atoms per mole} \\ \text{of molecular species } A \end{array} \right\}, \quad J = 1,2,\ldots,T, \text{ and } A = 1,2,\ldots,N \quad \text{(D.3)}$$

To develop an atomic species balance, we multiply Eq. D.1 by N_{JA} and sum over all molecular species to obtain

$$\frac{d}{dt}\int_V \sum_{A=1}^{A=N} N_{JA}\, c_A dV + \int_A \sum_{A=1}^{A=N} N_{JA}\, c_A \mathbf{v}_A \cdot \mathbf{n}dA = \int_V \sum_{A=1}^{A=N} N_{JA}\, R_A dV \quad \text{(D.4)}$$

On the basis of Axiom II, we see that the last term in this result is zero and our atomic species balance takes the form

Axioms I & II: $\quad \dfrac{d}{dt}\displaystyle\int_V \sum_{A=1}^{A=N} N_{JA} c_A dV + \int_A \sum_{A=1}^{A=N} N_{JA} c_A \mathbf{v}_A \cdot \mathbf{n}dA = 0, \quad J = 1,2,\ldots,T$ (D.5)

Here, we have indicated explicitly that there are T atomic species balance equations instead of the N molecular species balance equations which are given by Eq. D.1. When $T \leq N$ it may be convenient to solve material balance problems using atomic species balances.

In many applications of Eq. D.5, diffusive transport at the surface of the control volume is negligible and $\mathbf{v}_A \cdot \mathbf{n}$ can be replaced by $\mathbf{v} \cdot \mathbf{n}$ leading to

$$\frac{d}{dt} \int_V \sum_{A=1}^{A=N} N_{JA} c_A \, dV + \int_A \sum_{A=1}^{A=N} N_{JA} c_A \mathbf{v} \cdot \mathbf{n} \, dA = 0, \quad J = 1, 2, \ldots, T \qquad (D.6)$$

If the concentration is given in terms of mole fractions one can use $c_A = x_A c$ to express Eq. D.6 as

$$\frac{d}{dt} \int_V \left\{ \sum_{A=1}^{A=N} N_{JA} x_A \right\} c \, dV + \int_A \left\{ \sum_{A=1}^{A=N} N_{JA} x_A \right\} c \mathbf{v} \cdot \mathbf{n} \, dA = 0, \quad J = 1, 2, \ldots, T \qquad (D.7)$$

For some reacting systems these results, rather that Eq. D.1, provide the simplest algebraic route to a solution.

When problems are stated in terms of species mass densities and species mass fractions, it is convenient to rearrange Eq. D.5 in terms of these variables. To develop the forms of Eqs. D.6 and D.7 that are useful when mass flow rates are given, we begin by representing the species concentration in terms of the species density according to

$$c_A = \rho_A / MW_A \qquad (D.8)$$

The species density is now expressed in terms of the mass fraction and the total mass density according to

$$\rho_A = \omega_A \rho \qquad (D.9)$$

and when these two results are used in Eq. D.6 we obtain

$$\frac{d}{dt} \int_V \sum_{A=1}^{A=N} \left(\frac{N_{JA} \omega_A}{MW_A} \right) \rho \, dV + \int_A \sum_{A=1}^{A=N} \left(\frac{N_{JA} \omega_A}{MW_A} \right) \rho \mathbf{v} \cdot \mathbf{n} \, dA = 0, \quad J = 1, 2, \ldots, T \qquad (D.10)$$

This represents the "mass" analog of Eq. D.7, and it will be convenient when working with problems in which mass flow rates and mass fractions are given. We should note that Eq. D.10 actually represents a molar balance on the Jth atomic species since an examination of the *units* will indicate that

$$\int_A \left\{ \sum_{A=1}^{A=N} \frac{N_{JA} \omega_A}{MW_A} \right\} \rho \mathbf{v} \cdot \mathbf{n} \, dA \rightarrow \left\{ \begin{array}{c} moles \ per \\ unit \ time \end{array} \right\} \qquad (D.11)$$

To obtain the actual atomic species mass balance, one would multiply Eq. D.10 by the atomic mass of the Jth atomic species, AW_J.

At steady state, Eq. D.4 represents a system of homogeneous equations in terms the *net* atomic species molar flow rates leaving the control volume. For there to be a nontrivial solution, the rank $= r$ of $[N_{JA}]$ must be less than N. Consequently, there are at most r linearly independent atomic species balances (see Sec. 6.2.3 for details).

EXAMPLE D.1   PRODUCTION OF SULFURIC ACID

In this example, we consider the production of sulfuric acid illustrated in Figure D.1a. The mass flow rate of the 90% sulfuric acid stream is specified as $\dot{m}_1 = 100\,lb_m/h$, and we are asked to determine the mass flow rate of the pure sulfur trioxide stream, $\dot{m}_2$. As is often the custom with liquid systems the percentages given in Figure D.1a refer to mass fractions, thus we want to create a final product in which the mass fraction of sulfuric acid is 0.98.

In order to analyze this process in terms of the *atomic* species mass balance, we use the steady state form of Eq. D.10 given here as

$$\int_A \sum_{A=1}^{A=N} \left(\frac{N_{JA}\omega_A}{MW_A} \right) \rho v \cdot n dA = 0, \quad J \Rightarrow S,H,O \tag{1}$$

For the system shown in Figure D.1a, we see that Eq. 1 takes the form

$$\dot{m}_1 \sum_{A=1}^{A=N} \left(\frac{N_{JA}\omega_A}{MW_A} \right)_1 + \dot{m}_2 \sum_{A=1}^{A=N} \left(\frac{N_{JA}\omega_A}{MW_A} \right)_2 = \dot{m}_3 \sum_{A=1}^{A=N} \left(\frac{N_{JA}\omega_A}{MW_A} \right)_3, \quad J \Rightarrow S,H,O \tag{2}$$

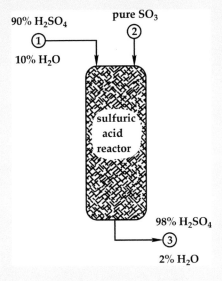

FIGURE D.1a   Sulfuric acid production

In the process under investigation, there are three atomic species indicated by S, H and O, and there are three molecular species indicated by H_2SO_4, H_2O, and SO_3. Since only sulfur and oxygen appear in the stream for which we want to determine the mass flow rate, we must make use of either a sulfur balance or an oxygen balance. However, sulfur appears in only two of the molecular species while oxygen appears in all three.

Because of this a sulfur balance is preferred. At this point, we recall the definition of N_{JA} given by Eq. D.3

$$N_{JA} = \left\{ \begin{array}{c} \text{number of moles of} \\ J\text{-type atoms per mole} \\ \text{of molecular species } A \end{array} \right\}, \quad J = 1,2,...,T, \text{ and } A = 1,2,...,N \tag{3}$$

which leads to

$$\begin{aligned} N_{JA} = 1, \quad J = S, \quad A = H_2SO_4 \\ N_{JA} = 1, \quad J = S, \quad A = SO_3 \end{aligned} \tag{4}$$

Letting J in Eq. 2 represent sulfur we have

$$\frac{\dot{m}_1 \left(\omega_{H_2SO_4} \right)_1}{MW_{H_2SO_4}} + \frac{\dot{m}_2}{MW_{SO_3}} = \frac{\dot{m}_3 \left(\omega_{H_2SO_4} \right)_3}{MW_{H_2SO_4}} \tag{5}$$

At this point, an overall mass balance given by

$$\dot{m}_1 + \dot{m}_2 = \dot{m}_3 \tag{6}$$

can be used to eliminate $\dot{m}_3$ in terms of $\dot{m}_1$ and $\dot{m}_2$. This allows us to solve for the mass flow rate of sulfur trioxide that is given by

$$\dot{m}_2 = \frac{\left[\left(\omega_{H_2SO_4} \right)_3 - \left(\omega_{H_2SO_4} \right)_1 \right] \dot{m}_1}{\left[\dfrac{MW_{H_2SO_4}}{MW_{SO_3}} - \left(\omega_{H_2SO_4} \right)_3 \right]} \tag{7}$$

In this case, we have been able to obtain a solution using only a single atomic element balance along with the overall mass balance. Given this solution, we can easily compute the mass flow rate of pure sulfur trioxide to be

$$\dot{m}_2 = 32.65 \, lb_m / h \tag{8}$$

In the previous example, we solved a problem in which there were three molecular species and three atomic species and we found that a solution could be obtained easily using an atomic species balance. When the number of atomic species is less that the number of molecular species, there is a definite computational advantage in using the atomic species balance. However, the physical description of most processes is best given in terms of molecular species and this often controls the choice of the method used to solve a particular problem.

EXAMPLE D.2   COMBUSTION OF CARBON AND AIR

Carbon is burned with air, as illustrated in Figure D.2a, with all the carbon oxidized to CO_2 and CO. The ratio of carbon dioxide produced to carbon monoxide produced is 2:1. In this case, we wish to determine the

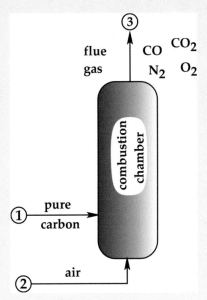

FIGURE D.2a   Combustion of carbon and air

flue gas composition when 50% excess air is used. The percent excess air is defined as

$$\left\{ \begin{array}{c} \text{percentage of} \\ \text{excess air} \end{array} \right\} = \frac{\left\{ \begin{array}{c} \text{molar flow} \\ \text{rate of oxygen} \\ \text{entering} \end{array} \right\} - \left\{ \begin{array}{c} \text{molar rate of} \\ \text{consumption of oxygen} \\ \text{owing to reaction} \end{array} \right\}}{\left\{ \begin{array}{c} \text{molar rate of} \\ \text{consumption of oxygen} \\ \text{owing to reaction} \end{array} \right\}} \times 100 \quad (1)$$

and this definition requires a single application of Eq. D.1 in order to incorporate the global molar rate of production of oxygen, R_{O_2}, into the analysis.

ATOMIC SPECIES BALANCES

The steady-state form of Eq. D.7 is given by

$$\int_A \left\{ \sum_{A=1}^{A=N} N_{JA} x_A \right\} c\mathbf{v} \cdot \mathbf{n} dA = 0, \quad J \Rightarrow C, O, N \tag{2}$$

and for the system illustrated in Figure D.2a, we obtain

$$\left\{ \sum_{A=1}^{A=N} N_{JA} x_A \right\}_1 \dot{M}_1 + \left\{ \sum_{A=1}^{A=N} N_{JA} x_A \right\}_2 \dot{M}_2 = \left\{ \sum_{A=1}^{A=N} N_{JA} x_A \right\}_3 \dot{M}_3 \tag{3}$$

We begin by choosing the *Jth* atomic species to be carbon so that Eq. 3 takes the form

Atomic Carbon Balance: $\quad\quad\quad \dot{M}_1 = \left[(x_{CO})_3 + (x_{CO_2})_3 \right] \dot{M}_3 \tag{4}$

and we continue with this approach to obtain the atomic oxygen and atomic nitrogen balance equations given by

Atomic Oxygen Balance: $\quad 2(x_{O_2})_2 \dot{M}_2 = \left[2(x_{O_2})_3 + 2(x_{CO_2})_3 + (x_{CO})_3 \right] \dot{M}_3 \tag{5}$

Atomic Nitrogen Balance: $\quad\quad\quad\quad 2(x_{N_2})_2 \dot{M}_2 = 2(x_{N_2})_3 \dot{M}_3 \tag{6}$

Directing our attention to the percentage of excess air defined by Eq. 1, we define the *numerical* excess air as

$$\beta = \frac{\left[(\dot{M}_{O_2})_2 \right] - \left[-R_{O_2} \right]}{\left[-R_{O_2} \right]} = 0.5 \tag{7}$$

At this point, we must make use of the mole balance represented by Eq. D.1 in order to represent the global net rate of production of oxygen as

Molecular Oxygen Balance: $\quad\quad\quad R_{O_2} = (\dot{M}_{O_2})_3 - (\dot{M}_{O_2})_2 \tag{8}$

This can be used to obtain a relation between the numerical excess air and the molar flow rates of oxygen given by

$$\beta (\dot{M}_{O_2})_2 = (1+\beta)(\dot{M}_{O_2})_3 \tag{9}$$

For use with the other constraining equations, this can be expressed in terms of mole fractions to obtain

$$\beta (x_{O_2})_2 \dot{M}_2 = (1+\beta)(x_{O_2})_3 \dot{M}_3 \tag{10}$$

This result, along with the atomic species balances and the input data, can be used to determine the flue gas composition

ANALYSIS

We are given that the molar rate of production of carbon dioxide is two times the molar rate of production of carbon monoxide. We can express this information as

$$\dot{M}_{CO_2} = \gamma \dot{M}_{CO}, \quad \gamma = 2 \tag{11}$$

and in terms of mole fractions this leads to

$$(x_{CO_2})_3 = \gamma (x_{CO})_3 \tag{12}$$

The final equation required for the solution of this problem is the constraint on the sum of the mole fractions in Stream #3 that is given by

$$(x_{O_2})_3 + (x_{N_2})_3 + (x_{CO})_3 + (x_{CO_2})_3 = 1 \tag{13}$$

In addition to this constraint on the mole fractions in Stream #3, we assume that the air in Stream #2 can be described by

$$(x_{O_2})_2 = 0.21, \quad (x_{N_2})_2 = 0.79 \tag{14}$$

Use of Eq. 12 in Eqs. 4–6 gives

Atomic Carbon Balance: $$\dot{M}_1 = (x_{CO})_3 (1+\gamma) \dot{M}_3 \tag{15a}$$

Atomic Oxygen Balance: $$(x_{O_2})_2 \dot{M}_2 = \left[(x_{O_2})_3 + \left(\gamma + \frac{1}{2} \right) (x_{CO})_3 \right] \dot{M}_3 \tag{15b}$$

Atomic Nitrogen Balance: $$(x_{N_2})_2 \dot{M}_2 = (x_{N_2})_3 \dot{M}_3 \tag{15c}$$

At this point, we note that we have eight unknowns and seven equations; however, one of the molar flow rates can be eliminated to develop a solution.

ALGEBRA

In order to determine the flue gas composition, we do not need to determine the absolute values of $\dot{M}_1$, $\dot{M}_2$, and $\dot{M}_3$ but only two dimensionless flow rates. These are defined by

$$\mathcal{M}_2 = \dot{M}_2 / \dot{M}_1 \tag{16a}$$

$$\mathcal{M}_3 = \dot{M}_3 / \dot{M}_1 \tag{16b}$$

so that Eqs. 15 and 10 take the form

$$1 = (x_{CO})_3 (1+\gamma) \mathcal{M}_3 \tag{17a}$$

$$(x_{O_2})_2 \mathcal{M}_2 = \left[(x_{O_2})_3 + \left(\gamma + \frac{1}{2}\right)(x_{CO})_3 \right] \mathcal{M}_3 \tag{17b}$$

$$(x_{N_2})_2 \mathcal{M}_2 = (x_{N_2})_3 \mathcal{M}_3 \tag{17c}$$

$$\beta (x_{O_2})_2 \mathcal{M}_2 = (1+\beta)(x_{O_2})_3 \mathcal{M}_3 \tag{17d}$$

From these four equations, we need to eliminate $\mathcal{M}_2$ and $\mathcal{M}_3$ in order to determine the mole fractions and thus the composition of the flue gas. Equations 17a and 17b can be used to obtain the following representations for $\mathcal{M}_2$ and $\mathcal{M}_3$

$$\mathcal{M}_3 = \frac{1}{(x_{CO})_3 (1+\gamma)} \tag{18a}$$

$$\mathcal{M}_2 = \frac{(x_{N_2})_3}{(x_{CO})_3 (1+\gamma)(x_{N_2})_2} \tag{18b}$$

Use of these two results in Eq. 17b leads to

$$(x_{N_2})_3 = \underbrace{\left\{ \frac{(x_{N_2})_2}{(x_{O_2})_2} \right\}}_{known} (x_{O_2})_3 + \underbrace{\left\{ \frac{(x_{N_2})_2 \left(\gamma + \frac{1}{2}\right)}{(x_{O_2})_2} \right\}}_{known} (x_{CO})_3 \tag{19a}$$

and when we eliminate $\mathcal{M}_2$ and $\mathcal{M}_3$ from Eq. 17d, we find

$$(x_{N_2})_3 = \underbrace{\left\{ \frac{1+\beta}{\beta} \frac{(x_{N_2})_2}{(x_{O_2})_2} \right\}}_{known} (x_{O_2})_3 \tag{19b}$$

These two equations, along with the constraint on the mole fractions given by

$$(x_{O_2})_3 + (x_{N_2})_3 + (1+\gamma)(x_{CO})_3 = 1 \tag{20}$$

are sufficient to determine $(x_{O_2})_3$, $(x_{N_2})_3$ and $(x_{CO})_3$. These mole fractions can be expressed as

$$(x_{O_2})_3 = \frac{1}{1 + K_3 + (1+\gamma)(K_3 - K_1)/K_2} \tag{21a}$$

$$(x_{N_2})_3 = \frac{K_3}{1 + K_3 + (1+\gamma)(K_3 - K_1)/K_2} \tag{21b}$$

$$(x_{CO})_3 = \frac{(K_3 - K_1)/K_2}{1 + K_3 + (1+\gamma)(K_3 - K_1)/K_2} \tag{21c}$$

$$(x_{CO_2})_3 = \gamma(x_{CO})_3 \tag{21d}$$

where the numerical values of K_1, K_2 and K_3 are given by

$$K_1 = \frac{(x_{N_2})_2}{(x_{O_2})_2} = 3.7619 \tag{22a}$$

$$K_2 = \frac{(x_{N_2})_2}{(x_{O_2})_2}\left(\gamma + \frac{1}{2}\right) = 9.4048 \tag{22b}$$

$$K_3 = \frac{(x_{N_2})_2}{(x_{O_2})_2}\left(\frac{1+\beta}{\beta}\right) = 11.2857 \tag{22c}$$

Use of these values in Eqs. 21 gives the following values for the mole fractions

$$(x_{O_2})_3 = 0.0681, \quad (x_{N_2})_3 = 0.7685, \tag{23}$$
$$(x_{CO})_3 = 0.0545, \quad (x_{CO_2})_3 = 0.1080$$

The sum of the mole fractions is 0.999 indicating the presence of a small numerical error that could be diminished by carrying more significant figures. On the other hand, the input data for problems of this type are never accurate to ±0.1% and it would be misleading to develop a more accurate solution.

In this appendix, we have illustrated how problems can be solved using atomic species balances instead of molecular species balances. In general, the number of atomic species balances is smaller than the number of molecular species balances, thus the use of atomic species balances leads to fewer equations. However, information is often given in terms of molecular species, such as the percent excess oxygen in Example D.2, and this motivates the use of molecular species balances. Nevertheless, the algebraic effort in Example D.2 is less than one would encounter if the problem were solved using molecular species balances.

D.1 PROBLEMS

D.1. A stream of pure methane, CH_4, is partially burned with air in a furnace at a rate of 100 moles of methane per minute. The air is dry, the methane is in excess, and the nitrogen is inert in this particular process. The reactants and products of the reaction are illustrated in Figure D-1. The exit gas contains a 1:1 ratio of $H_2O:H_2$ and a 10:1 ratio of $CO:CO_2$. Assuming that all of the oxygen and 94% of the methane are consumed by the reactions, determine the flow rate and composition of the exit gas stream.

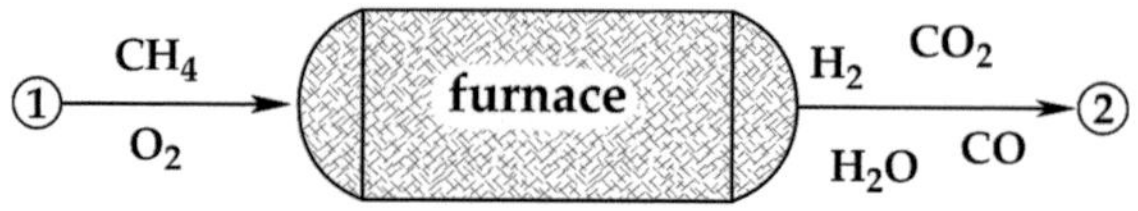

FIGURE D-1   Combustion of methane.

D.2. A fuel composed entirely of methane and nitrogen is burned with excess air. The dry flue gas composition in volume percent is: CO_2, 7.5%, O_2, 7%, and the remainder nitrogen. Determine the composition of the fuel gas and the percentage of excess air as defined by

$$\left\{ \begin{array}{c} \text{percent of} \\ \text{excess air} \end{array} \right\} = \frac{\left(\begin{array}{c} \text{molar flow} \\ \text{rate of oxygen} \\ \text{entering} \end{array} \right) - \left(\begin{array}{c} \text{molar rate of} \\ \text{consumption of oxygen} \\ \text{owing to reaction} \end{array} \right)}{\left(\begin{array}{c} \text{molar rate of} \\ \text{consumption of oxygen} \\ \text{owing to reaction} \end{array} \right)} \times 100$$

Appendix E: Conservation of Charge

In Chapter 6, we represented conservation of atomic species by

Axiom II:
$$\sum_{A=1}^{A=N} N_{JA} R_A = 0, \quad J = 1, 2, \ldots, T \tag{E.1}$$

and in matrix form, we expressed this result as

Axiom II:
$$
\begin{bmatrix}
N_{11} & N_{12} & N_{13} & \cdots & N_{1,N-1,} & N_{1N} \\
N_{21} & N_{22} & & \cdots & N_{2,N-1} & N_{2N} \\
N_{31} & N_{31} & & \cdots & & \\
\cdot & & & \cdots & & \\
& & & \cdots & & \\
N_{T1} & N_{T2} & & \cdots & N_{T,N-1} & N_{TN}
\end{bmatrix}
\begin{bmatrix}
R_1 \\ R_2 \\ R_3 \\ \cdot \\ \cdot \\ R_{N-1} \\ R_N
\end{bmatrix}
=
\begin{bmatrix}
0 \\ 0 \\ 0 \\ \cdot \\ \cdot \\ 0
\end{bmatrix}
\tag{E.2}
$$

If some of the species undergoing reaction are charged species (ions), we need to impose conservation of charge (Feynman, *et al.*, 1963) in addition to conservation of atomic species. This is done in terms of the additional axiomatic statement given by

Axiom II:
$$\sum_{A=1}^{A=N} N_{eA} R_A = 0 \tag{E.3}$$

in which N_{eA} represents the electronic charge associated with molecular species A. In terms of matrix representation, Axiom III can be added to Eq. E.2 to obtain a combined representation for conservation of atomic species and conservation of charge. This combined representation is given by

$$
\begin{bmatrix}
N_{11} & N_{12} & N_{13} & \cdots & N_{1,N-1,} & N_{1N} \\
N_{21} & N_{22} & \cdot & \cdots & N_{2,N-1} & N_{2N} \\
N_{31} & N_{32} & & \cdots & \cdot & \cdot \\
\cdot & \cdot & & \cdots & \cdot & \cdot \\
\cdot & \cdot & & \cdots & \cdot & \cdot \\
N_{T1} & \cdot & & \cdots & \cdot & \cdot \\
N_{e1} & N_{e2} & & \cdots & N_{eN-1} & N_{eN}
\end{bmatrix}
\begin{bmatrix}
R_1 \\ R_2 \\ R_3 \\ \cdot \\ \cdot \\ R_{N-1} \\ R_N
\end{bmatrix}
=
\begin{bmatrix}
0 \\ 0 \\ 0 \\ \cdot \\ \cdot \\ 0
\end{bmatrix}
\tag{E.4}
$$

Here, the elements in the last row of the $(T+1) \times N$ matrix take on the values associated with the charge on species 1, 2, ...N such as

$$
\begin{aligned}
N_{e1} &= 0, && \text{non-ionic species} \\
N_{e2} &= -2, && \text{ionic species such as } SO_4^= \\
N_{e3} &= +1, && \text{ionic species such as } Na^+
\end{aligned}
\tag{E.5}
$$

As an example of competing reactions in a redox system[*], we consider a mixture consisting of ClO_2^-, H_3O^+, Cl_2, H_2O, ClO_3^-, and ClO_2. The visual representation for the atomic / electronic matrix is given by

Molecular Species and Charge $\rightarrow$ ClO_2^- H_3O^+ Cl_2 H_2O ClO_3^- ClO_2

$$
\begin{array}{l}
\textit{chlorine} \\
\textit{oxygen} \\
\textit{hydrogen} \\
\textit{charge}
\end{array}
\begin{bmatrix}
1 & 0 & 2 & 0 & 1 & 1 \\
2 & 1 & 0 & 1 & 3 & 2 \\
0 & 3 & 0 & 2 & 0 & 0 \\
-1 & +1 & 0 & 0 & -1 & 0
\end{bmatrix}
\quad (E.6)
$$

and use of this result with Eq. E.4 leads to

$$
\text{Axiom II \& III:} \quad
\begin{bmatrix}
1 & 0 & 2 & 0 & 1 & 1 \\
2 & 1 & 0 & 1 & 3 & 2 \\
0 & 3 & 0 & 2 & 0 & 0 \\
-1 & +1 & 0 & 0 & -1 & 0
\end{bmatrix}
\begin{bmatrix}
R_{ClO_2^-} \\
R_{H_3O^+} \\
R_{Cl_2} \\
R_{H_2O} \\
R_{ClO_3^-} \\
R_{ClO_2}
\end{bmatrix}
=
\begin{bmatrix}
0 \\
0 \\
0 \\
0
\end{bmatrix}
\quad (E.7)
$$

At this point, we follow the developments given in Chapters 6–9 and search for the *optimal form* of the atomic/electronic matrix. We begin with

$$
A_e =
\begin{bmatrix}
1 & 0 & 2 & 0 & 1 & 1 \\
2 & 1 & 0 & 1 & 3 & 2 \\
0 & 3 & 0 & 2 & 0 & 0 \\
-1 & +1 & 0 & 0 & -1 & 0
\end{bmatrix}
\quad (E.8)
$$

and apply a series of elementary row operations to find the *row reduced echelon* form given by

$$
A_e^* =
\begin{bmatrix}
1 & 0 & 0 & 0 & 5/3 & 4/3 \\
0 & 1 & 0 & 1 & 2/3 & 4/3 \\
0 & 0 & 1 & 0 & -1/3 & -1/6 \\
0 & 0 & 0 & 1 & -1 & -2
\end{bmatrix}
\quad (E.9)
$$

---

[*] Porter, S.K. 1985, How should equation balancing be taught?, J. Chem. Education **62**, 507–508.

Use of this result in Eq. E.7 leads to

$$
\begin{bmatrix}
1 & 0 & 0 & 0 & 5/3 & 4/3 \\
0 & 1 & 0 & 0 & 2/3 & 4/3 \\
0 & 0 & 1 & 0 & -1/3 & -1/6 \\
0 & 0 & 0 & 1 & -1 & -2
\end{bmatrix}
\begin{bmatrix}
R_{ClO_2^-} \\
R_{H_3O^+} \\
R_{Cl_2} \\
R_{H_2O} \\
R_{ClO_3^-} \\
R_{ClO_2}
\end{bmatrix}
=
\begin{bmatrix}
0 \\
0 \\
0 \\
0
\end{bmatrix}
\qquad \text{(E.10)}
$$

We now follow the type of analysis given in Sec. 6.4 and apply the obvious *column / row* partition to obtain

$$
\underbrace{\begin{bmatrix}
1 & 0 & 0 & 0 \\
0 & 1 & 0 & 0 \\
0 & 0 & 1 & 0 \\
0 & 0 & 0 & 1
\end{bmatrix}}_{\substack{non-pivot \\ submatrix}}
\begin{bmatrix}
R_{ClO_2^-} \\
R_{H_3O^+} \\
R_{Cl_2} \\
R_{H_2O}
\end{bmatrix}
+
\underbrace{\begin{bmatrix}
5/3 & 4/3 \\
2/3 & 4/3 \\
-1/3 & -1/6 \\
-1 & -2
\end{bmatrix}}_{\substack{pivot \\ submatrix}}
\begin{bmatrix}
R_{ClO_3^-} \\
R_{ClO_2}
\end{bmatrix}
=
\begin{bmatrix}
0 \\
0 \\
0 \\
0
\end{bmatrix}
\qquad \text{(E.11)}
$$

Making use of the property of the identity matrix leads to

$$
\begin{bmatrix}
1 & 0 & 0 & 0 \\
0 & 1 & 0 & 0 \\
0 & 0 & 1 & 0 \\
0 & 0 & 0 & 1
\end{bmatrix}
\begin{bmatrix}
R_{ClO_2^-} \\
R_{H_3O^+} \\
R_{Cl_2} \\
R_{H_2O}
\end{bmatrix}
=
\begin{bmatrix}
R_{ClO_2^-} \\
R_{H_3O^+} \\
R_{Cl_2} \\
R_{H_2O}
\end{bmatrix}
\qquad \text{(E.12)}
$$

and substituting this result in Eq. E.11 provides the desired result

$$
\underbrace{\begin{bmatrix}
R_{ClO_2^-} \\
R_{H_3O^+} \\
R_{Cl_2} \\
R_{H_2O}
\end{bmatrix}}_{\substack{non-pivot \\ species}}
=
\underbrace{\begin{bmatrix}
-5/3 & -4/3 \\
-2/3 & -4/3 \\
1/3 & 1/6 \\
1 & 2
\end{bmatrix}}_{\substack{pivot \ matrix}}
\underbrace{\begin{bmatrix}
R_{ClO_3^-} \\
R_{ClO_2}
\end{bmatrix}}_{\substack{pivot \\ species}}
\qquad \text{(E.13)}
$$

As discussed in Chapter 6, the choice of pivot and non-pivot species is not completely arbitrary. Thus, one must arrange the atomic / electronic matrix in row reduced echelon form as illustrated in Eq. E.9 in order to make use of the pivot theorem indicated by Eq. E.13.

Use of Eq. E.1 with Eq. E.3 is a straightforward matter leading to Eq. E.4. Within the framework of Chapter 6, one can apply Eq. E.4 in a routine manner in order to solve problems in which charged species are present. In Chapters 6 and 7, we dealt with problems in which net rates of production had to be measured experimentally, and Eq. E.13 is an example of this type of situation. In this case, the net rates of production of the *pivot species*, CLO_3^- and CLO_2, must be determined experimentally so that the pivot theorem can be used to determine the net rates of production of the *non-pivot species* ClO_2^-, H_3O^+, Cl_2, and H_2O.

E.1   MECHANISTIC MATRIX

In our studies of reaction kinetics in Chapter 9, we made use of chemical reaction rate expressions so that all the net rates of production could be calculated in terms of a series of *reference reaction rates*. These reaction rates were developed on the basis of mass action kinetics and thus contained rate coefficients and the concentrations of the chemical species involved in the reaction. That development made use of *elementary stoichiometry* which we express as

$$\text{Elementary stoichiometry:} \quad \sum_{A-1}^{A=N} N_{JA} R_A^k = 0, \quad J=1,2,\dots,T, \; k=\text{I, II,}\dots,\text{K} \quad (E.14)$$

This result ensures that atomic species are conserved in each elementary kinetic step, and Eq. E.1 is satisfied by imposition of the condition (see Eqs. 9.46 and 9.47)

$$\sum_{k=\text{I}}^{k=\text{K}} R_A^k = R_A, \quad k = \text{I, II,}\dots,\text{K} \quad (E.15)$$

When confronted with charged species (ions) in a study of reaction kinetics, one makes use of *elementary conservation of charge* as indicated by

$$\text{Elementary conservation of charge:} \quad \sum_{A-1}^{A=N} N_{eA} \, R_A^k = 0, \quad k=\text{I, II,}\dots,\text{K} \quad (E.16)$$

Thus, charge is conserved in each elementary step of a chemical kinetic schema, and total conservation of charge indicated by Eq. E.3 is automatically achieved in the construction of a *mechanistic matrix*.

Appendix F: Heterogeneous Reactions

Our analysis of the stoichiometry of heterogeneous reactions is based on conservation of atomic species expressed as

Axiom II:
$$\sum_{A=1}^{A=N} N_{JA} R_A = 0 \tag{F.1}$$

We follow the classic continuum point of view[*] and assume that this result is valid everywhere. That is to say that Axiom II is valid in homogeneous regions where quantities such as R_A change slowly and it is valid in interfacial regions where R_A changes rapidly. We follow the work of Wood *et al*[†] and consider the $\gamma - \kappa$ interface illustrated in Figure F.1. The volume V encloses the $\gamma - \kappa$ interface

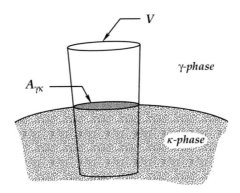

FIGURE F.1 Catalytic surface.

and extends into the homogeneous regions of both the γ-phase and the κ-phase. The total net rate of production of species A in the volume V is represented by

$$\int_V R_A \, dV \equiv \int_{V_\gamma} R_{A\gamma} \, dV + \int_{V_\kappa} R_{A\kappa} \, dV + \int_{A_{\gamma\kappa}} R_{As} \, dA \tag{F.2}$$

---

[*] Truesdell, C. and Toupin, R. 1960, The Classical Field Theories, in *Handbuch der Physik*, Vol. III, Part 1, edited by S. Flugge, Springer-Verlag, New York.
[†] Wood, B.D., Quintard, M. and Whitaker, S. 2000, Jump conditions at non-uniform boundaries: The catalytic surface, Chem. Engng. Sci. **55**, 5231–5245.

Here, the dividing surface that separates the γ-phase from the κ-phase is represented by $A_{\gamma\kappa}$ and the heterogeneous rate of production of species A is identified by R_{As}. This quantity is also referred to as the surface excess reaction rate[‡]. Multiplying Eq. F.2 by the atomic species indicator and summing over all molecular species leads to

$$\int_V \sum_{A=1}^{A=N} N_{JA} R_A \, dV = \int_{V_\gamma} \sum_{A=1}^{A=N} N_{JA} R_{A\gamma} \, dV$$

$$+ \int_{V_\kappa} \sum_{A=1}^{A=N} N_{JA} R_{A\kappa} \, dV + \int_{A_{\gamma\kappa}} \sum_{A=1}^{A=N} N_{JA} R_{As} \, dA, \quad J = 1, 2, \ldots, T \tag{F.3}$$

From Eq. F.1, we see that the left-hand side of this result is zero and we have

$$0 = \int_{V_\gamma} \sum_{A=1}^{A=N} N_{JA} R_{A\gamma} \, dV + \int_{V_\kappa} \sum_{A=1}^{A=N} N_{JA} R_{A\kappa} \, dV$$

$$+ \int_{A_{\gamma\kappa}} \sum_{A=1}^{A=N} N_{JA} R_{As} \, dA, \quad J = 1, 2, \ldots, T \tag{F.4}$$

At this point, we require that the homogeneous net rates of production satisfy the two constraints given by

$$\sum_{A=1}^{A=N} N_{JA} \, R_{A\gamma} = 0, \quad \sum_{A=1}^{A=N} N_{JA} \, R_{A\kappa} = 0, \quad J = 1, 2, \ldots, T \tag{F.5}$$

and this leads to the following form of Eq. F.4

$$\int_{A_{\gamma\kappa}} \sum_{A=1}^{A=N} N_{JA} R_{As} \, dA = 0, \quad J = 1, 2, \ldots, T \tag{F.6}$$

Catalytic surfaces consist of catalytic sites where reaction occurs and non-catalytic regions where no reaction occurs. Because the heterogeneous rate of production is highly non-uniform, it is appropriate to work in terms of the area average

$$\langle R_{As} \rangle_{\gamma\kappa} = \frac{1}{A_{\gamma\kappa}} \int_{A_{\gamma\kappa}} R_{As} \, dA \tag{F.7}$$

[‡] Whitaker, S. 1992, The species mass jump condition at a singular surface, Chem. Engng. Sci. **47**, 1677–1685.

so that Eq. F.6 takes the form

$$\sum_{A=1}^{A=N} N_{JA} \langle R_{As} \rangle_{\gamma\kappa} = 0. \quad J = 1, 2, \ldots, T \tag{F.8}$$

We summarize our results associated with Axiom II as

Axiom II (general):
$$\sum_{A=1}^{A=N} N_{JA} R_A = 0, \quad J = 1, 2, \ldots, T \tag{F.9}$$

Axiom II (γ–phase):
$$\sum_{A=1}^{A=N} N_{JA} R_{A\gamma} = 0, \quad J = 1, 2, \ldots, T \tag{F.10}$$

Axiom II (κ–phase):
$$\sum_{A=1}^{A=N} N_{JA} R_{A\kappa} = 0, \quad J = 1, 2, \ldots, T \tag{F.11}$$

Axiom II (γ–κ interface):
$$\sum_{A=1}^{A=N} N_{JA} \langle R_{As} \rangle_{\gamma\kappa} = 0, \quad J = 1, 2, \ldots, T \tag{F.12}$$

For a reactor in which only homogeneous reactions occur, we make use of Eq. F.9 in the form

Axiom II:
$$A R = 0, \quad R = \begin{bmatrix} R_A \\ R_B \\ . \\ . \\ R_N \end{bmatrix} \tag{F.13}$$

in which A is the atomic matrix. For a catalytic reactor in which only heterogeneous reactions occur at the $\gamma-\kappa$ interface, we make use of Eq. F.12 in the form

Axiom II:
$$A \langle R_s \rangle_{\gamma\kappa} = 0, \quad \langle R_s \rangle_{\gamma\kappa} = \begin{bmatrix} \langle R_{As} \rangle_{\gamma\kappa} \\ \langle R_{Bs} \rangle_{\gamma\kappa} \\ . \\ . \\ \langle R_{Ns} \rangle_{\gamma\kappa} \end{bmatrix} \tag{F.14}$$

The pivot theorem associated homogeneous reactions is obtained from Eq. F.13 and the analysis leads to Eq. 6.80 which is repeated here as

Pivot Theorem (homogeneous reactions): $\qquad R_{NP} = P\,R_P \qquad$ (F.15)

The pivot theorem associated with heterogeneous reactions is obtained from Eq. F.14 and is given here as

Pivot Theorem (heterogeneous reactions): $\qquad \left(\langle R_s \rangle_{\gamma\kappa}\right)_{NP} = P\left(\langle R_s \rangle_{\gamma\kappa}\right)_P \qquad$ (F.16)

The fact that the axiom and the application take exactly the same form for both homogeneous and heterogeneous reactions has led many to ignore the difference between these two distinct forms of chemical reaction.

In general, measurement of the net rates of production are carried out at the macroscopic level, thus we generally obtain experimental information for the global net rate of production. For a homogeneous reaction, this takes the form

$$R_A = \int_V R_A\,dA, \quad A = 1, 2, \ldots, N \qquad (F.17)$$

while the global net rate of production for a heterogeneous reaction is given by

$$R_A = \int_{A_{\gamma\kappa}} \langle R_{As} \rangle_{\gamma\kappa}\,dA, A = 1, 2, \ldots, N \qquad (F.18)$$

Here, we note that the global net rates of production for both homogeneous reactions and heterogeneous reactions have exactly the same physical significance, thus it is not unreasonable to use the same symbol for both quantities. Given this simplification, the global version of the pivot theorem can be expressed as

Pivot Theorem (global form): $\qquad R_{NP} = P\,R_P \qquad$ (F.19)

for both homogeneous and heterogeneous reactions.

References

Amundson, N. R. 1966, *Mathematical Methods in Chemical Engineering: Matrices and Their Application*, Prentice-Hall, Inc., Englewood Cliffs, NJ.

Aris, R. 1965, *Introduction to the Analysis of Chemical Reactors*, Prentice-Hall, Inc., Englewood Cliffs, NJ.

Aris, R. 1965, Prolegomena to the rational analysis of systems of chemical reactions, Arch. Ration. Mech. Anal., **19**, 81–99.

Bailey, J. E. and Ollis, D. F. 1986, *Biochemical Engineering Fundamentals*, Sec. 7.7, 2nd Edition, McGraw Hill Higher Education, New York.

Bird, R. B., Stewart, W. E., and Lightfoot, E. N. 2002, *Transport Phenomena*, 2nd Edition, John Wiley & Sons, Inc., New York.

Birkhoff, G. 1960, *Hydrodynamics: A Study in Logic, Fact, and Similitude*, Princeton University Press, Princeton, NJ.

Bjornbom, P. H. 1977, The relation between the reaction mechanism and the stoichiometric behavior of chemical reactions, AIChE J., **23**, 285–288.

Bodenstein, M. and Lind, S. C. 1907, Geschwindigkeit der Bildung der Bromwasserstoffes aus sienen Elementen, Z. physik. Chem., **57**, 168–192.

Bradie, B. 2006, *A Friendly Introduction to Numerical Analysis*, Pearson Prentice Hall, Englewood Cliffs, NJ.

Corliss, G. 1977, Which root does the bisection method find?, SIAM Rev., **19**, 325–327.

Denn, M. M. 1980, Continuous drawing of liquids to form fibers, Ann. Rev. Fluid. Mech., **12**, 365–387.

Dixon, D. C. 1970, The definition of reaction rate, Chem. Engr. Sci., **25**, 337–338.

Feynman, R. P., Leighton, R. B., and Sands, M. 1963, *The Feynman Lectures on Physics*, Addison-Wesley Pub. Co., New York.

Frank-Kamenetsky, D. A. 1940, Conditions for the applicability of Bodenstein's method in chemical kinetics, J. Phys. Chem. (USSR), **14**, 695–702.

Gates, B. C. and Sherman, J. D. 1975, Experiments in heterogeneous catalysis: Kinetics of alcohol dehydration reactions, Chem. Eng. Ed. Summer, 124–127.

Gibbs, J. W. 1928, *The Collected Works of J. Willard Gibbs*, Longmans, Green and Company, London.

Gleick, J. 1988, *Chaos: Making a New Science*, Penguin Books, New York.

Herzfeld, K. F. 1919, The theory of the reaction speeds in gases, Ann. Physic, **59**, 635–667.

Hinshelwood, C. N. and Askey, P. J. 1927, Homogeneous reactions involving complex molecules: The kinetics of the decomposition of gaseous dimethyl ether, Proc. Roy. Soc., **A115**, 215–226.

Horn, F. and Jackson, R. 1972, General mass action kinetics, Arch. Ration. Mech. Anal., **47**, 81–116.

Hougen, O. A. and Watson, K. M. 1943, *Chemical Process Principles*, John Wiley & Sons, Inc., New York.

Hurley, J. P. and Garrod, C. 1978, *Principles of Physics*, Houghton Mifflin Co., Boston, MA.

Kolman, B. 1997, *Introductory Linear Algebra*, Sixth Edition, Prentice-Hall, Upper Saddle River, NJ.

Kvisle, S., Aguero, A., and Sneeded, R. P. A. 1988, Transformation of ethanol into 1,3-butadiene over magnesium oxide/silica catalysts, Appl. Catal., **43**, 117–121.

Lavoisier, A. L. 1777, Memoir on Combustion in General, Mem. Acad. r. Sci. Paris, 592–600.

Levich, V. G. 1962, *Physicochemical Hydrodynamics*, Prentice-Hall, Inc., Englewood Cliffs, NJ.

Lindemann, F. A. 1922, The radiation theory of chemical action, Trans. Faraday Soc., **17**, 598–606.

Michaelis, L. and Menten, M. L. 1913, Die Kinetik der Invertinwirkung, Biochem Z., **49**, 333–369.

Mono Lake Committee, https://www.monolake.org/learn/aboutmonolake/savingmonolake/

Monod, J. 1942, *Recherche sur la Croissance des Cultures Bactériennes*, Herman Editions, Paris.

Monod, J. 1949, The growth of bacterial cultures, Ann. Rev. Microbiol., **3**, 371–394.

National Institute of Standards and Technology, SI Prefixes, http://physics.nist.gov/cuu/Units/prefixes.html.

Noble, B. 1969, *Applied Linear Algebra*, Prentice-Hall, Inc., Englewood Cliffs, NJ.

Peitgen, H.-O., Jürgens, H., and Saupe, D. 1992, *Chaos and Fractals. New Frontiers of Science*, Springer-Verlag, New York.

Perry, R. H., Green, D. W., and Maloney, J. O. 1984, *Perry's Chemical Engineer' Handbook*, 6th Edition, McGraw-Hill Books, New York.

Perry, R. H., Green, D. W., and Maloney, J. O. 1997, *Perry's Chemical Engineer' Handbook*, 7th Edition, McGraw-Hill Books, New York.

Polanyi, M. 1920, Reaction isochore and reaction velocity from the standpoint of statistics, Z. Elektrochem., **26**, 49–54.

Porter, S. K. 1985, How should equation balancing be taught?, J. Chem. Education, **62**, 507–508.

Prigogine, I. and Defay, R. 1954, *Chemical Thermodynamics*, Longmans Green and Company, London.

Reklaitis, G. V. 1983, *Introduction to Material and Energy Balances*, John Wiley & Sons, Inc., New York.

Ramkrishna, D. and Song, H.-S. 2008, A rationale for Monod's biochemical growth kinetics, Ind. Eng. Chem. Res., **47**, 9090–9098.

Ramsperger, H. C. 1927, Thermal decomposition of azomethane over a large range of pressures, J. Am. Chem. Soc., **49**, 1495–1512.

Reid, R. C., Prausnitz, J. M., and Sherwood, T. K. 1977, *The Properties of Gases and Liquids*, Sixth Edition, McGraw-Hill Books, New York.

Reppe, W. J. 1892–1969. https://www.chemeurope.com/en/encyclopedia/Walter_Reppe.html

Rodgers, A. and Gibon, Y. 2009, Enzyme kinetics: Theory and practice, Chapter 4 in *Plant Metabolic Networks*, edited by J. Schwender, Springer, New York.

Rouse, H. and Ince, S. 1957, *History of Hydraulics*, Dover Publications, Inc., New York.

Rucker, T. G., Logan, M. A., Gentle, T. M., Muetterties, E. L., and Somorjai, G. A. 1986, Conversion of acetylene to Benzene over palladium single-crystal surfaces. 1. The low-pressure stoichiometric and high-pressure catalytic reactions, J. Phys. Chem., **90**, 2703–2708.

Saeleczky, J. and Margolies, R., eds. 1998, *Shreve's Chemical Process Industries*, 5th Edition, McGraw-Hill Professional, New York.

Sandler, S. I. 2006, *Chemical, Biochemical, and Engineering Thermodynamics*, 4th Edition, John Wiley and Sons, New York.

Sankaranarayanan, T. M., Ingle, R. H., Gaikwad, T. B., Lokhande, S. K., Raja. T., Devi, R. N., Ramaswany, V., and Manikandan, P. 2008, Selective oxidation of ethane over Mo-V-Al-O oxide catalysts: Insight to the factors affecting the selectivity of ethylene and acetic acid and structure-activity correlation studies, Catal. Lett., **121**, 39–51.

Segel, I. 1993, *Enzyme Kinetics: Behavior and Analysis of Rapid Equilibrium and Steady-State Enzyme Systems*, Wiley-Interscience, New York.

Steding, D. J., Dunlap, C. E., and Flegal, A. R. 2000, New isotopic evidence for chronic lead contamination in the San Francisco Bay estuary system: Implications for the persistence of past industrial lead emissions in the biosphere, Proc. Natl Acad. Sci., **97** (21), 11181–11186.

Stein, S. K. and Barcellos, A. 1992, *Calculus and Analytic Geometry*, McGraw-Hill, Inc., New York.

Tanaka, M., Yamamote, M., and Oku, M. 1955, Preparation of styrene and benzene from acetylene and vinyl acetylene, USPO, 272–299.

Toulmin, S. E. 1957, Crucial experiments: Priestley and Lavoisier, J. Hist. Ideas, **18**, 205–220.

Truesdell, C. 1968, *Essays in the History of Mechanics*, Springer-Verlag, New York.

Truesdell, C. and Toupin, R. 1960, The Classical Field Theories, in *Handbuch der Physik*, Vol. **III**, Part 1, edited by S. Flugge, Springer-Verlag, New York.

Wegstein, J. H. 1958, Accelerating convergences of iterative processes, Comm. ACM, **1**, 9–13.

Whitaker, S. 1968, *Introduction to Fluid Mechanics*, Prentice Hall, Inc., Englewood Cliffs, NJ.

Whitaker, S. and Cassano, A. E. 1986, *Concepts and Design of Chemical Reactors*, Gordon and Breach Science Publishers, New York.

Whitaker, S. 1988, Levels of simplification: The use of assumptions, restrictions, and constraints in engineering analysis, Chem. Eng. Ed., **22**, 104–108.

Whitaker, S. 1992, The species mass jump condition at a singular surface, Chem. Eng. Sci., **47**, 1677–1685.

Wisniak, J. 2001, Historical development of the vapor pressure equation from Dalton to Antoine, J. Phase Equilibrium, **22**, 622–630.

Wood, B. D., Quintard, M., and Whitaker, S. 2000, Jump conditions at non-uniform boundaries: The catalytic surface, Chem. Eng. Sci., **55**, 5231–5245.

Wylie, C. R., Jr. 1951, *Advanced Engineering Mathematics*, McGraw-Hill Book Co., Inc., New York.

Ypma, T. J. 1995, Historical development of the Newton-Raphson method, SIAM Rev., **37**, 531–551.

Author Index

409

Subject Index